Lecture Notes in Energy

Volume 71

Lecture Notes in Energy (LNE) is a series that reports on new developments in the study of energy: from science and engineering to the analysis of energy policy. The series' scope includes but is not limited to, renewable and green energy, nuclear, fossil fuels and carbon capture, energy systems, energy storage and harvesting, batteries and fuel cells, power systems, energy efficiency, energy in buildings, energy policy, as well as energy-related topics in economics, management and transportation. Books published in LNE are original and timely and bridge between advanced textbooks and the forefront of research. Readers of LNE include postgraduate students and non-specialist researchers wishing to gain an accessible introduction to a field of research as well as professionals and researchers with a need for an up-to-date reference book on a well-defined topic. The series publishes single- and multi-authored volumes as well as advanced textbooks. **Indexed in Scopus and EI Compendex** The Springer Energy board welcomes your book proposal. Please get in touch with the series via Anthony Doyle, Executive Editor, Springer (anthony.doyle@springer.com).

More information about this series at http://www.springer.com/series/8874

John Erik Meyer

The Renewable Energy Transition

Realities for Canada and the World

Springer

John Erik Meyer
Canadians for a Sustainable Society
Parry Sound, ON, Canada

ISSN 2195-1284 ISSN 2195-1292 (electronic)
Lecture Notes in Energy
ISBN 978-3-030-29114-3 ISBN 978-3-030-29115-0 (eBook)
https://doi.org/10.1007/978-3-030-29115-0

This Springer imprint is published by the registered company Springer Nature Switzerland AG
The registered company address is: Gewerbestrasse 11, 6330 Cham, Switzerland

To Katie, for your hope, enthusiasm, and support. You and your friends deserve better than the tools we are passing on to you.

Preface

The growth of monetary systems, energy use, science, human numbers, and general human welfare has occurred concurrently over the past several centuries. The heady progress of human systems has allowed us to dominate the world around us and to ignore perhaps, the health of the seemingly invisible natural stocks and flows which support us. But changes in our environment are becoming more obvious and we have not yet developed a means to value it and prioritize our interaction with the natural world.

Modern society is awash in statistics, measures, and quantification, but public policy making has focused most of this data processing in the area of the commercial marketplace using fiat currency as the core metric. In the era of apparent endless abundance, quantification of things that were in endless supply made little sense. But now, limits to our energy supplies and resource bases are beginning to appear on the horizon. It behooves us to expand our singular focus on consumption to include a comprehensive representation of the natural assets and processes which make that consumption possible.

Doing so will require new tools and a new perspective on how we live, and what allows us to live, on this planet. Unlike the commercial marketplace which is constructed of trillions of money-based transactions annually, the natural world we interact with is a complex of energy transactions and stocks of mineral and biological reserves. These physical stocks and flows can only be represented by physical measures. When they enter the commercial marketplace, they acquire monetized valuation but beyond the human realm, where natural wealth is created and maintained, real physical measures are required to understand their origins and health.

Two points which are clear to the vast majority of people on this planet are that the climate is changing and that depletion of fossil fuel resources is inevitable. What the impacts will be of these trends, the time span over which they will occur, and who should bear the cost of mitigating these problems are points of contention currently roiling both domestic and international politics. Confusion surrounding responsibilities and appropriate actions are resulting in the failure of governments to make timely progress on these critical issues. But at the root of this lack of action is the ability of policy makers to quantify the full scale of the problem. Monetary metrics don't represent natural systems. Essentially, we can't address what we don't understand and, in modern society, we can't understand that which we can't measure.

Progress lies in the direction of being able to represent physical processes, in their entirety, in a manner that all those studying the issue can absorb. Metrics which capture the essence of the critical issues have been developed by researchers but not adopted by mainstream policy makers, most of whom cling to monetary metrics. For these policy makers, representative or not, the only full set of numbers they have to work with are monetary. Physical representations, although they might be much more accurate and informative, are currently narrow and may send very different signals than monetary metrics.

As national policy makers begin to pay attention to the physical world around us, commercial leaders are still locked into monetary metrics and use those metrics to plan for what they assume will be the continuance of the trajectory upon which the fossil fuel mother lode launched us. Options apparent to those using physical measures look entirely different. Hence, the need to differentiate between the two measurement systems, the responsibilities of those who use them, and the different time horizons each is working with. Public policy needs to be informed by a clear representation of the physical world around us because that is what both nations and the global community ultimately interact with. The recognition of the existence of different systems should allow an information based and non-polarized conversation to be established. The choice is not either / or.

This book was written with the intent of giving policy makers and those interested in history and our future options more insight into the limits of markets and monetary systems. The importance of energy to Canada and other countries can best be understood if presented in an historical context and also in relation to available resources and the energetic demands of the climate in northern regions.

Energy, climate, and resource education is fundamental to a society's ability to anticipate and adjust to a changing world. This book presents several new approaches to illuminate the problems and potential solutions in the energy transition but does not strive to deliver a silver bullet solution offering a sure means of success. Currently, our experience and models are in some way away from being capable of delivering this. Rather, areas of promise, areas in need of further development, and the habits and expectations which must change are highlighted to produce, if not a solution, perhaps a broader perspective. This broader biophysical perspective reveals a clear direction and several stepping stones along the path to the eventual development of a sustainable society.

The options humanity will have before it in the near future very much depend on the progress made in transitioning to renewable energy over the next two short decades. If significant progress is to be made, leaders need to have answers to several questions clear in their minds:

- How long do we have to change our consumption patterns before the climate is changed irreparably?
- How long will it be before fossil fuel resources decline in quality to the point of destabilizing energy flows and unsettling financial and production systems?
- How long will it take for a society to transition away from fossil fuels to the degree that both climate and energy supply threats are greatly reduced?

Simply asking these questions will better frame the issues and allow constructive discussion to grow into positive, coherent, and timely action.

Parry Sound, ON, Canada John Erik Meyer

Acknowledgements

"Charlie" Hall crossed the auditorium at a Biophysical Economics conference in 2018 and instead of requesting that I tone down my comments, asked if I might be interested in writing a book on energy in Canada. He then asked me, what the book would be about. After 10 s of frantic processing, I answered "what it will take for Canada to transition to renewable energy." I thought it was a subject in desperate need of airing. My thanks go to him, for this opportunity has been a great learning experience. My long-time friend Andrew Marshall was good enough to climb back into the editorial saddle and spend many hours correcting and improving my grammar and logic. David Hughes and Pedro Prieto freely offered the benefits of their deep practical experience and analysis. Teryl provided editing and enthusiastic support. Thomas Homer-Dixon assisted with research contacts and Vaclav Smil was good enough to provide comments and discussion despite his incredible publication schedule. Researchers, Michael Dale, Christian Breyer, Uwe Schneider, Graham Palmer, and Ferruccio Ferroni, and many others found time to answer questions and provide information.

Contents

About the Author

John Erik Meyer has been circling the issues of per capita resources, social equality, and leadership since high school. A BA in Economics and years of working with physical units and dollar accounting and costing systems brought the different abilities of these measurement processes into sharp relief. He is a patent holder and has had a number of articles published in Canada's major newspapers dealing with a range of topics from population, immigration, and the environment to the failings of GDP-based metrics for social policy formation. He has also presented a paper on Energy Currency at a conference of that name in Split, Croatia. He maintains a site www.theperfectcurrency.org which focuses on the concept of energy-based currency. Mr. Meyer is currently President of the NGO "Canadians for a Sustainable Society" and owner and designer at a small medium tech manufacturing company. He also dabbles in solar electric and heating systems as well as electric bikes and has designed and built a house which he expects to be (eventually) energy positive. His primary interests are the changes necessary to achieve a sustainable society, population cycles, biophysical economics, and the reasons for failed human social structures throughout history.

List of Figures

List of Tables

Chapter 1
Stored Energy Builds a Northern Nation

Abstract Times of stability in human history are merely grace periods in between different resource and climatic regimes. Adaptation to these underpinnings of human existence is a necessary survival skill. Now, as the human footprint extends over more of the planet, our ability to quantify, monitor, and assess our own impact will play a critical part in improving our ability to adapt and prosper. Canada, the nation, owes its existence and its degree of prosperity to fossil fuels to a greater degree than most nations. For thousands of years, its harsh northern climate facilitated the husbanding of natural resources beyond the reach of large populations and these resources could be exploited on a large-scale basis only with the work done by fossil fuels. Work was done and lifestyles changed by the harnessing of fossil fuel power and as these fuels fade into history, how much of the benefits they brought can be sustained by less dense and more variable renewable energy?

In the past 500 years, Canada has experienced two major transitions. The first was the displacement of indigenous peoples by Europeans starting in earnest by the early 1600s. The second was the dawning of the fossil fuel energy age in which humans, across the world, learned to exploit and apply the planet's huge energy fossil fuel stores accumulated over the past 400 million years.

Before the twenty-first century closes, we will have experienced two more transitions of at least equal magnitude; the destabilization of the exceptionally balanced and favorable climate the earth has experienced over the past 10,000 years and the transition from fossil fuels to renewable energy (Fig. 1.1).

The first European explorers to arrive in Canada encountered indigenous societies living in relative balance[1] with their environment with the population fluctuating with climatic and environmental conditions. The well-being of the Amerindians moved in lockstep with their environment as their ability to store energy was limited to the amount of food they could set aside. This amount rarely exceeded what was

[1] Amerindians having conquered the land thousands of years before wiped out large mammals such as mastodons, saber tooth tigers, and other megafauna.

J. E. Meyer, *The Renewable Energy Transition*, Lecture Notes in Energy 71,
https://doi.org/10.1007/978-3-030-29115-0_1

Fig. 1.1 Louis Parker, Iroquois Canoe

necessary to survive a normal winter. Amerindian populations were thus subject to variations in the food supply and living conditions dictated by climate.

The first documented[2] arrival of Europeans took place around 1000 AD[3] in the form of the Vikings. This group possessed marginally more advanced steel weapons and agricultural technology than the aboriginals who were at a Neolithic or Late Stone Age level of technological development with stone axes and stick and shell digging implements. Crops and animal husbandry gave the Vikings an advantage in their ability to survive in more marginal conditions. Both groups had mastered fire but the Vikings had become proficient in using it for iron production allowing them to work in both the land and wood more efficiently. Weaving gave them greater flexibility in clothing developed for specific purposes.

Iron working applied to warfare provided the advantages of swords and superior spears and arrows. Steel banded shields, helmets, and chain mail made them more durable in battle while their ship technology facilitated their ability to cross greater spans of water more rapidly and in larger numbers.

The Vikings, and the Amerindians they encountered, shared roughly the same stature and level of health. The Scandinavians were seafarers, hunters, fishermen, and farmers and at least as capable of living off the land as their indigenous counterparts. Given equal numbers, the Vikings, with their small technological advantages, might have gradually displaced the indigenous groups of the Americas but

[2] Saint Brendan The Navigator, the Irish monk, circa 500 AD is held by some to have visited North America but there is no archeology to support this possibility.

[3] Across Atlantic Ice: The Origin of America's Clovis Culture Book by Bruce A. Bradley and Dennis Stanford.

they did not have the critical mass of population to hold on to their initial settlement at L'Anse aux Meadows in Newfoundland.

The pressure of growing population had pushed the Vikings out from the fjords and fields of Norway, Sweden, and Denmark with enough force to successfully settle in Britain, France, and through the river systems of western Russia. These destinations were between 3 and 10 days sailing away. But in crossing the North Atlantic, first to Iceland (7 days) and then to Greenland (add 3 weeks), their numbers dwindled until upon finally reaching the verdant shores of North America (add 6 weeks), they could barely field one or two long ships full of explorers and farmers at the end of 10-weeks journey spent in an open boat on the North Atlantic. This was not enough to prevail against the much more numerous native tribes.

Five hundred years later, the tables had tilted decidedly against the fortunes of the native societies of the Americas. Europeans had developed distinctly advanced technology both in terms of weapons and agriculture. And in transportation and navigation, they had been learning quickly. Columbus' first voyage took 8 weeks to cross the Atlantic. His second voyage was much quicker and a typical voyage of the day might have been 6 weeks. John Cabot sailed from Bristol to North America in a scant 33 days.

And when the European ships arrived, they carried probably four times as many people and perhaps double the cargo of the Viking ships. The Mayflower carried 102 passengers, 30 crew, and 180 tons of cargo. Europeans had developed the means to transport significant numbers of people, large amounts of goods and livestock in their larger ships across the Atlantic and thus the fate of the Amerindian societies was sealed.

Why Migrate?

The means to cross the ocean in force was one thing but above all, Europeans had the driving need to do so as their numbers were growing rapidly while their resource base was in decline, creating dismal living conditions and a desperate, restive populace.

Every species experiences population cycles in which their numbers grow rapidly in a favorable environment and then, with the depletion of a key resource or a change in climate, population declines inevitably occur. Although well understood in most societies for millennia, this process was finally explained mathematically by Malthus on one of whose shoulders was firmly planted the right foot of Charles Darwin. The population declines and social chaos of the 1600s[4] were vivid illustrations of the end portion of a population cycle and likely informed Malthus's work 150 years later.

[4] Global Crisis deals with the demographic, social, and economic repercussions of the cooling of the climate in the mid-1600s.

Conditions in Europe and across most of Europe and the world were cold and miserable in the 1600s as so elegantly and painstakingly depicted in such grim detail in Geoffrey Parker's "Global Crisis" as crop failures led to the collapse of trading networks and widespread civil disruption. In that century, a growing population in Europe had encountered years of colder and wetter climate which spawned civil unrest, famine, and war. During one particularly hard winter it was even possible to walk from Europe to Asia over a frozen Bosporus. The end of the latest population cycle, which had begun in the late 1300s, saw population declines on the order of 30% in Europe and Asia along with extreme social stress. The level of social chaos is illustrated well by the incomprehensible 30 Years War.

Desperate people make willing migrants, which explains the commitment made by the colonists to endure a long dangerous and miserable ocean voyage to travel to a little known and harsh land. In the early days of the Virginia colony, mortality rates approached 50% with many people being married five or six times before their early deaths. Yet, conditions being what they were in Europe, people still came.

Stored Energy Emerges

Not only did the Europeans of the 1500s have greater transport capacity and agricultural technology than their Viking predecessors, they had much more advanced military technology. Significantly, this technology was based on the first application of human created and stored energy. Gun powder constituted a chemical energy storage system and it gave the Europeans an immense advantage on the field of battle.

Gun powder marked the earliest use of manufactured, concentrated and stored energy and provided an indication of the patterns which would develop in our modern energy-based society. It was unlike any technology humans had developed before. The gunpowder template changed the way we worked with tools and the environment around us. It had the following revolutionary characteristics:

- Gunpowder was neither harvested nor mined directly, but was produced in several different chemical processes all requiring energy inputs of their own by very inefficient means. The energy potential in the gunpowder that resulted was vastly less than the process energy which went into its making but that initial pulse of energy (on the order of 1000 kW or 1500 horsepower) for a third of a millisecond was something humans had never been able to create before.
- Gunpowder was manufactured, accumulated, and stored; it was a stored energy which could be applied at any time in the future under the complete control of the operator.
- It stored energy in very dense and transportable packages.
- If stored properly, it could hold its energy in excess of 50 years.
- When needed, it could be used almost immediately.
- It could do a very large amount of work, well beyond the capacity of humans.
- It required tools to properly apply it and these tools could evolve quickly.

In sum, gunpowder offered as much energy as was needed, wherever it was needed and whenever it was needed. This was a world changing process for humanity and a portent of things to come with the discovery and application of fossil fuels.

In terms of applied energy, the first blunderbuss firearms delivered seven or eight times the force that a good bow drawn by a strong man could. This allowed even a minimally skilled combatant of any stature to deliver a larger projectile further and faster than any warrior could hope to achieve with a bow. The difference in accuracy between the bow and early firearms is an open question, but firearms developed quickly whereas native bows were a technology which had essentially plateaued. Bows could be shot much more rapidly but this was not enough to overcome the strengths of the new firearms technology.

Although this quote originated in another time and place, it neatly sums up the critical difference technology makes on a faraway battlefield when confronted with superior indigenous numbers: "Whatever happens, we have got The Maxim gun, and they have not." Belloc (1898).

When a battle was joined in the Americas, the Europeans fielded smaller numbers of fighters and they were likely smaller, less fit, and less healthy than their indigenous opponents who had the added advantage of vastly greater knowledge of the landscape. Their low quality, starchy diet of grain, and a life of frequent food shortages left Europeans of the 1500s to 1700s as much as 15 cm shorter than they are today and much smaller than the healthy, broad-based diet hunter-gatherers who stood against them (Steegmann and Haseley 1988; Steckel and Rose 2002). This is not to imply that hunter-gatherers had an idyllic life. Their varied and more protein dense diet may have been healthier, but starvation, accidents, and violent death associated with the nomadic hunting life and near-continuous low-level warfare with neighboring tribes were constant threats. Still, some hunter-gatherer groups in Canada likely were healthier than their more sedentary Mayan and Aztec distant cousins who had advanced to corn and grain-based farming societies with dense, disease-prone population centers (Milton 2000; Armelagos and Cohen 2013).

Controlled energy was very important in developing key tools to allow a foothold to be gained, but once established, the agricultural technology and disease were the essential components in the success of the European displacement of the Amerindians.

Societies don't run on gunpowder but since it can dictate the outcome of critical events, the energy used to make it was well spent even if the energy output was a small fraction of the energy input. Now 500 years on, our societies no longer run on human, animal, and wind power combined with the energy embedded in crops and wood, but on stored energy created in a process taking several hundred million years. And since raw energy use is now so ubiquitous, understanding how much energy is required to produce more energy (variously called energy efficiency, Energy Returned on Energy Invested (EROI) or net energy) is of critical importance.

Although early battles were far from one-sided, the ever growing numbers of Europeans arriving, combined with their technological advantages, pushed the native populations back. Much can be made of the tactical technological difference between the two groups, but perhaps the greatest strategic weapon the Europeans possessed was disease. European diseases decimated Amerindian populations which had no defense against them (Beaujot and Kerr 2015). "In 1862 a ship infected with smallpox docked in Victoria, BC and the disease spread quickly to the crowded First Nations encampments on the city's outskirts. Over a 2 year period, records indicated a 60% decline in the Aboriginal population of this coastal region of British Columbia.") Often, European colonists simply occupied deserted or sparsely populated lands with little or no opposition.

In the words of historian Olive Dickason "The earliest European accounts of the New World all spoke of the 'great multitudes of people'; it was later, when colonization was gaining momentum, that the large stretches of territories were found unoccupied, and the notion of 'empty continent' gained currency" (Romaniuc 2014).

The population dynamics during the colonization of Canada are brought into relief by the point made by Roderic Beaujot in Population Change in Canada "If we accept Dickason's figure of a pre-contact population of about 500,000, it would take almost three centuries after Jacques Cartier first sailed into the Gulf of St. Lawrence in the 1530s for the European population to reach the half million mark. …. That is, early European settlement was accompanied by major depopulation of the First Nations population."

Disease hollowed out the native population with whole villages dying and their lands becoming overgrown completely beyond the view of the early settlers. The Europeans spread across the continents from east to west and as they did so, often moved into abandoned lands which were once densely populated, and often terraformed as the landscape was and food environment was optimized to best suit the tribes. Europeans walked unwittingly through once well-managed systems of environmentally sustainable agriculture, a human engineered and managed system of tree farms for walnuts, chestnuts, and hazelnuts and fields where their prime meat source, deer, could flourish (Mann 2005, 2006).

This pattern of obliviousness was similar to the British observations in Australia of aboriginal land management. The British did not know it at the time, but the bewildering and infuriating aboriginal penchant for walking naked through the countryside and setting fires had a purpose fused into their culture by over 50,000 years of experience. Managing the fire cycle with small fires allowed productive trees and wildlife to flourish, rather than having the buildup of large amounts of brush and trees combust in huge fires destroying all useful trees and game.

Anyone wandering the English countryside randomly setting fires, naked or not, would be arrested. In eighteenth century Australia, they were often simply shot. It was not a linguistic issue or a matter of translation. Even an absolutely clear explanation of the importance of managing the fire cycle to balance the environment would most likely not have been processed by English brains raised in a society from a verdant and temperate land, which held man's absolute dominance of nature

Fig. 1.2 JD Kelly thecanadasite.com John Cabot—The Europeans arrive in larger ships and stay

to be paramount. Nor would the concept, even if understood, have been readily accepted by the administrative infrastructure of an empire whose mission was to make the world England.

The farming and animal husbandry of the Europeans allowed them to establish large and stable settlements in which their numbers rapidly grew and from which they spread across the Americas. Amerindians were simply uprooted, driven off, and displaced as their traditional hunting grounds were turned into inhospitable and increasingly densely populated landscapes.

In short, the Amerindians of the new World suffered a decline in population in the range of 95% in the period between 1500 and 1650. During this "Great Dying," between 40 and 90 million lives were lost due to warfare, slavery, displacement, disease, and outright genocide. They were replaced by a burgeoning population of European settlers who, like all human populations, when placed in a rich environment, reproduced rapidly (Fig. 1.2).

Immigration was a side note as domestic population growth reached over 3% in many colonies meaning that their population doubled every 25 years. This population pressure pushed the frontier through Lower Canada, and in the USA, over the Appalachians and across the Midwest, finally ending in California, Oregon, and Washington.

The quest for good farmland spread the population over the continent. As Europeans displaced native Amerindians, colonized the Americas, and exploited the verdant resource base, they applied their advanced technology to build sophisticated societies and they laid the groundwork for the dawning of the age of abundant energy. The advent of fossil fuels greatly facilitated the deepening of this extensive development.

The colonization of the Americas was not a new or unique process. About 4000 years before, the Beaker people from central and Western Europe conducted a colonization of the British Isles which ended in a 95% reduction in the population of the indigenous Neolithic people over a period of 150–200 years.[5] In that conquest from 2750 to 2500 BC, energy technology was not a likely factor but an edge in tools and agriculture were key as were the diseases the Beaker people (nee Celts?) carried with them (Reich 2018). Although the scale of the conquest of the Americas was absolutely exceptional in world history, its pattern was not. The ebb and flow of various tribes and peoples across time and the face of the planet almost certainly followed this pattern. The success or aggression of Europeans in their out-migration was due to their mobility and their technological edge (Crosby 2004).

By the late 1700s, European society blossomed in the eastern part of Canada and additionally had access to the whole of western and northern portions of the country. Indigenous people had been displaced from the best lands and their

---

[5] There are so many similarities with the European colonization of the Americas, perhaps this is actually a template for the large number of colonizations which have taken place down through human history.

1. The migrating people were more technologically advanced. This would have given them an edge in combat as well as being able to make better use of the lands they acquired giving them a population advantage.
2. In a time span of 150–200 years, the indigenous population had been reduced by 95%.
3. Upon the arrival of Europeans in the Americas, the native population was decimated by disease, war, displacement, slavery, and outright genocide.
4. Disease was the primary instrument of population decline.
5. The indigenous population almost disappears from the record of the most desirable lands and is pushed into much more marginal regions.
6. After several hundred years of decline the indigenous population stages a revival as they blend into migrant society and adapt to their new circumstances and technology.
7. With more advanced agricultural technology, the fields and forests in Britain might have appeared to the Beaker as the "virgin" lands of the Americas (particularly Eastern North America) did to early colonists.
8. The rapid growth of the Beaker people could be due to migration but also to a high rate of domestic growth as they exploited a rich new land. The annual population growth rate of the new American colonies approached 3% and more in the frontier areas. In short order, migration was no longer required to build the population base.
9. At 3% annual growth (doubling every 25 years), a population of 10,000 would grow to 2.5 million in 200 years.

Is this a template for some of the great migrations? Indigenous populations suffer decline due to climate/resource difficulties and migrants with superior technology and strengthening numbers arrive as the climate is improving. Either through technological advantage or sheer numbers or disease, the migrants fairly decisively brush aside the indigenous peoples and either wipe them out completely or marginalize them. Their own numbers, through higher birth rates then increase geometrically as they apply their advanced technology to their new resource base.

numbers had fallen precipitously. In addition to displacement, the equilibrium between humans and the environment had been dramatically altered by the advent of trade.

In the Amerindian culture, animals provided food, clothing, and a wide range of materials. There was no need to take more than what was required. In a trading economy, animals left alive constituted "money left on the table," and it made sense to take the whole even if only a small portion like the hide or pelt was used. This destructive commercial mechanism is alive and well today in the form of elephant ivory poaching and shark "finning."

The culture of mining resources that Europeans brought with them led to the catastrophic decline of a number of key species, most famously the buffalo. This once abundant staple of the Plains Indians was reduced from a population of 80 million to near extinction by 1880. Similarly the eastern caribou, beaver and the northwestern otter almost vanished and "the overkill of whales and walruses in the 1870s led to widespread starvation in the 1870s in the Arctic and sub-Arctic" (Romaniuc 2014). It is easy to understand why the Europeans were so ready to embrace the New World: they had essentially worn out their old one.

The first great Canadian transition, the march to dominance of Europeans had taken place in two centuries. The public infrastructure was now in place which could readily adapt to the second great Canadian transition—the rise of the fossil fuel energy society. This new society had a vastly altered social structure and a view of the relationship between humans and the land diametrically opposed to the balance so central to the culture of the Amerindians it replaced.

The Americas before contact were a world almost completely without harvested and stored energy. From Tierra del Fuego to the tip of Ellesmere Island, the Amerindians depended on the renewable resources of the land, crops, forests, wildlife, and fish for their sustenance and maintenance of their societies. Like all pre-fossil fuel societies ranging from hunter-gatherers to agricultural societies with large urban centers, balance with nature was a well-known principle that was infused into decision-making (Hayden 1972).

Those societies (predominantly in Central America) which ignored that basic rule grew quickly and collapsed even more quickly.

These sophisticated societies, which grew from the large populations with time to learn and experiment, developed large cities, advanced mathematics, and a wide range of technologies. They also fostered stratified social structures and as their populations grew at the expense of the health of the land, they set themselves up for agricultural crises which detached elites could neither comprehend nor successfully deal with. It followed that war, famine, and population crashes became a part of the historical record as it was in all other parts of the world.

No matter what the era, solar energy and latitude generally dictate the richness of the resource base and the effort required for survival (Fig. 1.3).

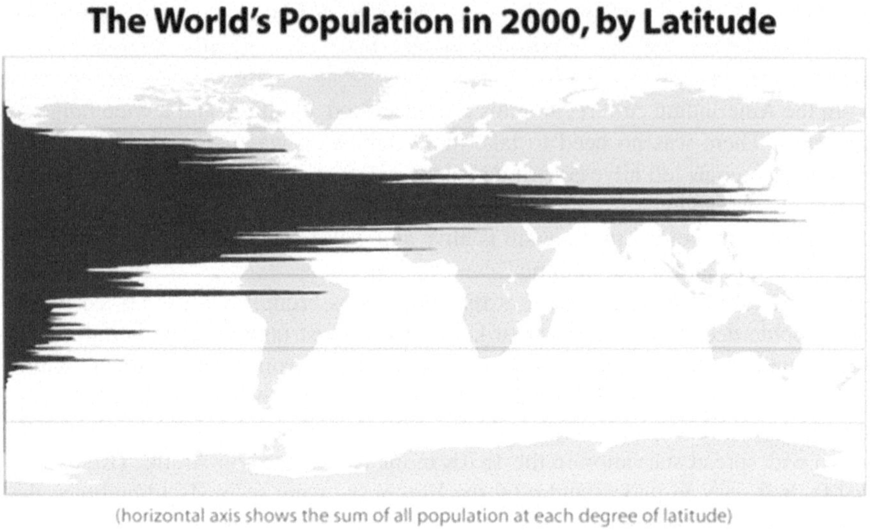

Fig. 1.3 World population by latitude from Graham Palmer

The Rise of Energy as a Separate Commodity

Fossil fuels, with their millions of years of stored solar energy, changed the game. They allowed us to make our own rules and establish our ever-lasting (to this point) preeminence over nature. As humans learned to develop cheap and abundant stored energy resources, our interface with nature changed. How much additional benefit was conferred on society by the use of the new fuels? That varied dramatically depending on the region.

In areas richly endowed with favorable climate and rich soils such as Central America, the use of fossil fuels underwrote a dramatic increase in both population and consumption. But these increases were not remotely as large as those experienced in more marginal lands like Canada and its far north.[6] There, fossil fuels did not simply augment the existing favorable environments but allowed large communities to prosper in regions where even bare survival was previously next to impossible.

When the explorers and colonists first set foot in Canada, they were awed by the extent of the resources, from fish to forests to wild game and minerals. It had been thousands of years since Europe represented such an unspoiled wilderness so full of potential. By the mid-centuries of the last millennia, those resources had largely been consumed. Populations had increased and resources had been degraded, leaving Europe crowded with armies of the poor and undernourished. In comparison, Canada understandably seemed like the "untapped treasure trove of unlimited natural resources" early explorers claimed it to be.

[6] For certain crops this may not hold true as noted by Charles Hall, "For maize, the yield is 4+ times more at 40° latitude than at the equator. The main reason is that in the growing season, the daylight time is twice the night time, allowing much more time when plants do not burn up their profit in nighttime respiration in warm temperatures."

At least, these resources were "Unlimited" for several generations. As the very limited farmland in the most southern regions of Canada were developed and filled up and as accessible forests were clear cut and animals overharvested, the burgeoning population found opportunities fewer and further between. Subsequent, ever larger generations of Canadians thus encountered limits which made foreign shores more attractive and from the early 1800s to the mid-1900s, except for a few years around 1910, Canada was a nation of emigration.

Canada was chock full of mineral, timber, and other resources but was a largely inhospitable land and the area that would support a community in comfort, even with European technology of the day, was very limited. So harsh were conditions that of the initial pulse of 15,000 French colonists to arrive in Quebec in 1600s, fully 5000 opted to return to the abysmal conditions in France they had fled just several years earlier. With a life expectancy of 30–35 years, if a woman was fortunate enough to survive until 20, she still stood a 35% chance of dying before she reached 45 (Beaujot and Kerr 2015).

Development was slow but the arrival of fossil fuels unlocked access to the vast resource base by making transport and resource processing possible and by providing food supplies and comfortable accommodations not possible before.

The Energy Lever

A similar pattern unfolded in the rest of the Americas but the "leverage" fossil fuels provided in the more marginal lands like Canada was vastly greater than it was in the more temperate lands.

At the time of European contact, Canada's population was between 250,000 and 500,000 people including 2000 Inuit in the Arctic spread over 10 million square kilometers. That is a population density of 1 person per 20 or 40 km^2. In the Arctic, there was one Inuit per 700 km^2.

The Americas as a whole were, on average, much richer and accommodating lands where the 80 million inhabitants lived on 43 million square kilometers meaning each inhabitant required just over half a square kilometer to survive. In the richest areas of Central America, the land was capable of supporting 120 people per square kilometer.

This dense population was fed in Mayan times by crop yields of corn of close to 1.3 metric tons per hectare. Today the average yield in Mexico is 3.25 tons/ha or 2.5 times the Mayan output.[7]

[7] – Mayan output https://www.uwlax.edu/urc/jur-online/PDF/2001/S_Fischbeck.pdf
- Guijarral terrace area was calculated to be 196, 747.0 m^2 or 19. 7 ha.
- This area multiplied by a potential corn yield of *1.3 metric tons per hectare* would yield.
- Approximately 25.6 metric tons of corn per planting. A conservative estimate of 66% occupancy of residential structures at Guijarral yielded a population density of approximately *124 people per square kilometer*, with an average occupancy of five persons per structure.
- Modern yield in Mexico—The USDA attaché's most recent annual report on the grains sector in Mexico forecast a 2016–2017 maize crop of 22.6 million tons from 6.95 million hectares.
- This equals *3.25* tons/ha and that is (3.25/1.3) *2.5 times* the Mayan output In Quebec, yield is 9 tons/ha or 2–8 times the Iroquois yield in upstate NY Iroquois yields footnote (22–76 bu/acre (1155–4127 kg/ha) In ideal New York fields.

In the better soils of Quebec and Ontario today, the yield, with higher levels of fertilizer, irrigation, and mechanization, is almost three times the current output per hectare in Mexico. This is seven or eight times the output of the ancient Mayans. How? Oil and oil-enabled technology.

The importance of fossil fuels—the new and greater abilities it gave the Canadian society—can be seen clearly in a comparison of the growth of human populations in the richest versus the harshest regions of the Americas. From a precontact population of 250,000, Canada's population has grown 150 times to 38 million. In the rest of the Americas, the population grew from a precontact level of 80 million to 1 billion—an increase of 12 times.

Population Density of Regions at Various Times in History Pre-1492 (Fig. 1.4 and Table 1.1)

The Cahokians built the most sophisticated society north of Mexico at the confluence of the Mississippi and Missouri Rivers, which is still over 400 km south of the most southerly point in Canada. The coast of British Columbia and its southern interior was the most densely populated region in Canada supporting perhaps as many as 250,000 people, over half of the indigenous people living in what is now Canada (Beaujot 2019).

The Iroquois were the largest tribe in northeastern North America and in the richest areas south of the Great Lakes, and the village size may have reached the 3000 person level. North of this in Canada, the average village was much smaller with several longhouses with no more than 150 people each and usually far fewer (Roots Native American 2011).

The only society to even approach the Cahokian was that of the Anasazi in the south western United States. The Anasazi built clay brick buildings and constructed scores of kilometers of irrigation canals through the 80,000 km^2 occupied by their 30,000 strong population.

The Vikings in Greenland had a large hunting range as well as domesticated cattle. Their settlement, begun in 985 AD was viable only in the warmest period of the last 2000 years and once temperatures declined slightly, the Greenland settlements, which may have had a peak population of 5000, declined and finally disappeared by the early 1400s. Vikings buried their dead but a trading ship, probably from Iceland, visiting the area in the mid-1400s, found the remains of a lone Viking male on a stone beach. Perhaps he was the last of his society and the one for whom no burial was possible.

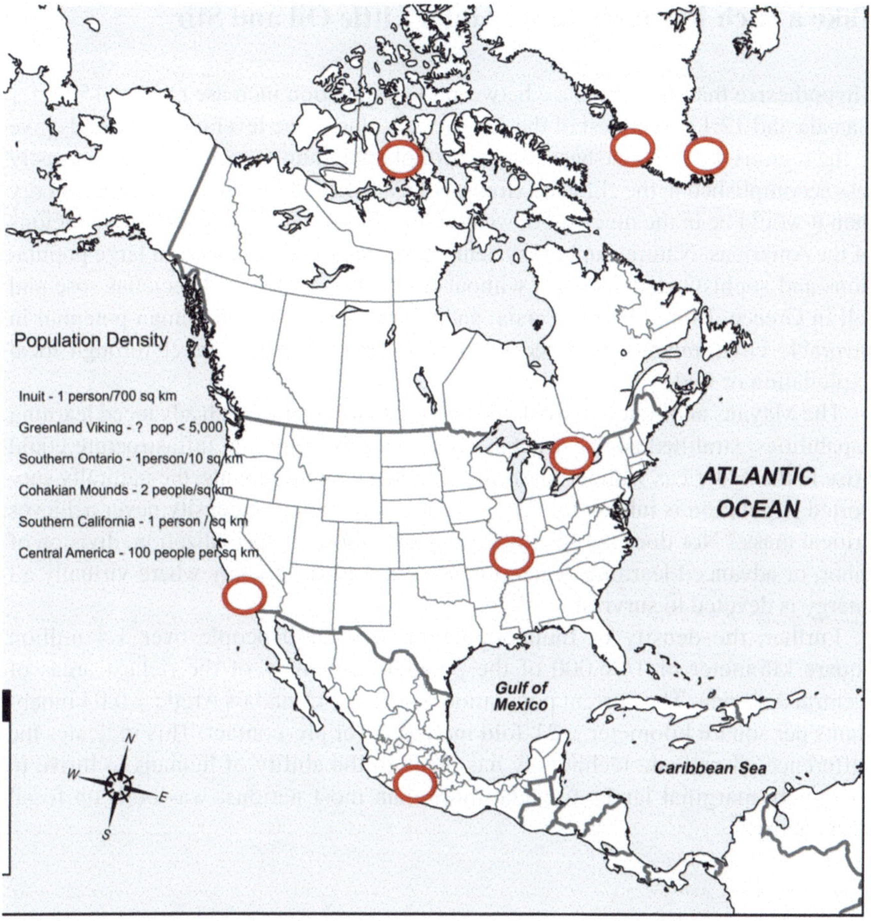

Fig. 1.4 Population density by latitude

Table 1.1 Population density

Region/people	Population density
Inuit	1 person per 700 km^2
Greenland Viking	1 person per 100 km^2?
Southern Ontario	1 person per 10 km^2
Cahokian	2 people per km^2
Southern California	1 person per km^2
Central America	100 people per km^2

Take a Rich Resource Base, Add a Little Oil and Stir

I hypothesize that the difference between the population increase ratio of 150:1 for Canada and 12:1 for the rest of the Americas, is due to the leverage fossil fuels give to the regions with the harshest climate. Simply put, human society would be vastly less accomplished in the climate extremes of Canada without abundant cheap energy than it would be in the much more welcoming climes of the more temperate regions of the Americas. Natural support systems in verdant lands can sustain large populations and sophisticated societies without fossil fuels. Advanced societies rose and fell in Greece, Rome, Egypt, Persia, and China and illustrated human potential in favorable environments with access to necessary resources, either through local exploitation or trade.

The Mayans and Aztecs have demonstrated that societies with advanced learning capabilities, stratified social structures and extensive physical infrastructure could arise in the Americas without fossil fuels. In harsh environments, the naturally supported population is inherently very low and the population density never achieves critical mass.[8] Nor does the free time necessary for craft specialization, division of labor, or advanced learning develop in a semi-nomadic society where virtually all energy is devoted to survival.

Further, the density of Inuit population was 2000 people over 1.4 million square kilometer or 1/86,000 of the population density of the richest areas of Central America. The current population density of Canada's Arctic = 0.03 inhabitants per square kilometer, a 21-fold increase over pre-contact. This indicates the difference energy and technology has made to the ability of humans to thrive in extremely marginal lands. Canada, more than most nations, was built on fossil fuel energy.

[8]The furthest north in the Americas the conditions for a highly developed society existed appeared to be at the confluence of the Mississippi and Missouri Rivers near present day St. Louis, MO. There the Cahokian Indians created a city which, in 1250 AD, was larger than London, England. The Cahokian civilization collapsed several centuries prior to the arrival of Europeans but evidence of their achievements can still be found in the in the form of their "Indian Mounds." These are very large and geometrically accurate pyramid-like structures with a very complex construction. The "mounds" are 30 m tall and cover almost 6 ha.

They are located 500 km further south than Canada's largest city, Toronto and over 1100 km south of the 49th parallel and 2500 km south of the most southerly Inuit range. To illustrate the environmental extreme this represents, 2500 km is about the distance from Toronto to Cuba.

This region featured moderate winters, access to large bison herds, forests and allowed agricultural production on a large scale. The collapse of this society may have been due to crop failure. But whatever the cause, it failed to the extent that when the first European explorer and Spanish explorer Hernando de Soto discovered them in 1540, no local natives knew who had built the mounds or what had happened to the builders.

The Most Energy Efficient Culture?

All civilizations, no matter the level of sophistication, grow to fill the environmental potential of the land and end up walking a fine line between environmental limits and rapid resource collapse leading to societal decline. But the true masters of living on the edge are the Inuit, the people who have survived for thousands of years at the limit of where human life is still possible. This highly adapted culture had 52 words for ice and snow (Encyclopedia Canadian 2019).

Picture the precontact Inuit band as an energy system in the high, treeless Arctic. The band members have no access to imported energy or energy stored by nature. The only energy available is that which is embedded in the flora and fauna around them in the form of the small amounts stored in fish, seals, caribou, and the very sparse vegetation that presents itself for a few weeks in the summer. Firewood might be available in the form of driftwood from thousands of kilometers away found washed ashore.

For the Inuit, lifespans were short but highly variable depending on the region and climatic period in which they lived. The average was possibly 35–40 years, and those short lives experienced very little leisure time. The band was sustained by constant effort working with the very limited resources and material they had available to make just the basic tools and clothing they needed to survive.

Although the winter diet of the Inuit contained no plant matter, food sources were complex ranging from seals to fish to whales and polar bears. With only one person for every 700 km^2, the density of available food and material resources was obviously very low so whatever was available, was fully utilized.

As an example of the most elementary energy system, let's assume that the seals were the sole source of food and that the average band was 15–18 people with four or five hunters. The Harp seal might average 130 kg in weight with their meat yielding about 1000 cal/kg.

Being a highly energetic people in a harsh environment, the average daily calorie consumption of the Inuit was approximately 3100 kcal or (3.4 kWh) compared to about 2500 kcal or 2.9 kWh for the average twenty-first century Canadian. Assuming that the seal yielded 100 kg of meat, and thus 116 kWh of energy, one seal would sustain one Inuit for 34 days.

If he were living on his own, a hunter would have to kill only one seal a month to survive for a short period. But long-term survival of the hunter requires much more than simply harvesting food. Clothing, tools, and shelter had to be made and maintained. Meals had to be prepared and children had to be born, raised, and educated. Long-term survival requires a community with a broad range of skills.

In the lone hunter model, one seal lasts for 1 month so the total expenditure of energy of the hunter exactly equaled the energy from his harvest. The ratio of the energy expended by the hunter to the energy he harvested would therefore have been 1:1.

But in a sustainable society, the hunter's effort had to provide a substantial surplus to generate sustenance for other members of the band. In the long term, if there was no band, there would be no hunter. Hunters, spouses, children, and maybe aged constituted a working, enduring community. Assuming there were three or four "dependents," if that term can be applied to the critical support structure, the math changes radically. Now a hunter has to harvest five seals, with an embedded energy content of 500 kWh, a month to maintain the band. The effort expended, the calories burned by the hunter would only be a fraction of his total harvest.

Energy Return on Energy Investment: EROI, the Critical Metric

The math works out to five seals (~500 kWh) harvested for an expenditure of 30 days of effort (~100 kWh) for a ratio of around 5:1. The term used in energy measurement is Energy Return on Energy Invested (EROI, or "EE-RRR-O-I" for short). This is the ratio between the energy harvested and the energy used in the harvesting process. It is an absolutely critical metric for both our early hunting society and for our advanced and complex societies which are built on very high levels of energy consumption.

For the hunter to provide for the rest of the band, he had to harvest five times the energy he used himself so the net energy he produced was 4, (5 minus his own consumption of 1).

With four band members per hunter and the resultant total need of the five persons depending on the output of one hunter, it can be said that in the harshest environment on the planet, the minimum EROI for survival of a people was 5:1. This energy balance allowed for survival but no accumulation of infrastructure or learning. Such a society could endure in perpetuity and, given a climate stable within a certain range, we would find the same hunter-gatherer groups with essentially the same tools, living in the same area, just as the previous thousands of generations had.

The populations would have been too small and their tools too rudimentary for them to substantially affect their environment which would remain stable as well, fluctuating only with the climate. So, with extremely limited access to a very large but low-density resource base, the Inuit could indeed endure but they would never have the free time or the energy stores or the access to the necessary broad resource base to develop the population size and learning infrastructure that would allow them to ultimately put a human on the moon. They were a truly resilient culture but trapped by resources and climate in eternal technological stasis.

The number of seals harvested annually by 2000 Inuit would have been on the order of 18,000. The precontact Harp seal population in the Arctic was in the neighborhood of 12 million making the Inuit harvest sustainable for as long as the Arctic climate endured.[9]

But relocate this tight, disciplined and focused group from their very sparse resource base into an environment rich with resources requiring far less daily demands in terms of shelter, clothing, and food consumption. Central America, presented such an environment with consistent and moderate climate and huge stocks of readily accessible soil, minerals, forests, and wildlife. The available resources there were vastly higher and the effort needed to survive in terms of shelter and clothing was vastly lower.

Migration was not required. Houses were permanent and any effort put into infrastructure and more sophisticated tools and learning would endure and accumulate. Not only would our transplanted Arctic band now have the ability to develop their technology continuously but they would also have the ability to increase their population. They could then eventually overwhelm their resource base by extracting ever greater harvests of crops from the soil, wood from the forests, and minerals from the richest deposits.

At some point, their exploitation of the resource base would begin to produce diminishing returns for the effort they made and the society would become more focused on basic survival activities than on the non-essential religion, learning, and art. When the climate swung into an unfavorable cycle, stress would increase on the already stretched resource base exacerbating the resource–demand tension of a society depending on a diminished resource base to support a large population.

The inevitable result was higher mortality rates, war, and social decline. Societies are based on the wealth of their resources, which can be roughly divided into the sectors of climate, energy, minerals, biomass (plants, animals, and fish), and geography.

The rest of Canada south of the Artic featured a highly diverse web of bioregions which varied greatly in their ability to sustain human populations. The west coast was a rich region with large food resources from the ocean, tremendous stands of forest in the hinterland, and a moderate climate. This resource base allowed year round food supplies to be maintained while abundant wood resources assured the cooking and heating fires would never go out. Permanent settlements could be established which could support relatively sophisticated societies in terms of build-

[9]– Seal notes
- 1 harp seal = 130 kg or 100 kg for 100,000 cal or 116 kWh
- Seal meat averages 1000 kcal/kg.
- 1 food calorie = 0.00116 kWh
- kWh per seal = 130,000 kcal $*$ 0.00116 = 150 kWh,
- Average calorie consumption per day of Inuit = 3100 (3.6 kWh).
- Therefore 1 seal = 42 days of food.
- Average consumption = 9 seals per person per year.
- Precontact Inuit population = 2000.
- Total seals harvest annually = 18,000.
- 1 barrel of oil = 1628 kWh
- 1 seal = 150 kWh = 0.09 barrels of oil.

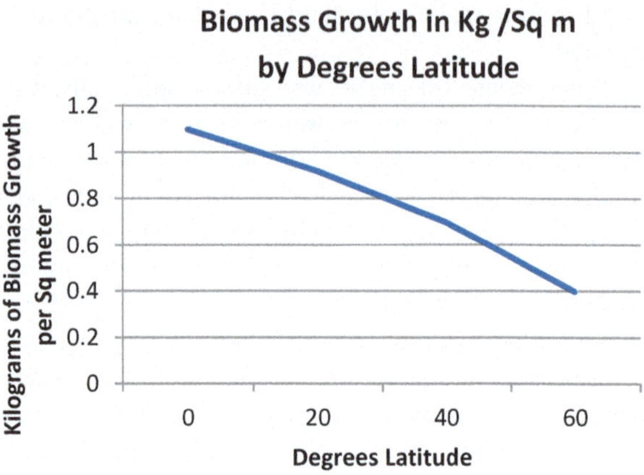

Fig. 1.5 Latitude, productivity, and species richness (Gillman 2014)

ings, art, and culture. Amerindians occupying the Pacific shore thus had the available time to accumulate knowledge and develop advanced food storage and preservation technology (Fig. 1.5).

The ability of a region to support a human population depended on the amount of food it could produce. This was strongly influenced by latitude with higher latitudes generally producing much more sparse vegetation. For comparison, in Canada, the average timber biomass added to the forest per hectare is about one-third that of the USA and 1/30 that of the Amazon jungle.

The agricultural land along the St. Lawrence, while not as rich as the waters and forests of the Pacific northwest, was much more hospitable than the vast area north of it inhabited by a few Amerindians "whose numbers were proportionate to the scantiness of the soil" (MacKenzie 1911).

The mighty Iroquois tribe which straddled the USA/Canadian border around the Great Lakes may have counted 60,000 souls in its society over tens of thousands of square kilometers of some of the best land in Canada. Although they practiced slash and burn agriculture, moving their village perhaps every 20 years, their numbers were small enough that the soil and forests were easily able to retain their health.

Across Canada, the availability of natural resources sustained widely different and fluctuating levels of population but in no area did it reach the point at which technology or science could flourish.

Necessary Conditions for a Sophisticated Society

Sophisticated societies, featuring writing, organized religion, public service, and organized learning, have developed in many parts of the world over the past 6000 years but these were necessarily agriculturally based. The richer the soil, the denser the population and the greater the division of labor and the stratification of the social order. Thomas Homer-Dixon in his book "The Upside of Down," showed how energy flowed through the Roman Empire. It was an empire limited by the ability of the agricultural system to support human labor. And the Roman agricultural system, which found it easier to enslave rather than innovate, depended on slaves whose average life expectancy was a mere 26 years. Although slaves were first used en masse in agriculture, they were soon applied to supporting the lifestyle of their masters in every conceivable manner. From artisans, to domestic servants to engineers and educators, it took scores of slaves to support the lifestyle of comfortable Roman citizens and thousands to maintain the opulent lifestyle of Rome's most powerful leaders.

About 1500 years later, when energy itself became readily available in its raw form, society was able to take a leap into the technological era in which the amount of work that could be done was multiplied by a factor of over 100. The limitation of biomass output was left behind as we learned to burn the earth's stores of fossil fuels. First coal, then oil and natural gas launched our civilization. Coal and oil energy became not only cheaper than slaves (Saraydar 1989) but more capable. The economic need for human slaves disappeared and their place was taken by what Andrew Nikiforuk termed "energy slaves."[10] And it seems that as we move "forward," humans are adding digital "thinking slaves" to complement their physical energy slaves (Nikiforuk 2014).

When Europeans arrived with their advanced technology and were able to exploit coal deposits in the Maritimes, the process by which technology could be spread and infrastructure developed across the country had begun.

Coal Fueled the Railway, the Railway Built Canada

A few notes.

- On November 7, 1885, the last spike was driven by Canadian Pacific director Donald Smith in to complete the first Canadian transcontinental main line at Craigellachie B.C. in the Eagle Pass.
- William Van Horne took the occasion to make his famous 15-word speech "All I can say is that the work has been well done in every way." Long on doing and

[10]
- In 1824, steam power did the work of 750,000 men
- Coal added the work of three billion slaves by the late 1800s.
- A man's daily labor was the equivalent of burning 4 lb of coal.
- The granite in the great pyramid of Egypt which took thousands of slaves and decades to build could be raised by the burning of the same amount of coal used by one large foundry in a week.
- Admiral Rickover—every American effectively used the equivalent of 244 manufacturing slaves.
- David Hughes—healthy individual on a bike riding an 8 h day with normal holidays would take over 7 years to output the energy contained in 1 barrel of oil.
  - A slave riding 16 h a day with no holidays would take 3.8 years.
- The average North American consumes 24 barrels of oil. Employs 89 virtual energy slaves.

short on talking, obviously a career in politics was out of the question for Van Horne.

- By that time the population had grown to 4.5 million.
- This was 10–15 times the size of the original Amerindian population which was then 107,000.
- Before the fossil fuel revolution, the time to travel from Montreal to the Pacific coast was measured in months, likely with a period of overwintering at some point along the route. The travel time of the first train trip from Montreal to the Pacific terminus of Port Moody was 5½ days.
  - The train would have been capable of pulling close to several million pounds (1000+ tons) of freight
  - The capacity of a six man voyageur freight canoe was 3 tons.
- It was by far the longest railway ever constructed at the time. It had taken 12,000 men, 5000 horses, and thousands of tons of dynamite to construct the Lake Superior section alone.[11]
- From Ottawa River to Pacific required 2500 miles or 4000 km of new line
- In total, it took over 30,000 workers to build the CPR 3200 km in 4½ years

After oil gushed out of the ground in Southwest Ontario, the full potential of Canada began to be exploited. Oil lit the fuse of the meteoric rise of technology, development, science, and population both in Canada and the world. Oil thrust human society into the lofty position it now occupies and will now have to endeavor to maintain.

When the first European colonization of Canada began in earnest, the demographic makeup consisted of scattered bands of Amerindians living sustainably in a late stone-age technological stasis. Two hundred and fifty years later, the first great transition was complete as a country with a unified government, a large and rapidly growing population of Europeans and coast to coast transportation infrastructure stood poised to exploit the rich resources of this vast land.

With the advent of fossil fuel energy, life in Canada was able to move beyond the economics of mere survival. Lady Agnes MacDonald, the wife of Canada's first Prime Minister, the railroad builder Sir John A. Macdonald, clearly no wilting lily, was so enthralled by the majestic views in the Rockies she insisted on riding the cowcatcher on the front of the locomotive for many hours.

---

[11] – Dynamite—a later development related to gunpowder but differing in its energy density (much higher) and in its speed of reaction—it was a "high velocity" explosive. Again, it required large amounts of energy to make but it offered energy in very concentrated form which could be focused to apply a huge amount of force in a very short period of time. Given its speed of detonation, dynamite was good for shattering, blowing things apart whereas its much slower cousin, gunpowder, was ideal for propelling projectiles at high speed over long distances.
  - For comparison, gunpowder contains 3 MJ/kg, dynamite contains 7.5 MJ/kg, and gasoline contains 47.2 MJ/kg (though gasoline requires an oxidant, so an optimized gasoline and O_2 mixture contains 10.4 MJ/kg).
  - TNT has a detonation (burn) rate of approximately 5.8 km/s (ten times faster than the fastest gunpowder) black powder, or smokeless gunpowder has a burn rate of 171–631 m/s.

Fig. 1.6 Members of the royal tour of 1901 during a stop in Glacier, British Columbia. (Library and Archives Canada). "Here are more details to last week's story of Prime Minister J.A. MacDonald and Lady Agnes MacDonald's trip on the Canadian Pacific Intercontinental Railway to the British Columbia coast. At one stop at Lake Louise, Lady Macdonald was studying the broad front of the locomotive with the cowcatcher and decided that it would be exciting to ride from that vantage point of the locomotive while going through the Kicking Horse Pass. The Prime Minister was not keen on the idea, but the railway superintendent thought it best she be accompanied and so he rode with her after arranging secure seating for both of them. She was fascinated by the journey, and later the Prime Minister accompanied her on the cowcatcher, although he said he preferred the comfort of the observation car. Lady Agnes MacDonald wrote of her experiences on the railway trip through the Rockies in an article entitled, "By Car and Cow Catcher" in Murray's Magazine." The article, and a photo of the MacDonalds, may be seen in "Tales of the Canadian Rockies" by Brian Patton.

The thrill of effortless movement on a massive vehicle across a beautiful, and now unthreatening world, opened up a new aspect of life for a large part of the population; travel for pleasure. Survival in the natural world had given way to a level of dominance never before possible in human experience (Fig. 1.6).

References

Armelagos G, Cohen MH (2013) Paleopathology at the origins of agriculture.
Beaujot (2019) private email communication
Beaujot R, Kerr D (2015) Population change in Canada, Oxford, Canada
Belloc H (1898) The Modern Traveller.
Crosby AW (2004) Ecological imperialism: the biological expansion of Europe, 900–1900
Encyclopedia Canadian (2019). https://www.thecanadianencyclopedia.ca/en/article/inuktitut-words-for-snow-and-ice
Gillman LN (2014) Latitude, productivity and species richness
Hayden B (1972) Population control among hunter/gatherers
MacKenzie (1911) Chevalier de La Salle – MacKenzie, vol II, 1911:343
Mann CC (2005) Native Intelligence, Smithsonian Magazine. https://www.smithsonianmag.com/history/native-intelligence-109314481/
Mann CC (2006) 1491, Knopf, New York
Milton K (2000) Hunter-gatherer diets—a different perspective
Nikiforuk A (2014) The energy of slaves. Greystone, Vancouver
Reich D (2018) The Beaker phenomenon and the genomic transformation of northwest Europe
Romaniuc A (2014) Aboriginal population of Canada. University of Alberta, Edmonton
Roots Native American (2011). https://nativeamericannetroots.net/diary/1081
Saraydar E (1989) The conflation of productivity and efficiency in economics and economic history
Steckel RH, Rose JC (2002) Skeletal health in the Western Hemisphere from 4000 BC to the present
Steegmann AT Jr., Haseley PA (1988) Stature variation in the British American colonies

Chapter 2
Canada's Energy History

Abstract Sophisticated lifestyles require large amounts of energy in northern regions. Indigenous peoples clung to survival through the consumption of energy embedded in the food they ate and the wood they were able to gather. This, and the rudimentary tools they were able to use to build shelter and fashion clothing enabled them to live for thousands of years at roughly the same level of technical development their ancestors had when they crossed over the land bridge from Asia during the end of the last Ice Age. But the level of energy and resources available, combined with the high demands of northern climates did not allow these peoples to accumulate tools and processes past the development stage of hunter-gatherers. Their cousins, who journeyed far to the south to Central America, were able to prosper, as indicated by their enormous population growth, and develop very sophisticated cities and stratified cultures on the entirely different energy/resource potential of the region and its low energy requirement.

Implications of Northern Geography

Canada's landmass covers close to 10 million square kilometers over which a great deal of sun shines, wind blows, and rain falls. These, along with their by-products, vegetation and wildlife, were the energy stocks and flows which the Amerindians and early settlers exploited to survive in this challenging environment.

The world's longest coastline encloses a panoply of geoclimatic regions. Canada is slightly less than three-fifths as large as Russia and marginally smaller than Europe. In total area, Canada is larger than either the USA or China but ranks fourth in land area once our world's largest stock of fresh water lakes and rivers are deducted.

Wind and solar radiation are diffuse energy sources, but the stored energy of biomass is much more concentrated, and, flowing water in rivers vastly more dense still. Despite the immense energy flows over the surface of the country, the current population of 38 million is still relatively sparse on a world basis. However, when viewed from the perspective of where the great majority of population lives, the most southerly 10% of the country, Canadian population densities come much closer to the world average.

J. E. Meyer, *The Renewable Energy Transition*, Lecture Notes in Energy 71,
https://doi.org/10.1007/978-3-030-29115-0_2

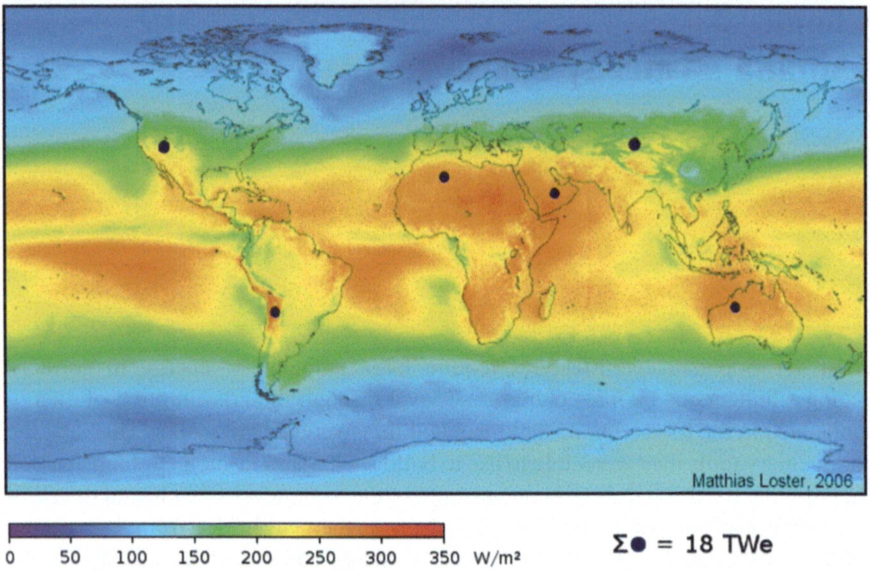

Fig. 2.1 Solar radiation potential map (Loster 2006)

Even the southernmost concentrations of population are possible only through the heavy use of fossil fuels. The northernmost settlement in Canada, and in the world, is Canadian Forces Station, Alert, on the northern tip of Ellesmere Island, just 800 km south of the North Pole. It owes its existence to the use of large amounts of oil for heating, electricity, and transport of food and supplies. Survival of humans, with the ratio of available harvestable energy to minimum energy requirements, would not be possible at that latitude.

Canada has a large land area but its northern location dictates that the solar energy we receive per square meter is much less intense than that received by countries closer to the equator (Fig. 2.1).

Above is the average Watts per square meter that reach the earth's surface. Clearly the northern regions receive far less solar energy than more equatorial land masses. Not only is the average solar energy much less but also the seasonal variability of that energy is far higher than regions closer to the equator.

Below we see the differences in temperature variability between Igloolik, latitude 69° (3° above the Arctic Circle) and Guadalajara, Mexico, at 20° latitude. In Igloolik, solar energy drops to essentially zero for months at a time. Guadalajara requires minimal energy for heat, year round, while its energy level remains very even throughout the year (Fig. 2.2).

Below is a Solar Capacity Chart for the two towns. Capacity factor represents the ratio between the rated maximum output of a solar panel installation and its actual output over time. If a typical rooftop photovoltaic installation was rated at 10 kW maximum output, given an hour of strong sun, it would produce 10 kWh of energy. If that system had perfectly strong sun for 6 h a day (1/4 of the 24 h day), and total

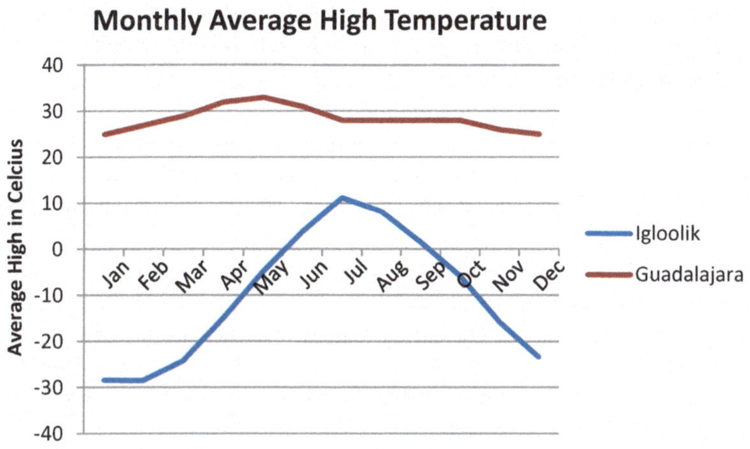

Fig. 2.2 Monthly average high temperature, Igloolik, Guadalajara (Degreedays 2019)

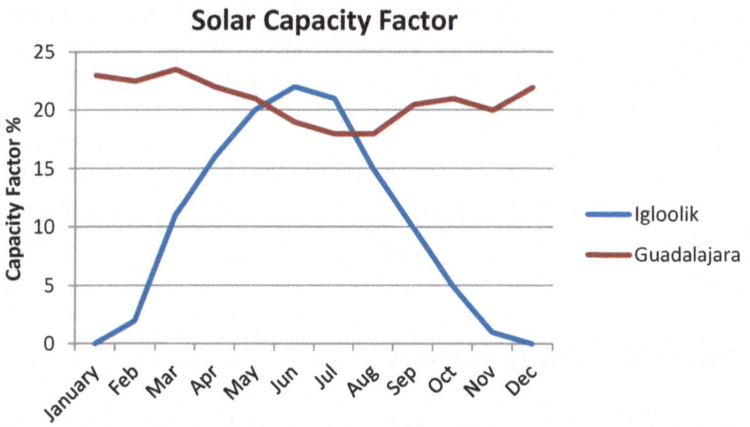

Fig. 2.3 Solar capacity factor by month, Igloolik, Guadalajara (NREL 2019)

darkness for the remaining 18 h it would produce one-fourth of the energy it had the capacity to produce for a capacity factor of 25%.

"Capacity Factor" represents the percentage of power an energy source, in this case solar PV panels, can actually produce in an installation compared to what it could produce under ideal circumstances. Here it is possible to see the difference latitude makes as the capacity factor for Guadalajara is very high and consistent through the year. Conversely, the northern location can generate significant energy for only 6 months a year and almost none at all for 4 months a year when energy is needed most, due to the extreme cold. The energy requirements needed for a healthy, comfortable life are also vastly lower all year round at the temperate latitudes than in the extreme north (Fig. 2.3).

Igloolik represents the northernmost Inuit precontact settlement while Guadalajara is at the same latitude as the most densely populated Mayan and Aztec regions. The average monthly temperatures of the two locations gives an indication of the extreme differences in energy required for humans to exist. The growing season in the far north is a matter of a few weeks making any form of agriculture impossible; hence most food energy is derived from the sea. In contrast, plants and animals can grow all year round at more southern latitudes with the main restriction being the water cycle.

Still, given its land mass, Canada receives a great deal of solar energy and added to the power embedded in the winds and flow of the rivers, it can be said that we live in an energy rich environment. The ability to take advantage of that energy is dependent on the level of technology available. Before the scientific leaps of the last two centuries, harvesting energy directly from the environment was minimally possible through waterwheels and windmills.

Energy for basic survival can be harvested from renewable sources in a wide variety of ways. Whatever the source, the end products for humans are food, heat, and light. Heat and some food can be released directly from the stock of biomass by burning wood and plant material. Food, heat, and light can be pulled from the sea in the form of fish, seals, and whales. Food can be harvested from grazing animals, as they have harvested the energy from plant matter and represent a stock of concentrated energy and nutrients.

But the native populations did not have the technology to harvest energy directly from the primary source, the sun, or from the next-level energy systems of moving air and water. They lived off the stocks of energy in the form of plants and animals, not the flows of energy carried by the wind and water.

Survival and Energy in Canada

The history of energy use, and extraction in Canada, post European contact, is one of the exceptionally high energy use demanded by the country's harsh climate and a plethora of energy resources with widely divergent densities and levels of accessibility.

Virtually every region in Canada requires less energy to survive than the Arctic, and yet the Arctic, with its vast expanses of tundra, possesses the least rich energy resources. Small wonder it supported only a tiny population of humans. Few trees of a useful size grow above the 60th parallel and, even in the modern age, few fossil fuel sources have been developed.

Certainly there were none available to the Inuit who were initially known as "Eskimo," meaning "eaters of raw meat." Since it has been shown that the process of cooking meat allows the body to absorb 30% more nutrients (Carmody 2016), clearly, if there was firewood available, the Inuit, like all hunter-gatherers, would have cooked their food. But this was not always possible given the resources at hand.

In the Arctic, most of the energy sources were not even stationary, (i.e., embedded in the form of trees and vegetation to be burned as external heat), but were either walking or swimming in the form of caribou, musk-oxen, polar bears, fish seals, walrus and whales which had to be ingested. Unlike the Giant Moa of New Zealand, which allowed humans to simply stroll up and club them, animals in the Arctic had become wary of humans over many millennia and needed to be stalked. This required the expenditure of a great deal of energy, except in the rare cases of polar bears hunting humans. In these instances, the "food source" came to the "hunter."

Still, the Inuit endured and developed one of the highest grade forms of energy available to any society: seal oil. With an energy density similar to gasoline, (roughly 10 kWh/L) seal oil offered heat, light, and sustenance to the small bands of Inuit in an easily transportable form, ideal for their lean, nomadic life. Medicinally, it could be used as a salve for wounds and skin irritations as whale oil was for the British troops in World War I, where it was liberally applied to bare feet to prevent trench foot. And as a currency for trade, it offered compactness, consistency, and scalability with high intrinsic value—virtually the perfect currency.

Seal oil and the closely related whale oil took effort and energy to produce although, as in the example of gunpowder and colonists, Inuit society did not run on it, but rather used it for specific, vital tasks. Seal oil presaged the advent of crude oil in its power and wide range of uses. If seal oil had been easy to obtain in large quantities, the history of the north might have been very different. Although highly constrained by one of the poorest resource bases on the planet to which to apply their pure energy, given large amounts of it, it is quite possible the Inuit would have been able to slowly develop their way out of their fragile subsistence existence.

Other native groups in Canada had the luxury of forests from which to draw firewood and although severe in many regions, the environment was still less harsh than in the Arctic. Some native groups had access to outcrops of coal in the Maritimes and oil sludge or tar in Alberta and Southwestern Ontario. How much of these were used though is an open question as the smoke would have made using these fossil fuels in open fires in lodges or teepees an unsavory prospect.

Hearths and chimneys minimized this problem for early settlers as far as heat was concerned, and stoves may have allowed these fuels to be used for cooking. The Hopi Indians of the American Southwest built permanent adobe houses featuring chimneys and this tribe was known to use coal. According to historian William Ascarza, "The earliest miners in the area were the Hopi, who mined 100,000 tons of coal between 1,300 and 1,600 A.D. for firing pottery and heating fuel. The Spanish explorer Agustín de Vetancurt described the mission at a village called Awatobi in 1697, noting that the Hopis burned coal, giving off noxious fumes."

The Quest for Raw Energy: Settlers Look Past Fish, Forest, and Fur

In the early days of European colonization, the first raw energy harvesting took place via windmills and water wheels. It was at this point in the Americas that raw energy was finally established as something apart from the animal, fish, and biomass forms which had previously been its only sources. But vegetation and animals were beginning to lose their monopoly on energy supply in the human–energy relationship.

Wind and Water

Historically, the waterwheel was almost certainly in use by 4000 BC in China and the Middle East and vertical models were created during Roman times. The initial uses included irrigation and milling grain but uses later expanded to drinking water supply, lumber milling, pumping of water for drainage and air for forge bellows, as well as the powering of textile mills. Wind and water drove the first means by which mechanical energy from the environment replaced that of humans and animals.

In terms of New World timeline, the waterwheel preceded coal slightly but it was used almost exclusively for agricultural purposes, whereas coal and steam were mainly applied to heat, transportation, and heavy industry. The waterwheel had the advantage of being completely renewable, and after the initial capital investment of dam, sluice and wheel mechanism was made, using mostly local materials, the facility could do work the work of many men and oxen with little additional input for many decades.

The first Canadian water-powered grain mill commenced operation in 1607 in Port Royal in the prime farm country of the Annapolis Valley in Nova Scotia. It would have been needed for a number of months a year, for as long as a supply of grain was available, but it could not have operated during severe low temperatures or when water levels were low. Although the energy for a water wheel did not have to be transported to the mill, all of the raw material the mill was processing did have to be transported. Predictably, areas near a good source of water power developed far more quickly than those without ready access to water power.

Where water could not be depended upon as a source of power, in many areas, wind could drive the crude machinery that evolved in early tech-savvy agrarian societies. The first windmills originated in China and Persia turning their pumps, grinding their grain, and sawing their timber. In terms of absolute power and uninterrupted operation, watermills had a distinct advantage over windmills, but not every community had access to large flows of water rapidly descending steep slopes or cliffs.

Also, wind can blow on any one of 365 days a year, whereas the flow of water, particularly in wintery climates can fluctuate seasonally. Early windmill designs featured sails on a vertical axis, like a children's playground merry-go-round, which required space and dictated they be close to the ground. By the early middle ages, from the Eastern Mediterranean to the British Isles, horizontal-axis windmills, with their whole surfaces facing the wind, had emerged. These took up far less space and also elevated the blades or sails far above the ground where the wind currents were stronger and more uniform. The horizontal shaft to which the blades were attached then drove a vertical shaft which transmitted power from the top of the tower to the grinding or pumping mechanisms on the lower floors of the building.

Windmills were built into the fortifications of early Quebec City. However, no people had the critical level of need for the work windmills could do than the Dutch. The modern Netherlands features a land area 26% below sea level while only 50% rises higher than 1 m above sea level. In this environment, starting in the 1300s, the windmill was aggressively developed by the Dutch who used them to pump water from their marshy low-lying lands and claim large swaths of superbly rich soil for farming.

> The Dutch essentially affixed the standard post mill to the top of a multi-story tower, with separate floors devoted to grinding grain, removing chaff, storing grain, and (on the bottom) living quarters for the windsmith and his family. Darrell Dodge, historian.

Wind power, as demonstrated by the Dutch, was ideal for the application of pumping dry, large tracts of land as it was not a time or season sensitive application. The work could be done on the wind's timetable with little accommodation needed for human or crop schedules. The same meshing of capability and need occurred on millions of farms across North America where small steel windmills pumped water for decades on their own schedule and without a drop of fossil fuels or a minute of human effort. These simple trestle towers topped by wooden or steel paddle-type blades were ubiquitous in North American farmlands.

> Between 1850 and 1970, over six million mostly small (1 horsepower or less) mechanical output wind machines were installed in the U.S. alone. The primary use was water-pumping and the main applications were stock watering and farm home water needs. (Dodge 2014).

Waterwheels and windmills spread out through rural areas while coal followed the railroad to the mines and factories, which had large powerful steam engines or furnaces that required a continuous supply of very energy dense fuel. They needed to run on demand, for as long as required without the daily or seasonal fluctuations in power inherent with water and wind power generation. And coal could generate the heat required for the increasing number of industrial processes which involved an ever larger list of mineral components and advanced alloys.

Above ground, there are dispersed energy flows. Below ground, there are stores of dense. Wind and water cemented the agricultural base of Canada but fossil fuels lit the fuse of dramatic industrial, technological, and social development which sustained a larger and growing population.

Buried Energy: Coal

> In Roman Britain, the Romans were exploiting all major coalfields (save those of North and South Staffordshire) by the late 2nd century AD. While much of its use remained local, a lively trade developed along the North Sea coast, supplying coal to Yorkshire and London. This also extended to the continental Rhineland, where bituminous coal was already used for the smelting of iron ore. It was used in hypocausts to heat public baths, the baths in military forts, and the villas of wealthy individuals. Excavation has revealed coal stores at many forts along Hadrian's Wall as well as the remains of a smelting industry at forts such as Longovicium nearby. (Wikipedia 2019a; CoalScience 2018).

With the arrival of coal savvy Europeans in the Americas, its development proceeded swiftly. But the rise of coal in the Americas was driven by more than the need to heat dwellings and cook food. Coal enabled new work to be undertaken, new processes to be developed, and new products to be produced and transported.

The Hopi Indians took pottery to a higher level with the use of coal since it offered greater consistency and higher temperatures with less work. But it was only with the arrival of the metal-working Europeans and their established coal-fired forge technology, that the demand for ever increasing quantities of coal drove the Americas and Canada into the fossil fuel age.

Fish, fur, and timber led to the initial European settlement of Canada, followed in short order by farming. Technology was a marginal factor as Europeans spread into the interior through sweat and sacrifice, working to overcome nature with essentially hand tools. But coal allowed improved tools to be developed, which then morphed into powerful machines and newer, larger, and faster methods of transport. Coal fueled the railway networks, which allowed access to every major resource center across the land by more people and more equipment, while providing an outlet for timber, minerals, and food stuffs to large overseas markets. Coal opened the door to continental transport and industrialization and intercontinental trade.

There are four major types of coal denoting their energy content but there are only two types of uses—metallurgical and thermal. Anthracite: The highest grade coal, black and lustrous, is often referred to as hard coal or metallurgical coal. Bituminous coal has a high heating (Btu) value and is the most common type of coal used in electricity generation in the North America with sub-bituminous coal is dull with a higher heating value than lignite. Emissions per unit of energy output increase as the grade of coal decreases.

Peat is not actually coal, (it may be a precursor) but was often used in some northern regions, notably Scotland, as it was the best fuel available. Canada has a huge amount of peat of highly varying qualities mixed in to its 1.2 million square kilometers of muskeg. Peat, being just an accumulation of biomass at various stages of decay and must be dried before use. It can be thought of as coal in its early stages. Add several score million years of extremely high pressure and temperature and peat ultimately transforms into coal.

Although it had been used in early Roman times for heat for water, and buildings as well as for smelting in iron production, coal's potential for transport was not unlocked for another 1800 years. When coal was finally applied to transport, the Americas and Canada in particular were opened up to an industrial level of development.

Coal was first mined in Canada 1639 at Grand Lake, New Brunswick. In 1720, French soldiers began to mine at Cow Bay (Cape Breton, Nova Scotia) to supply the fortress at Louisbourg. Cape Breton later supplied coal to Boston and other American ports, and to the militia in Halifax. In Cape Breton, locals could gather lumps of "sea coal" (yes, some coal floats) washed up on shore from underwater or shoreline seams. A small bag of 10 kg could be quickly filled and its 80 kWh of energy was enough to keep a modest home with a reasonably efficient stove toasty for an entire winter's day.

By 1870, 21 collieries (a coal mine and its connected buildings) were operating in Cape Breton. Commercial coal mining in New Brunswick began in 1825 and, except for some early exports, most of the province's coal production was used locally. In western Canada, coal was first mined on Vancouver Island in the mid-nineteenth century. The building of the transcontinental railways through Alberta and BC allowed coal mines to be developed on the banks of the Oldman River near Lethbridge, at Banff, Drumheller and Edmonton.

By 1867, coal production had reached an annual total of 3 million tons. This translates to a huge energy lever being applied to every undertaking necessary for human endeavor. How much of a productivity boost did this give to the then Canadian population of 3.5 million? Immense.

- 3 million tons = 3 billion kg
- At 8 kWh per kg, 3 billion kg × 8 = 24 billion kWh
- At 3.1 kWh per person per day = 8 billion person days of labor
- This equals the effort of 22 million people working all year long.

The Canadian population, 3.5 million strong, now had at its disposal, the energy and the work that 22 million additional people could do.

What was the attraction of coal over wood or other methods of energy acquisition? In a few words, energy density, EROI (E-Roy) and effort. Coal came out of the ground ready to use. A lot could be produced in a short period of time with simple tools and the workforce did not have to pick up and move every few weeks looking for new sources. Its EROI was extremely high and once opened, a seam could deliver years or decades of high energy output with low effort input consistently year round. This made its EROI, *energy returned for energy invested*, extremely high.

Furthermore, coal was easy to transport and it didn't go bad or change from one day or season to the next. Coal gave us the ability to do more work, more quickly than ever before. But additionally, coal, unlike gun powder, enabled people to do things and develop tools and processes that were undreamt of before its advent.

It must be kept in mind that not all of the energy in a process could be applied to the end use. Much of it went to the mining and transportation of the coal. The machinery to mine, distribute, and apply the coal's energy, say a power shovel, locomotive, or smelter, had to be built and, of course, coal was a heat process so it had to be burned, and the efficiency of early steam engines did not remotely approach 10%. The Inuit hunter could not survive without his band nor could the energy worker produce energy for long without a complex support infrastructure.

In order to benefit from coal's energy potential we have to feed it through a wide variety of processes to produce the food, shelter, and tools we require. Therefore the spectacularly high EROIs of miners at the coal face or riggers at the oil wellhead, drop by a huge factor once all of the processes involved in delivering the final product are considered. This is called the life-cycle. If a steam engine had an efficiency rating of 10%, it meant that 90% of the potential energy in the coal went up the chimney as wasted heat.

Despite the system losses in delivering energy to the critical end use, the energy content of fossil fuels was so large and there was so much of it easily available in the infancy of the energy society, these system losses were trivial compared to the final output.

By 1911, western Canada had taken the lead in coal production. In 1947, the year that oil and natural gas were first produced commercially near Leduc, Alberta, coal supplied one-half of Canada's energy needs. The Drumheller, Alberta area alone produced 2 million tons of coal while employing 2000 workers. This averages out to 4 tons per person's day of work or 32,000 kWh per day.

Just after World War II, coal consumption peaked as the rapid conversion of coal's traditional markets to the super-fuels, oil and gas caused the coal mining industry almost to disappear. Beginning in about 1950, almost all coal used for domestic heating, industrial energy and transportation energy was replaced by petroleum products and natural gas.

But as the grid demand grew to a size beyond the ability of water power to supply, the huge reserves of coal found new relevance, creating the highest quality energy of all—electricity. Coal use peaked again in Canada in 1997 and declined after Ontario implemented its Green Energy plan and replaced coal-fired electricity generators with smaller, more flexible and cleaner natural gas generators. By 2040, it is expected that all thermal coal use in the generation of electricity will have been phased out with only metallurgical coal for steelmaking, used in the future.

But coal created the building blocks of a united and well developed, stable nation by the work it did and by the processes and transportation it enabled. Railways were vital to the development of Canada as a politically viable nation and to the exploitation of its natural resources and railways first ran on coal. In Canada, the history of coal and railways moved in lockstep with national development.

Canada's Arteries: The Railway

The power of coal was first applied to transport in the 1820s when the mail from Britain to America was entrusted to ocean-going steamships resulting in a crossing time of only one-third of that possible by the fastest schooners. Time has always been money both in terms of efficiency and competitive advantage and coal, in a world of sail and animal transport, was capable of making great chunks of it. Coal-fired steam engines expanded through the country both geographically and in the breadth of tasks they were used to perform. In 1830, a steam engine was used to hoist blocks of granite from the St. Lawrence River up the cliffs to build the Quebec Citadel (The Canadian Encyclopedia 2015). The viability of all manner of industries, rail and shipping transport was dramatically boosted by the power and heat coal could deliver on a reliable basis.

The first railway in the world opened between Liverpool and Manchester, England, in September 1830. Despite the death of an MP, who was run down by the locomotive at the opening ceremony, the 50 km L&M Railroad ignited a fever of track laying around the world. A short 6 years later, the potential of the railway for Canada was realized by a group of Montreal businessmen.

The Champlain and St. Lawrence Railroad opened in 1836 connecting La Prairie to St Johns on the Richelieu River 12 km away. It is notable that Canada's first railway went north and south, designed to make the trip from Montreal to New York easier by effectively connecting the St. Lawrence River to the Hudson River via Lake Champlain. This was a potentially dangerous precedent for Canadian sovereignty. Fortunately, 50 years later, the trans-Canada railway running east to west was completed establishing the viability of Canada as a Coast to Coast nation.

The Champlain and St. Lawrence railroad spawned a support industry and baggage cars were quickly built in a Montreal factory, the start of the nation's core industrial/transport complex. But it would not be correct to say that Canadians started to "ride the rails" at that point since the tracks were squared pine timbers. These were not replaced by iron rails until the 1850s.

Expensive for material transport, the short railroad thrived on "tourism" driven by relatively low costs and the prospect of effortless travel in which the journey was as much fun as getting there. Even in its infancy, fossil fuel transport clearly demonstrated that travel no longer needed to be an arduous and potentially dangerous undertaking.

In eastern Canada, short rail lines sprouted in all directions. Although serious construction of railways did not occur in Canada until the 1850s, once begun it dominated public policy by stimulating trade and industry, building cities, ferrying settlers westward, uprooting the First Nations, and stitching together an improbable country.

Not only did the railways open the west to the material flows, it allowed the establishment of law and order. Famously, the North West Mounted Police (eventually to become the Royal Canadian Mounted Police) was established to end the predations of the reviled American whisky traders upon the heavily marginalized First Nations at Fort Whoop-Up near modern day Lethbridge, Alberta. Before the railway in 1874, the first detachment of 275 men took 9 weeks to cover the 1700 km from Winnipeg to southern Alberta. Three years later, upon completion of Canadian Pacific Railway, the entire trip from Toronto would have taken under 4 days.

CPR Travel Times in 1928

Coal and the railways turned the far margins of the Canadian wilderness into a huge, accessible, civilized, treasure house of natural resources. Oil would extend this access and even make the once hostile environment, a pleasure in which to live and work (Fig. 2.4).

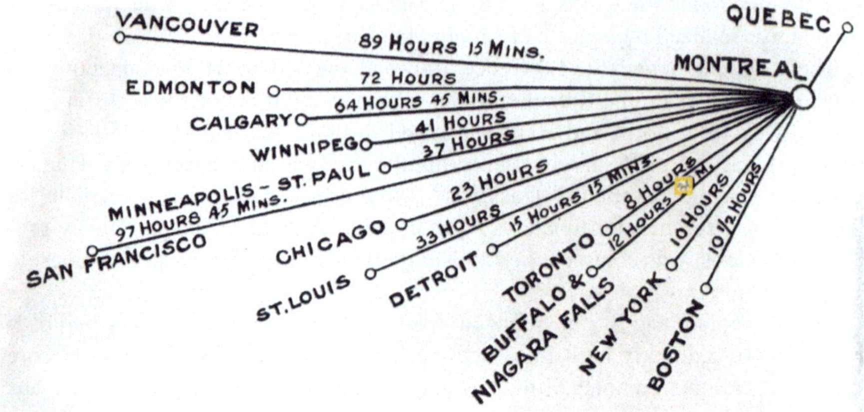

Fig. 2.4 Train travel times

Oil: The Uber-Commodity

To the point of oil discovery in the mid-1800s, most high technology being used in the Americas had come from Europe or through Europe from China or the Middle East. But with the discovery of oil, technological advances were poised to stream out of Canada and the USA. Europe and Great Britain had to import their oil but Canada and the USA were sitting on it and were quick to make good use of it.

Oil supplies became available in North America first in Southwestern Ontario and subsequently, Pennsylvania, (most Americans incorrectly attribute the first oil well to Colonel Drake), very close to the budding manufacturing heartlands of both countries. Coal-fired steam engines could do immense amounts of work but they were large and unwieldy. Oil/gasoline engines could be scaled down to much smaller sizes allowing a far greater variety of applications including mobile ones where the high power-to-weight ratios of these compact engines could be used to great effect. If coal built the major arteries of the nation, oil enabled the capillaries of small roads, rivers, and small lakes allowing automobiles, aircraft, and small boats to access the furthest reaches of Canada's great expanse.

A great deal has been written about the importance and global history of oil, so only the bare framework of its origins in Canada will be covered. However, the experiences of the first oil pioneers are quite remarkable. Two seminal discoveries, Oil Springs, Ontario and Leduc, Alberta, launched Canada into the oil age. These nation-building events were just as different from each other in scale and location as were the experience levels of the people who discovered them.

Conventional Oil: Ontario

From the archives of the Oil Museum of Canada—Lampton County Museums, we have the account of the first gusher in North America.

Fig. 2.5 First oil well in Canada (author)

> On January 16, 1862, a man named Shaw struck oil and the noise of this well exploding caused the earth to shake. The country's first oil gusher blasted crude oil to the treetops from 60 meters below ground at a rate of 3000 barrels a day and flowed unchecked for four days. No one knew how to turn off this gusher. The scene was described as "rushing up with a will—not giving time for the workmen to gather up their implements—filling the well in 15 minutes and shooting up a column of oil some 20 feet in the air. Everything obtainable was procured, but all that could be obtained, capable of holding oil, was insufficient for the wonderful supply, and hundreds of barrels of oil were flowing around the well, over the road into the creek." The oil flooded the hollow where the well was located, the Oil Springs fields were a foot deep in oil and it flowed down Black Creek into Lake St. Clair. The flow was stopped on the Monday by wrapping a flax-filled leather bag around the bottom of a 2 ½″ pipe and stuffed it down the 3″ bore of the well. The flow was diminished, but still gushed 20 feet high. Another bag and ¾″ pipe were forced inside the first, crimping the flow more, but still the well produced 500 barrels a day and filled 35-gallon barrel in one and ¾ minute. The flow was stopped when the ¾″ pipe was extended 20 feet above the ground.

Compare this to the energy output of the Inuit hunter who spent, on average, almost a week stalking each seal. By the above account, in energy terms, a seal equivalent was squirting out of the pipe every 2–5 s. The world had changed (Fig. 2.5).

Above is a picture of the first oil well in Canada located in Oil Springs, Ontario. The two large pots behind it constituted the first oil refinery in Canada.

Although still pumping today at a rate of about 30,000 barrels annually, the Southwestern Ontario fields around Oil Springs did not have the scale at any time to power the nation. Conventional oil fields discovered in Alberta certainly had the scale but did not have the "gusher eureka" moment that sparked rapid development in Ontario. Rather, development was slow and halting and it took three tries before the oil producing potential of Alberta became clear.

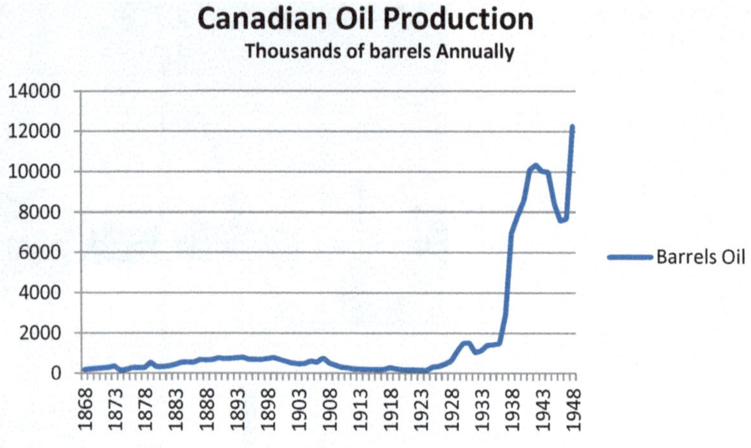

Fig. 2.6 Canadian oil production to 1948 (StatCan 2017)

Alberta

Waterton, Alberta was the location of the first oil "seeps" in the west where the Amerindians were recorded using the black goo as a medicinal salve for wounds and sore joints. These random deposits received scant attention until 1886 when Mr. John G. Brown began to fill sacks with the tar-like material and squeeze out the oil. The father–son team of William and Oliver Aldridge first dug a trench and then a well to collect the oil and, with Brown, sold it as lubricant and lamp fuel. The medicinal application was kept alive by the treatment of cattle infected with mange.

An up and down series of developments followed. Waterton wells petered out quickly and attention turned to Turner Valley which presented the complexity of "wet gas" on top of oil formations to the neophyte developers. The fields there first produced natural gas from 1914 to 1923 and then oil and gas from 1924 to 1936. During this period, although Turner Valley was Canada's largest oil field, natural gas was the primary product. Finally from 1936 to 1946, oil production peaked (Fig. 2.6).

Through the 1940s new major strikes were not occurring, but as production declined in the existing wells, the high drama of the Leduc Discovery unfolded. Note the huge increase in technological sophistication that 85 years of development had produced from Oil Springs to Leduc. From the Alberta Heritage Site we have the following account.

Leduc Discovery Day (Figs. 2.7 and 2.8)

> In 1946, Imperial Oil ran a major seismic survey across central Alberta. The results suggested a large, potentially oil-bearing geological anomaly similar to the Devonian formation previously found around Norman Wells, Northwest Territories. Imperial Oil selected a site on the farmstead of Mike Turta near the hamlet of Leduc and drilled an exploratory well.

Figs. 2.7 and 2.8 Opening of first Leduc well (Alberta History 2019)

The Leduc exploratory well was to be Imperial Oil's last attempt to find oil in Alberta. There had been no major oil discoveries in the previous twenty-five years, and Imperial Oil had drilled 133 consecutive dry wells. The company had decided to concentrate only on Alberta's natural gas reserves, but on April 9, 1946, the company's technical personnel reluctantly decided to drill hole No. 134.

In November 1946, veteran tool push Vern Hunter arrived and began preparation work for what would become Leduc No. 1. The rotary drilling rig, Wilson No. 2, was transported piece-by-piece to the site by train and truck. When erected, the 47-m (154-ft.) tall steel derrick was 12 m (39 ft.) taller than Edmonton's tallest building at the time. Despite his own personal skepticism, Hunter and his thirty-man crew began drilling on November 20.

Hunter drilled carefully and kept meticulous records. His analysis of periodic core samples began showing indications of oil, catching the immediate attention of Imperial Oil executives. In January 1947, Leduc No. 1 drilled into a layer of wet gas (natural gas containing gasoline) and a layer of porous rock laced with oil. A well test resulted in a small plume of oil, which the *Edmonton Journal* mistakenly and prematurely reported as a gusher. The well reached 1,544 m (5,066 ft.) and testing showed that the well would produce. Hunter told the company that he could bring Leduc No. 1 into production on February 13, 1947.

February 13, 1947, nearly proved to be anticlimactic. About 500 spectators, including local farmers, residents of Edmonton, journalists, executives, government officials and politicians gathered in bitter cold and waited, and waited. Equipment froze, and one piece failed, causing a long delay. Finally, at 3:55 p.m., Nathan E. Tanner, Alberta's Minister of Lands and Mines, turned a valve, and the flare line was set alight in an impressive column of flame and smoke as oil began to flow from Leduc No. 1.

Vern Hunter later recalled the events of the day:

It was about five o'clock when I got there. It was dark and cold. Well, it took us until about two o'clock that afternoon, working frantically. We got the other crews there too, you know. There were three shifts there. … Walker Taylor, my boss, was very good. There was people swarming around the rig. They had started coming at ten o'clock in the morning. Walker Taylor took all the dignitaries and the press into the boiler house. He was feeding them coffee and donuts, getting them out of my hair.

About two o'clock in the afternoon, we got everything all rigged up and ready to go again, then started swabbing, and of course there was nothing but mud coming at first and everybody was quite disappointed. "Look at what's coming out nothing but mud." But pretty soon the mud started to have some gas mixed with it and it really started to blow about three or four in the afternoon. It didn't go up the derrick. We had it under control, we had it piped out the flare line and up the pipe. It flared hundreds of feet.

Gas and oil and mud all mixed together. It threw a mushroom cloud just like an atomic bomb and then smoke rings. There were two or three very big smoke rings floating across the sky.

That was about three o'clock in the afternoon. We only let it blow for fifteen, twenty minutes. It was coming in, blowing, like, whoo! You could hear it all right. What gas came out with the oil was lit—one of the roughnecks went out with a burning rag to light it—and there was a clear flare.

(Alberta History 2019)

Leduc Field Booms

Right from the start, Leduc No. 1 produced up to 1,000 barrels a day. Geologists eventually identified the Leduc formation as a dual zone field, with multiple layers of oil overlapping each other at different depths. Although Imperial Oil had secured the most desirable drilling sites, a flurry of activity saw other large companies, and several smaller local companies, take leases and drill wells. By the end of June 1947, Imperial Oil had drilled twelve wells at the same depth as Leduc No. 1, and eleven deeper wells, all of which were productive.

Legacy of the Leduc Era

The Leduc discovery established Alberta as a major oil producer and opened up the province to extensive exploration and development. Leduc moved Canada from the net oil importer side of the ledger to the oil exporter side where it has stayed for the past 70 years and will surely remain for the duration of the oil era.

In terms of economic and social development, the oil industry in Alberta of 1957 employed 16,000 people at well above average wage rates. It added $625 million in royalties over a decade to the provincial coffers and turned Calgary into the oil business's headquarters. All this was with the production of 137 million barrels of oil with billions more to come.

In 1950, about three new wells were started every day. By the end of the year, 1057 wells had been drilled, raising the total number of producing wells in the province to nearly 2000, including the 726 at Redwater, 519 at Leduc, and 328 at Turner Valley.

Alberta and Canada were officially in the energy business and more discoveries added to the base established by Leduc. There was the Redwater field spread over an area 32 km by 6 km which supported over 900 wells which produced 30% of the provinces total output and justified the construction of a 50-km pipeline to Edmonton

Although oil seemed boundless in 1957, the reality of mining finite resource deposits began to sink in, in the subsequent years such as 1959 when Shell Canada drilled 41 wells, with only five producing. During the next 5 years, only two small discoveries were made.

But the Pembina field, with an area of 1500 km^2 and supporting 6000 wells, discovered in 1953, was being developed using a radically different method: rock fracturing. This new method was required because even if there is abundant oil in underground geological formations, it needs to be able to flow for extraction to be possible. The permeability or porosity of the stone must be sufficient for the oil to move.

In a conventional well, a vertical pipe is drilled down into the (porous rock) reservoir and either natural geological pressure squeezes the oil through the rock membrane and up the pipe to the surface or this process is assisted with mechanical pumping. Pumping is typically required as a field ages and natural pressure eases.

Since the sandstone formations of the Pembina field did not have the necessary permeability to allow conventional oil extraction, a new method had to be developed. Permeability was induced via fracturing of the sandstone by pushing fluid at extreme pressures down the borehole and its horizontal extension breaking up the surrounding rock and thus allowing the captive oil to flow and be extracted.

By 1956, Pembina was producing 100,000 barrels of crude per day.

Oil Sands 97% of Canadian Oil Reserves and the third largest oil field in the world.

The oil laden sands of Northern Alberta were well known to the both the native and settler populations. Although they obviously contained a great deal of oil, no one had found a way of developing it.

In 1717, Waupisoo of the Cree brought samples of the oil sands to the Hudson's Bay Company trading post at Fort Churchill while the great explorer, Alexander MacKenzie described the sands in 1790.

The Canadian government was well aware of the oil sands by the late 1800s and by 1900 had mapped out their extent. However, understanding exactly what they were and what they could produce would take another 50 years.

> In 1929, Karl Clark noted that Alberta's most significant oilfield at that time, in Turner Valley, was producing only one-quarter of the oil requirements for the province. While it was not known if there were any other conventional oil deposits in the province, the size, location, and potential of the oil sands resource was obvious.

Born in Georgetown, Ontario, Dr. Karl Clark, the father of oil sands extraction, joined the Scientific and Industrial Research Council of Alberta and began experimenting with ways of separating and recovering the oil from the tar sand using hot water and a chemical reagent. His research with two pilot plants sponsored by the Alberta Government, one at Clearwater in 1930, the second at Bitumount in 1949 proved conclusively that the hot water process could yield clean dry oil capable of being upgraded to commercial requirements. Commercial viability, however, would come much later (Fame 2019) (Figs. 2.9 and 2.10).

It was during World War II, when the energy demands of war production grew intense, both the Federal and Alberta governments determined to exploit this large but problematic resource. By 1949, the first separation plant was working but investors found the new discoveries of conventional oil far more attractive. The potential of the oil sands may have been unlocked, but it would be many years before the resource was tapped.

Suncor constructed the Great Canadian Oil sands plant and production began in 1967. Syncrude shipped its first barrel from the Athabaska Tar Sands in 1978.

Imperial Oil began production at the first commercial in situ project, in Cold Lake, in 1985, with production exceeding 140,000 barrels per day by 1989. Before that time, natural gas liquids and their by-products dominated production; Imperial Oil was largely responsible for increasing the production of bitumen fivefold from 1984 to 1996.

Fig. 2.9 Early research into processing of oil sands (Alberta History 2019) (*Scientist Karl Clark of the Alberta Research Council (above) pioneered a method for separating bitumen from sand. This process was key to the eventual development of large-scale oil sands mining projects*)

Fig. 2.10 Modern oil sands processing plant (O'Connor 2015)

Extent and Composition

Now well charted and understood, the Alberta oil sands are the third largest oil field on the planet and cover some 140,000 km^2 in the northern half of Alberta and stretch from the Saskatchewan border westward. The main deposits are centered on Fort MacMurray and also around Peace River in the western half of Alberta but to date most of the development has taken place in the area closer to Fort MacMurray (Fig. 2.11).

The oil sands are a large seam of oil impregnated sand varying in thickness from 15 to 150 m. At the eastern extent, this seam is near surface level but as the seam stretches to the west, its depth increases. In effect, the layer of oil-bearing sand is tilted downwards from east to west. Oil sands can be found at the surface at their eastern extent but the depth of the overburden increases to over 600 m at the western edge (Fig. 2.12).

The "pay dirt" in the east can be mined by removing the overburden and trucking out the oil-bearing sand for processing. Open-pit mining with its (in)famous images of trucks and excavators stripping vast areas of vegetation and top soil and extensive tailing ponds is only practical in 5500 km^2 of the oil field. The remaining 135,000 km^2 must be harvested by means of "in situ" developments which involve piping steam into the underground seam to the point at which the heated oil will flow, allowing it to be pumped to the surface.

Environmentally and certainly visually, in situ is a vastly less damaging to the local ecosystems. But this method requires more energy for a given output of oil and it also has lower yield than mining. The mining method involves processing the oil sand through heated and pressurized tanks and pipes which remove in excess of 90% of the oil from the sand. The maximum practical depth of open pit mining though is 70–100 m. Beyond that, access to the resource has to be made by drilling (Fig. 2.13).

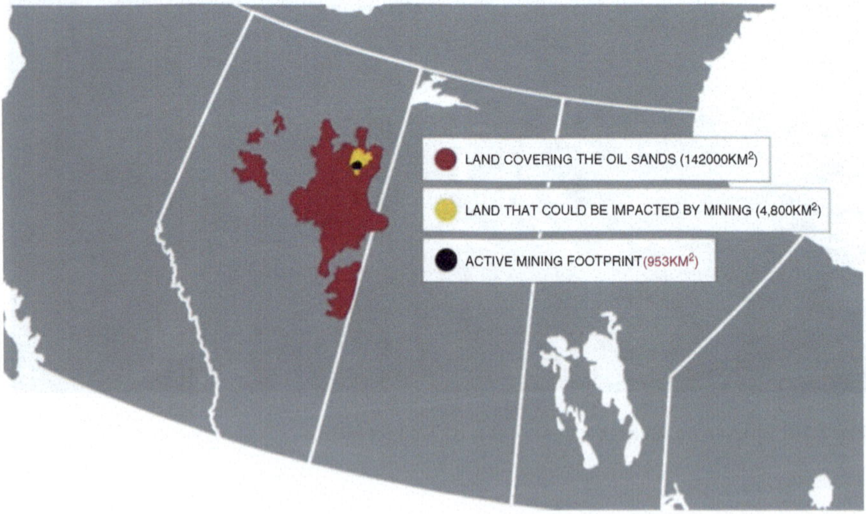

Fig. 2.11 Oil sands extent in Alberta (CAPP 2019)

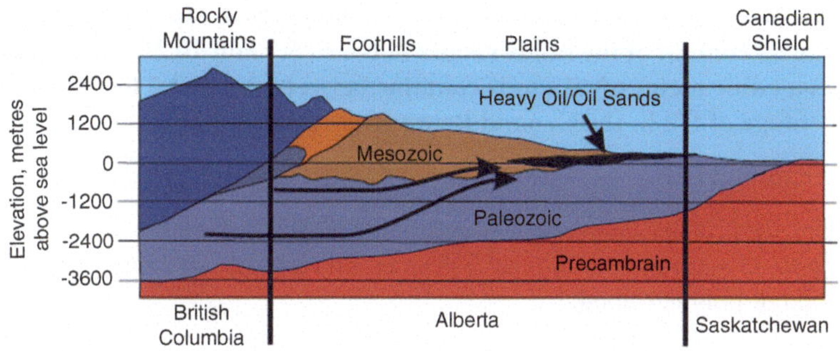

Fig. 2.12 Oil sands geologic profile (Schmitt 2013)

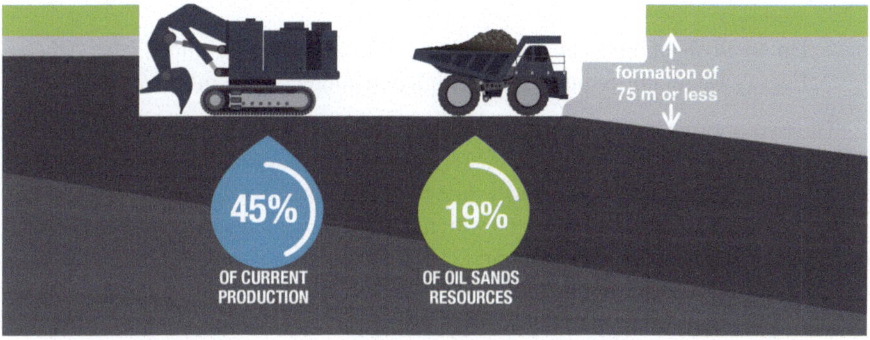

Fig. 2.13 Oil sands depth (Natural Resources Canada 2019)

How much oil is in the oil sands? It approaches 2 trillion barrels. The accessible oil—oil that can be taken out of the ground—is on the order of 166 billion barrels, giving Canada the third largest reserves in the world. For perspective, the world currently uses about 30 billion barrels of oil annually, while Canada and the USA consume 900 million and 8 billion barrels, respectively.

In conventional oil fields, only a fraction of the oil in the ground can be brought to the surface. This ranges from 25 to 30% "primary recovery" (Nolan 2019). For recovery rates higher than that, "enhanced methods" must be employed to force more of the oil to the surface. These include water or polymer floods where huge amounts of water are pumped underground behind or under the oil-bearing rock forcing the oil to migrate toward the wellhead. This is the method now used in the largest oil field in the world, the Ghawar in Saudi Arabia with its 48 billion of barrels of oil (Reguly 2019). In order to maintain the output of 3.8 million barrels per day from this aging field, the Saudis have been injecting 7 million gallons of sea water daily for over 50 years to maintain the pressure in the field and promote oil flow through its incredibly permeable rock (up to 35% porosity). The water "cut" is now over 33% for this aging field meaning, more than one barrel of water comes out of its wells for every 2 barrels of oil. Simply put, as oil fields age, more energy is required to maintain the energy output (Masnadi and Brandt 2017).

In the three major oil sands deposits, Athabasca, Cold Lake, and Peace River, the oil content of the sands ranges between 5.7 and 10.2% by weight. On average, a cubic meter of oil sands contains between 1 and 1.2 barrels of oil.

Percentage of Oil in the Ground Being Recovered

- **Conventional Oil**
  - 25% for primary recovery of conventional oil (drilled wells)
  - Up to 33% using conventional enhanced oil recovery methods such as water flood or polymer flood.
- Oil Sands
  - Up to 35–40% bitumen using Cyclic Steam Stimulation (in situ method)
  - Up to 50–60% of bitumen using Steam Assisted Gravity Drainage (in situ method)
  - Up to 90% of bitumen from mining. For mining operations, the Alberta Energy Regulator requires a very high percentage of the oil be removed.

The two approaches to harvesting oil from the oil sands depend on how deeply the sands are buried. Mined oil sands requires large amounts of process water to separate the bitumen from the sand, while the in situ process requires less water but greater volumes of steam. The two in situ methods are called steam-assisted gravity drainage (SAGD using two pipes) and cyclic steam stimulation (CSS) which uses one pipe for both steam application and oil collection. The resulting bitumen product can either be sent to an upgrader for conversion to synthetic crude oil or diluted and sold directly to refineries (Fig. 2.14).

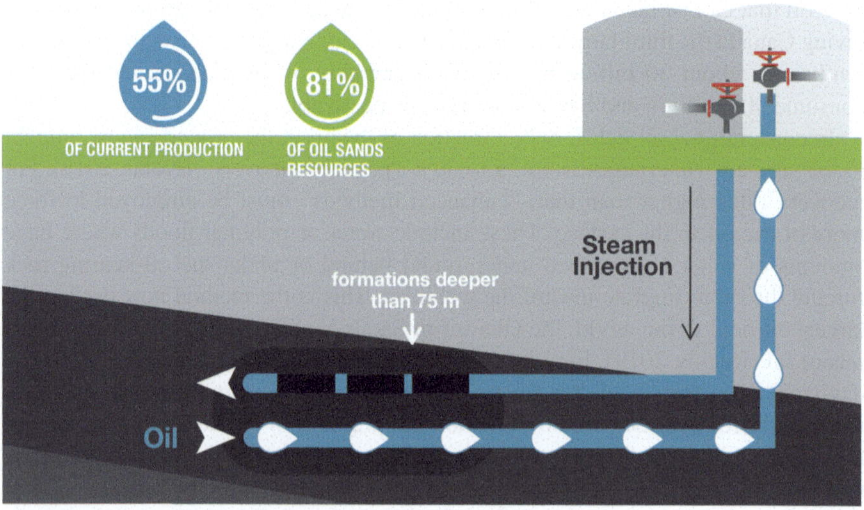

Fig. 2.14 Oil sands in situ method (Natural Resources Canada 2019)

The in situ method involves heating the underground seam and pumping out the oil that will readily flow leaving over 40% of it forever buried. The in situ method of extraction appears to be far less complex than the mining operations and certainly results in far lower land remediation costs. However, its EROI is lower because of its high heat inputs and lower yields, which is to say more oil may be coming out but a lot more energy goes into producing it.

The mining areas of the oil sands have EROIs in the neighborhood of 4.5 or 5 to 1. Getting 5 barrels output for every barrel (or equivalent energy) used in the process is a very low return by conventional oil standards. The world EROI average has fallen dramatically but in 2019 is still in the vicinity of 17:1. In situ can boast an EROI of no better than 3.5:1, probably the lowest level of any oil producing field currently deemed financially viable.

What Is "Bitumen"?

One of the easiest ways of understanding bitumen is to compare it to its cousin, conventional crude oil. Whereas conventional crude oil flows freely, bitumen does not. At room temperature it looks like cold molasses, and must be either heated or diluted before it flows. Like all petroleum, both conventional crude and bitumen are made up of hydrocarbons (i.e., organic compounds containing only carbon and hydrogen). However, compared to conventional crude oil, bitumen contains more carbon than hydrogen, as well as many more impurities, such as nitrogen, sulphur, and heavy metals. In order to produce synthetic crude, these impurities must be removed and the carbon–hydrogen imbalance corrected.

Dilbit is bitumen diluted with one or more lighter petroleum products, typically natural gas condensates such as naphtha. Diluting bitumen (with diluent) makes it much easier to transport, for example in pipelines. But bitumen has a higher percentage of carbon (higher energy content than oil) which explains the explosive power seen in the tragic Lac Megantic disaster which killed 47 when 74 rail cars filled with bitumen crashed and burned in the center of this small town in Southern Quebec.

Simply put, the oil sands process requires more complex processing than conventional oil. More processing means higher energy input.

- The oil (bitumen) must be either dug up and heated or heated underground for separation from the sand.
- The bitumen must be processed in "upgraders" before it can be turned into conventional oil or have diluent added for transport to a refinery.

This complexity creates the following problems:

• Given the high energy requirements of any of the oil sands extraction methods, carbon and other greenhouse gas emissions are 30% or more higher than those for conventional oil—up to 55 kg of carbon emissions per barrel.

- High consumption of natural gas—28 m^3 for every barrel of oil produced.
  - One barrel of oil contains ~1700 kWh of energy and 28 m^3 of natural gas contains ~300 kWh.
- Concentrated regional air pollution;
- Transferring toxic materials into the soil, water, and air.
- Loss of ecological habitat and biodiversity.
- Water pollution.
- High water use for mining—from two to five barrels (300–750 L) per barrel of oil.
- Forest destruction—over 800 km^2 of boreal forest has been destroyed.
- Long-term legacy issues of toxicity and seepage from tailings ponds.
- Oil sands do not have a high enough EROI to "breed" themselves—they are not self-sustaining.

Mines Don't Live Forever

At the current level of production of 4.2 million barrels per day, the lifespan of the oil sands would be over 100 years. If the production expands to the 5.6 million barrels per day in 2025 as expected by the Canadian Association of Petroleum Producers, the lifespan declines to 80 years. If the wildest growth dreams of some come true with production hitting a Saudi Arabia-like level of 10 million barrels per day, the lifespan will come down to 45 years. For reference, in early 2019, US oil production from fracking sites produced 7 million barrels per day with 3 million more barrels coming from conventional fields for a total production of 10 million barrels. Output from US fracked wells is expected to decline by 2025 even with the drilling of thousands of new wells annually.

As the share of production shifts increasingly westward to deeper deposits accessible only by in situ methods, the EROI is likely to drop (full discussion of EROI in Chap. 7). Certainly technology improvements will slow the decline but as we move from richer mined resources to lower quality resources, depletion eventually triumphs over technology.

Energy Source or Energy "Product"?

Production of oil from the oil sands is completely dependent on the availability of much higher EROI natural gas. Without a much higher EROI energy source as feedstock, the oil sands would not be a viable source of energy.

Certainly oil from the oil sands could be produced, refined, and shipped but the effort and energy input in doing so would completely consume the energy produced from the oil, leaving nothing for social infrastructure, schools, education, health, or security. The province of Alberta could not survive on an EROI of 4:1 and therefore the much energetically cheaper natural gas, whose EROI is in excess of 15:1, is used to power oil sands development. If natural gas were in short supply, oil sands production would quickly come to a halt. But unless exports of natural gas are vastly increased, Canada has more than enough to pull out every last barrel of oil from the oil sands and turn it into gasoline or diesel fuel.

Hence oil from the oil sands is more of an energy product than a source of energy. We are converting the energy from natural gas into a form of energy that is usable by our transportation system as it is currently configured.

Oil production is subsidized energetically by natural gas and therefore the oil sands must be viewed as a source of oil and not as a viable stand-alone source of net energy. Natural gas has a major role to play in the transition to renewable energy whereas oil does not. By using large quantities of natural gas to produce oil from the oil sands, we are, in effect, spinning gold into lead.

Technology vs Depletion: Ultimately Depletion Wins

- Changing EROI of oil sands
  - Tar sands (typically from Colorado) are impregnated sands that yield mixtures of liquid hydrocarbon and require further processing other than mechanical blending before becoming finished petroleum products. Oil sands (typically from Alberta and Venezuela) are deposits of bitumen; viscous oil that must be rigorously treated in order to convert it into an upgraded crude oil before it can be used in refineries to produce gasoline and other fuels. Bitumen is about 10–12% of the actual oil sands found in Alberta. The remaining 80–85% is mineral matter, including clay and sands, and around 4–6% water.

    http://ffden-2.phys.uaf.edu/102spring2002_web_projects/m.sexton/
  - Excellent site—oil skeptic

    http://energyskeptic.com/2017/eroi-of-canadian-natural-gas-a-peak-was-reached-despite-enormous-investment/
  - Subsidies

    It is critical to understand subsidies of one energy type by another because if the lifespan of the high EROI energy is significantly shorter than that of the low (end product) energy then the lifespan dictates that of the end product.

    For example, if the EROI of natural gas was 15 and the oils sands 3, and there was a 50 year supply of gas and a 300 year supply of oil, the oil could be produced only for as long as the natural gas held out. Once the natural gas declined precipitously, the oil development—upgrading—would cease.

- Oil sands and fracking should not be seen as energy sources as were their rich oil field predecessors, but as the final stage of an energy process which ultimately yields a critical and irreplaceable (to this point) goodgasoline and diesel that keep planes in the sky, ships on the ocean, tractors in the fields, and our current car and truck fleet on the road.
- Now the oil sands is a precursor to what might come—clean and more available energy applied very inefficiently to yield a critical energy for which there is simply no substitute.
- In the future, possibly electric energy will be used to run tractors to produce ethanol or biodiesel which will keep some planes in the air or fuel some critical task requiring heavy machinery or a dense and light fuel.

Newfoundland and Labrador

Oil was located off the Newfoundland coast in 1979 after 13 years of exploratory drilling. The Hibernia field with estimated reserves of 884 million barrels of oil is 300 km east of St. John's and produced its first barrel in 1997. Four oilfields are producing crude oil in Newfoundland and Labrador offshore areas Hibernia, Hebron, Terra Nova, and White Rose (Heritage Newfoundland and Labrador 2019) (Fig. 2.15).

Fig. 2.15 Off-shore oil field infrastructure (Heritage Newfoundland and Labrador 2019)

Of the smaller oilfields, Terra Nova oilfield began operating in 2002 and is estimated to hold 354 million barrels of recoverable oil, while White Rose entered production in 2005 and holds approximately 286 million barrels of oil. Estimates put the Hebron field at 700 million barrels of oil.

Because oil reserves are nonrenewable production is finite. The lifespans of these fields are estimated to be as follows:

- Hebron—25 years
- Hibernia—20 years
- White Rose—15 years
- Terra Nova—10 years

However, petroleum companies are exploring offshore sites on the Grand Banks, the Labrador shelf, and areas off Newfoundland's west and south coasts. The importance of oil to the provincial economy can't be overstated. To 2007, production totaled 867 million barrels of crude oil, worth $46 billion and amounting to 35% of the provincial GDP.

The oil and gas sector directly employed 1.3% of the province's labor force and almost all were Canadian. Support and short-term construction jobs indirectly employed 3.8% of the provincial workforce in 2007. At their peak in 2012, oil royalties provided 31% of the total provincial government budget.

Ontario

From its start in Oil Springs, Ontario, Canadian oil production has grown immensely. Initially oil production served as a replacement for whale oil used in lamps but the prime oil-bearing whale species were hunted to near extinction and a substitute had to be found. Prices for a barrel of oil fluctuated wildly depending on transport capability from $2.50 a barrel to 10 cents to $10 to 75 cents in the 1860s. In today's money, this would have covered the range from $2.20 to $220.00 a barrel.

But society did not yet run on oil as coal was providing the needed heat and mechanical motion adequately. In addition, there simply was not enough oil in the Southwestern Ontario fields to supply much more than the lighting application as demonstrated by their total cumulative production over 150 years of 28 million barrels. This equals roughly a week of production from the Alberta of 2019. Annual production is still around 30,000 barrels annually with an average well producing 0.6 barrels per day (Fairbank and Cochrane 2000).

From the initially impressive gusher in Oil Springs, 150 years ago to the huge industry it has become today, oil dominates the Canadian economy. According to the BP Energy Statistical Review, Canada produced 4.8 million barrels of oil per day in 2017 and consumed 2.4 million barrels for a net export of 50% of production (BP 2018).

Natural Gas

Natural gas is the ugly duckling story of fossil fuels. When it was first stumbled upon in early coal and oil fields it was nothing more than a nuisance. And in coal mines, a dangerous nuisance at that, since it could seep out of the coal seams and explode with catastrophic results. Consequently, natural gas was simply "flared off" at the wellhead whenever it got in the way of oil and coal operations.

Natural gas is the cleanest burning, lowest carbon emission fossil fuel and it can be found in abundance in the Prairie Provinces and northern British Columbia. Despite its past initial lowly position on the economic totem pole, natural gas is an excellent "bridge" fuel to smooth the transition to renewable energy. Gas-fired electrical generators can react very quickly to the surges and drops in power production from solar and wind sources and therefore can serve as the ideal base load. In addition, new dual stage (gas plus waste heat) turbines can increase the efficiency of turning fossil fuel into electricity from the 30 to 40% of coal and oil plants to 50%. Conversely, coal and nuclear plants require long periods of time to ramp up or decrease their output. In the future, the ugly duckling will spread its wings as the swan of low carbon energy systems.

Natural gas is often embedded in oil which is under tremendous pressure hundreds or thousands of feet below the earth's surface. Once that pressure is released by the drilling process, the gas forms and rushes to the surface much like the bubbles in a bottle of carbonated drink.

After natural gas forms, it will tend to rise toward the surface through porous spaces in the rock because of its low density compared to the surrounding rock. Most of the natural gas deposits we find today occur where the gas happened to migrate into a highly porous and permeable rock underneath an impervious cap rock layer, thus becoming trapped before it could reach the surface and escape into the atmosphere.

There are two general categories of natural gas deposits: conventional and unconventional. Conventional natural gas deposits are commonly found in association with oil reservoirs, with the gas either mixed with the oil or buoyantly floating on top, while unconventional deposits include sources such as shale gas, tight gas sandstone, and coalbed methane. There can also be pure conventional natural gas fields which are not associated with any form of oil (Fig. 2.16).

Accidental Industry

The difficulty of locating natural gas beneath the surface of the Earth did not mean that gas was untapped as a resource; it did mean, however, that its discovery and use were often a matter of chance. During the nineteenth and early twentieth centuries, many discoveries of natural gas were purely accidental, often occurring as a result of drilling or mining for some other desired resource. In 1859, natural gas was

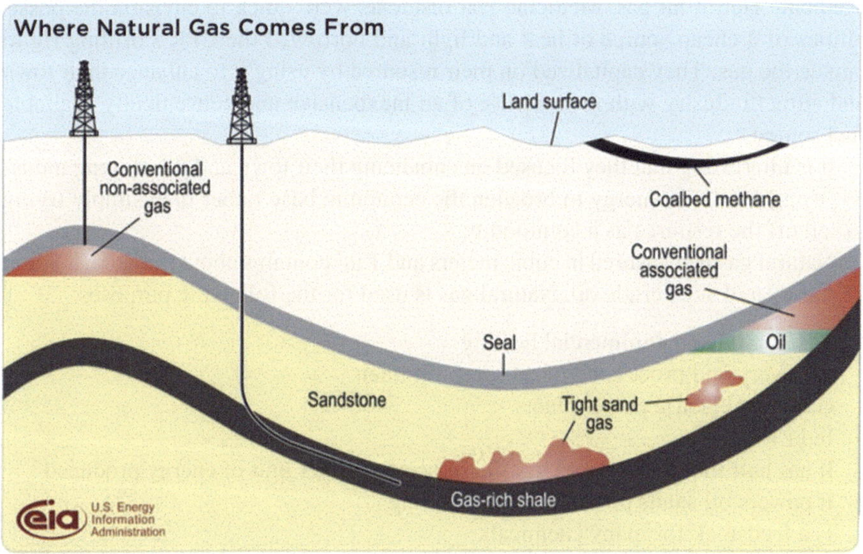

Fig. 2.16 Natural gas geologic profile (EIA 2019)

discovered in New Brunswick and flared off as a waste product. Other discoveries relied upon the keen observation skills required to detect a seep; evidence might be bubbles in a stream or funny tasting water or even flames quietly burning on a hillside.

The earliest documented use of natural gas in Alberta resulted from accidental discoveries made while searching for something else. As Canadian Pacific Railway (CPR) trains steamed their way across western Canada, railway workers needed steady supplies of water and coal to keep their locomotives running. In 1883, CPR drillers pierced the Earth near the Langevin (renamed Carlstadt in 1910 and Alderson ca. 1915) CPR railway Siding No. 8, 55 km northwest of Medicine Hat, seeking water; they found gas instead. This event was considered noteworthy enough for publication in December of 1883:

> At Langevin, 4th siding west of Medicine Hat, a rather singular phenomenon has presented itself. The well-borers have reached a depth of 1,120 feet without finding water, but a gas which rushes out of the tube, which, on taking fire emits a flame sufficient to light up the surrounding country. (Calgary Herald Mining and Ranch Advocate and General Advertiser 1884)

The discovery may have been accidental, but the outcome was fortunate as the gas was able to be put to good use at the nearby section house for cooking and heating. A second and less opportune accident illustrated the continued ignorance about the power of natural gas. Reported in a January 1884 edition of the *Calgary Herald* was an explosion that destroyed a building and caused several injuries.

In the year 1890, the CPR made another inadvertent contribution to the birth of the gas industry when, in digging for coal along the South Saskatchewan River near

Medicine Hat, it hit gas. Medicine Hat residents were quick to envision the possibilities of a cheap source of heat and light and borrowed the CPR's drilling rig to pursue the gas. They capitalized on their resource by using it to enhance their town and attract industry with the promise of an inexpensive and conveniently available fuel source.

It is interesting that they focused on enhancing their town and developing industry, using available energy to broaden the economic base rather than simply trying to sell off the resource as a commodity.

Natural gas is measured in cubic meters and 1 m^3 contains about the same amount of energy as 1 L of crude oil. Natural gas is used for the following purposes:

- Residential and commercial heating
- Recovery and processing of oil sands bitumen
- Gas-fired electric power plants
- Industrial heat
- It has half the greenhouse gas emissions of coal per unit of energy produced
- It powers oil sands production and refining
- Is a feedstock for many chemicals

The Look and Feel of a Kilowatt-Hour

Kilowatt-hours are used throughout this book as the basic measure of energy. The Watt is the most common measure of electrical energy and the kilowatt-hour is the most common measure of electrical power most people are familiar with. And it will become even more familiar as electricity replaces fossil fuel in more and more applications. Energy is produced and consumed in many different ways and it is sometimes hard for human perception to reconcile the magnitude of energy involved as we tend to overestimate the energy in motion and noise and underestimate it in heat transfers. Here is a list of applications of energy that may appear to use different amounts but actually do use the same amount.

- The means representing the power output of an automobile engine will switch from horsepower to kiloWatt. One horsepower equals 746 W or about three-fourth of a kiloWatt.
- The size of the gas tank will no longer be measured in liters or gallons but in kilowatt-hours.
- These are very easy number to work with once, they become familiar. A 10-kW electric motor running at full throttle for an hour consumes 10 kWh of electricity.
- Electricity can be used to produce heat as well as motion but the different levels of power required for both are somewhat counterintuitive.
- A small electric heater can sit quietly in a corner of a medium-sized room in a well-insulated house and use 1 kW to keep that room comfortable when the outside temperature is 0 °C. After an hour, it will have consumed 1 kWh of electrical energy.

- That same kilowatt hour can be used to push a car down the road. A moderately efficient EV sedan might use 0.2 kWh to cover a kilometer at a speed of 100 km/h. Consequently, the same seemingly small amount of electrical energy (1 kWh) used to heat a room for an hour is enough to push a large vehicle at high speed for 5 km.
- If the battery capacity of the vehicle was 40 kWh, the range of the vehicle at 100 km an hour with its consumption rate of one-fifth of a kWh/km, would be 200 km.
- A medium-sized house fan could be kept going at full speed for 10 h by that same kilowatt-hour of electricity. A passive application like heating air or water requires more energy than one would suspect compared to the energy required to move something from place to place.
- An automobile which uses 10 L of gasoline for every 100 km it travels is using 1 kWh for every kilometer it travels.
- An Olympic level cyclist can produce close to 400 W of mechanical power over an hour with sprints up to 1000 W. An average fit adult can maintain an output of 100 W for an hour while a very hard working manual laborer can sustain an output of 75 W over an 8 h day.
- One kWh would be enough to allow an efficient digital amplifier, driving very efficient loudspeakers, to produce deafening sound pressure levels for an entire 24 h period.

Electricity: The "Man-Made" Fuel

Electricity had been a curiosity for thousands of years but by the seventeenth century the new human learning process called "science" was beginning to focus on it. The 1740 invention of glass friction generators in Leyden, Germany may have inspired Benjamin Franklin's famous experiments and by 1820, Hans Christian Oersted was demonstrating the relationship between electricity and magnetism. In 1820, Michael Faraday and Joseph Henry invented a primitive electric motor, and in 1831 demonstrated the principle of the generator.

In 1870, Belgian Zenobe Gramme, created a steady direct current generator suitable for powering motors and electric lights which opened the imaginations of researchers to electricity's potential to light and power the world. By 1877, the streets of many cities across the world were electrically illuminated.

Advances in alternating current (AC) technology spawned new opportunities for power generation. Hydropower, for example, marked several milestones between 1890 and 1900 in Oregon, Colorado, Croatia (where the first complete multiphase AC system was demonstrated in by NiKola Tesla 1895), at Niagara Falls, and in Japan. Sir Charles Parsons built the first steam turbine generator with a thermal efficiency of 1.6% in 1884. By the 1910s, the coal-fired power plant cycle boosted net efficiency to about 15%. The mechanical production of electric power began the

Second Industrial Revolution driven by inventions by Thomas Alva Edison, Nikola Tesla, and a host of other scientists and tinkerers spread all over the world.

In the same way that fossil fuels opened the interior of Canada to development and enabled an expanding universe of industrial processes, electricity took human endeavors in many different directions. It allowed small but powerful motors to move inside small homes and shops to perform many different tasks. It enabled instantaneous communications across vast distances and it allowed human mobility to extend under water and into space. Electrical devices could be scaled down to extremely small sizes and it made possible the development of sophisticated computers and scientific testing apparatus.

Electricity also made possible the creation of a new and useful metal, aluminum. Aluminum does not occur in pure form but is refined from bauxite by an electrolytic reduction process which produces 1 kilo of aluminum from every 4 kg of bauxite. Although the amount of energy required in this process has been reduced by 50% over the past five decades, most aluminum smelter of any size are located close to large hydro or coal plants as they are voracious consumers of electricity. Norway, Iceland, China, and Quebec all house large aluminum smelters. Aluminum is strong for its light weight, highly electrically conductive, and it does not corrode beyond a very thing surface oxidation. Its applications range from aircraft skin to electrical wiring.

Electricity was a fuel that caught on quickly. It was quiet, clean, and cheap and it could do all manner of work in businesses, homes, and farms. It only took 10 years, from 1947 to 1956 to connect 75% of the farms in Manitoba to the grid (Canada Hydro 2008).

Electricity allowed human learning to explode and it could be generated by any energy source capable of turning a wheel from fossil fuel engines, to waterwheels to windmills.

Hydro Electricity

Rivers in Canada discharge approximately 105,135 m^3/s (cubic meters per second) or 3300 km^3 of flow to the oceans annually, which amounts to about 8% of world river discharge. Northward drainage accounts for almost half (48.6%) of this flow. As far as power generation is concerned though, large flows are of little use if they are not tumbling down steep slopes or over high cliffs. The smaller tributaries of the Amazon River produce more usable energy rushing down the eastern slopes of the Andes than does the huge, very broad river at its mouth as it slowly merges with the Atlantic Ocean.

Electricity, which is an energy carrier rather than a source of energy, is the purest form of energy humans have yet developed and a most ideal source of electricity can be found in Niagara Falls. Here, a single river focuses the water from a drainage area of 409,000 km^2 in a 400-m-wide channel plunging over a 51 m high cliff in the heart of the country's most densely populated and industrialized area.

Four of the five Great Lakes are upstream from Niagara Falls. Lake Erie, Lake Huron, Lake Superior, and Lake Michigan constitute a combined water storage volume of over 21,000 km^3 and account for 20% of the world's fresh water volume. Only Russia's extremely deep Lake Baikal holds more water. These huge storage reservoirs, shared between the USA and Canada, are able to maintain an average flow over the Falls of 5800 m^3/s.

The energy density here is immense. By the time the water, streaming over the crest of the Falls at 30 km/h, hits the water in the gorge below it has accelerated to over 100 km/h and contains approximately 2.7 GW of mechanical energy. Harnessing this power is done by channeling the flow of the river around the Falls to the Sir Adam Beck generating station which has the capacity of 2 GW.

In contrast, the immense Amazon River flows at 209,000 m^3/s dispersed through a 200 km wide estuary, equivalent to a 1.6-km-wide and 60-m-deep channel, where the water pushes into the Atlantic Ocean at the speed of 6–8 m/h. By the time the Amazon reaches the ocean its power is still extremely large but very diffuse (Niagara Falls Information 2019; Wateroffice 2019.

Electrical power in Canada got its start in Montreal in 1878 and by 1883 Victoria, BC had electric street lighting with the first hydroelectric generator coming into operation in 1884. A Conservative politician, Adam Beck, pushed for the creation of a public electric utility and in 1910 the publicly owned network was switched on. During World War I, this utility brought the largest hydroelectric power plant in the world at that time on line: Niagara Falls.

EROIs for hydro power projects are generally very high and are dependent on the size of the project and the variability of the waterflow it is driven by. Due to the relatively small infrastructure needed Niagara must be the absolute pinnacle in this regard although no EROI studies of it have been completed to date (Powermag 2017; Hydro Power 2018).

Hydroelectric capacity by province is a factor or rainfall and elevation changes (Fig. 2.17).

More rain falls on Quebec and runs off high rocks than any other province in Canada. There is potential for new hydro dams in the far north, but most of those in the southern half of Canada, where the rivers run south and east and west, which is to say close to the more densely populated regions, have been exploited already.

Because such a larger percentage of its electricity is supplied from hydro, the lowest cost source, electricity costs in Quebec are the lower than in any other region of North America and lower than most in the world. Manitoba has low electricity costs as well since it can supply a large majority of its electricity needs from hydro sources.

Ontario produces double the amount of hydroelectric power that Manitoba does but, given its large population, must supplement this with nuclear and natural gas plants as well as wind and solar farms. Once the last nuclear plant in Ontario is decommissioned in 2037, the outlook for the supply of electricity in the province becomes clouded. Nuclear generating plants currently supply 60% of the provinces electrical energy, ten times the amount of energy generated by the province's present wind and solar farms.

Canada ranks fourth in the world for hydropower development, with over 79 GW of installed capacity, including pumped storage from 475 major generating stations.

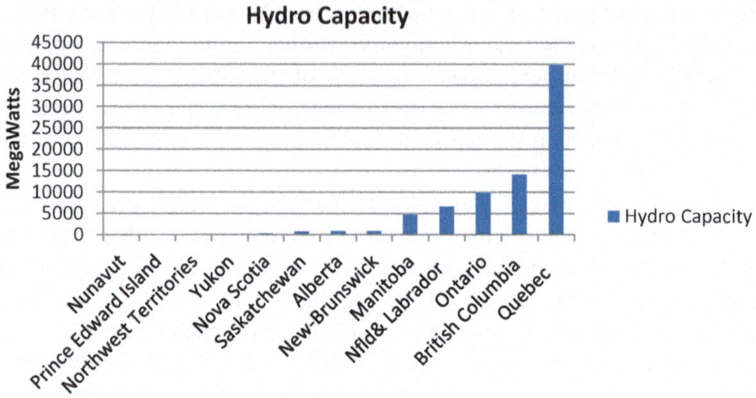

Fig. 2.17 Hydroelectric capacity by province (NEB 2019)

In terms of the key per capita hydroelectric power output, Canada ranks third behind Norway and New Zealand. Total annual generation from all hydropower facilities in Canada reached an estimated 380 TWh in 2016, with only China and Brazil exceeding this figure. Hydropower currently accounts for 62% of the country's electricity power mix.

A number of large projects are underway which will expand capacity by another 3 GW. Hydroelectricity will be vital in underpinning our transition to a renewable energy future.

Niagara Falls is an ideal energy source featuring 2000 MW of clean energy capacity with nearly unlimited storage (Great Lakes) in the heart of a prime agricultural, industrial, transportation, and population region. It produces 2149 GWh annual output or enough energy for 170,000 homes at a rate of use of 35 kWh per day.

Nuclear Energy

Ancient Greek philosophers first developed the idea that all matter was composed of invisible particles called atoms. The word atom comes from the Greek word, atomos, meaning indivisible. Uranium was discovered in 1789 by Martin Klaproth, a German chemist, and named after the planet Uranus. By 1900, physicists knew the atom contains large quantities of exploitable energy.

British physicist Ernest Rutherford was called the father of nuclear science because of his contribution to the theory of atomic structure. In 1904, he wrote: "If it were ever possible to control at will the rate of disintegration of the radio elements, an enormous amount of energy could be obtained from a small amount of matter." (Zohuri 2018).

After World War II and the first use of nuclear weapons, science focused on applying atomic energy to the production of electricity. In 1954, the world's first nuclear plant at Obninsk, Russia, came on line followed by Calder Hall plant in England in 1956 (Euro Nuclear 2016).

"Energy too cheap to meter." That was part of the promotion behind the launch of nuclear energy in Canada. Essentially, nuclear generated electricity was expected to be the source of near infinite clean energy at close to no cost. Events unfolded differently and the nuclear energy industry now faces a very uncertain future.

The full quote from the 1954 speech of Lewis Strauss, then chairman of the United States Atomic Energy Commission, to the National Association of Science Writers, not only shows his enthusiasm for nuclear energy but epitomizes the optimism of the 1950s. At that time, the USA and the world were riding the wave of surging energy supplies and the accompanying innovations were changing and improving life on an almost weekly basis.

> Our children will enjoy in their homes electrical energy too cheap to meter. It is not too much to expect that our children will know of great periodic regional famines in the world only as matters of history, will travel effortlessly over the seas and under them and through the air with a minimum of danger and at great speeds, and will experience a lifespan far longer than ours, as disease yields and man comes to understand what causes him to age.

In short, Strauss foresaw abundant food, effortless travel, greatly extended (and presumably healthy) lifespans, and near godlike learning. A great deal of what he predicted has come true but there have been unforeseen consequences.

Not only have the costs of nuclear been much higher than estimated, but the great unknowns of disposal of radioactive waste, the cost of the process of decommissioning aged nuclear plants, and the fear of further nuclear disasters have brought the construction of new plants in the western world to a near halt. On a parallel track, one must also take note of the possibility of the proliferation of nuclear weapons as a collateral result of widespread nuclear power plants.

An indication of the general level of naivety ("peak naivety"?) on environmental and downstream consequences during the giddy 1950s and 1960s was the 1958 suggestion of one American oil company of detonating nuclear bombs in the oil sands so the heat would melt the oil and it could gather in the craters ready to be pumped out. The unintended consequences and legacy costs from a 2019 perspective would be almost too much to contemplate. The Canadian federal government quashed the plan (Unger and Thistle 2013).

In the early days of nuclear power, there were many theorized applications for this new, almost limitless, quiet, and clean energy source. Nuclear submarines, of course, became a reality. They were capable of higher speeds, could run more quietly and could stay submerged many times longer than their diesel/electric battery counterparts.

One of the most widely read magazine of the 1950s and 1960s was Popular Mechanics which ran a story on nuclear-powered lawn mowers. These would do away with the smelly, noisy two-stroke engines which were used in applications needing a small, powerful energy source. Unsurprisingly, no nuclear-powered lawn mowers ever made it to market. Beyond power, radiation in the form of X-rays was used in many different ways, including a shoe fitting apparatus which allowed the wearer and prospective customer to put on the new shoes and look down through the viewing lens through the shoe material and see how the bone structure in his feet fit into the shoe. Children loved it as they could see the bones in their feet wiggling and parents could get a clear answer as to whether or not the shoes fit well.

These retail X-ray machines disappeared quickly from stores, possibly victims of their own success, since children loved them and the increasingly nervous sales staff may have been concerned for both the children and their own welfare, spending 8 h a day in the vicinity of a poorly isolated radioactive device.

The largest airplane built before the 1980s was the Convair 36H designed to serve as a heavy bomber in the US Air Force. The 36H had 6 huge piston engines to get it up to cruising altitude but once there, the motive force would be provided by hot air generated from an onboard nuclear power plant. Like their underwater war machine cousins, the Convair would be able to stay in the air indefinitely. The airplane featured an 11-ton radiation shield for the cockpit. It was made obsolete by progressively more powerful jet engines in the late 1950s and the program was grounded in the late 1960s over public safety concerns (Winchester 2005).

Despite well-publicized problems and false starts, there are 450 nuclear power plants in operation in 31 countries which have been producing over 10% of the world's electricity reliably and safely for decades. These plants generate a base load that displaces capacity from coal- and gas-fired plants, resulting in lower air pollution and greenhouse gas emissions as well as solid dependability of supply for regions with limited access to hydro or fossil fuel resources.

There are currently nearly 50 plants under construction in 15 countries but only 2 live projects in North America (WorldNuclear).

Canada Goes Nuclear

There are 19 nuclear reactors in 4 different plants in Canada, 3 in Ontario and 1 in New Brunswick which generate 15% of our electricity. These plants are the Candu design which features the ability to operate under full power while the fuel rods are being changed. Although the mining of uranium is energy intensive, Candus use unenriched uranium which requires much less energy than enriched uranium to make. The controlled nuclear fission process generates steam which drives the turbine and generator. Only one-third of the energy released by the fission process is able to be applied to turning the turbine, the other two-third lost as heat through the cooling water flows drawn from lakes or rivers (Garland 2019). Perhaps this waste heat could be used in geothermal storage applications in the future (more in later chapters) (Fig. 2.18).

A Candu reactor (the Pickering nuclear station has six reactors) has an output of 880 MW. For comparison, natural gas generating plants have capacities ranging from 100 to 1000 MW. The Sir Adam Beck generating station at Niagara Falls has a 2000 MW capacity. The largest windfarm in Ontario has a capacity of 200 MW

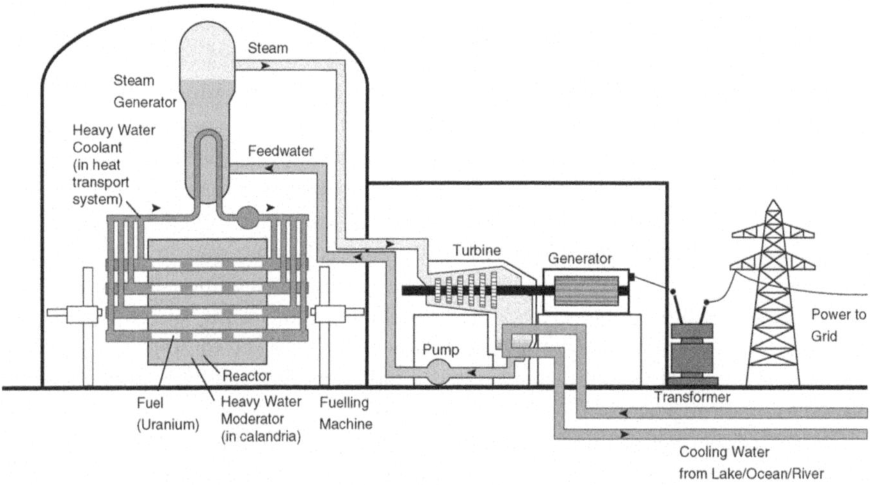

Fig. 2.18 Operation of a Candu Nuclear Reactor (UNENE 2019)

while Alberta and Quebec each have one windfarm approaching 300 MW (Wikipedia 2019b).

Nuclear power plants offer the advantages of a small footprint (under a square kilometer) for a large output of continuous power. Candu designs must be located beside large bodies of water, effectively increasing their footprint but even so, they are not at the mercy of the vagaries of wind, sun, or river flow for their energy nor do they require large constant inputs of coal or natural gas. The EROI for nuclear energy is also high although the decommissioning energy costs are largely unknown at this time. EROI estimates range from 5:1 to 15:1 but there appear to be no recent studies available (Hall 2019).

The advantages of the Candu design are they are inherently safer than most other nuclear facilities and they can use a wider range of cheaper fuels including natural, unprocessed uranium. Their ability to utilize thorium as well as uranium expands potential resources of feedstock by a factor of 3.

In the long term, nuclear energy is not renewable as it depends on mining uranium or thorium whose deposits are large but limited. However, nuclear energy can serve very well as a bridge technology supporting the transition from fossil fuels to 100% renewable sources (World Nuclear 2019).

Canada's first plant was commissioned in 1971 and the last one in 1993. There are no new nuclear plants in the planning process and the existing ones are scheduled for decommissioning over the next 18 years. Unless there is a major reversal in public sentiment by 2037, there will be no operating nuclear plants in Canada and the load will have to be picked up by new natural gas plants and renewable wind and solar.

Wind Energy

Mechanical energy from wind has been used in Canada for centuries but over the past 25 years, commercial electrical energy wind farms have sprung up across the country and now number 295 with a rated capacity of 12 GW. Hydro-Québec Research and Natural Resources Canada installed experimental vertical-axis turbines in the 1970s and 1980s; however most of these are no longer in operation. Horizontal-axis wind turbines proved to be more efficient, and in 1986 Hydro-Québec constructed the 65 kW Kuujjuaq wind turbine as a demonstration project. All of Canada's provinces have adopted wind energy including BC, which began construction of its first wind farm near the town of Chetwynd in 2008. Canada has substantial potential for more land-based wind farm sites, and has only recently turned to offshore projects.

In 1993, Alberta was able to claim the country's first wind farm (Cowley Ridge) aptly named "Optimist Wind Energy Wind Farm Project." In addition to the other 294 installations, wind farm output contributed 6% of Canada's electricity production or about 1.5% of our total energy output. Their 30 TWh annual output is comparable to four Candu reactors running at maximum capacity with no downtime. Additionally there are over 6400 smaller wind generators tied into the electrical grid.

The wind map of Canada shows many good locations but the best and those nearest existing transmission lines have already been developed. Locations in the far north are abundant but the grid connection costs remain prohibitive and the environmental/wildlife impacts are potentially large.

By way of comparison with the energy density in flowing water, a 30 km an hour wind has approximately 1/10,000 the energy that a similar cross section of water flowing over Niagara Falls would. But wind occurs in many places while high volume, high velocity water flows occur in only a very few.

The Canadian wind industry is growing rapidly at 15% annually since 2013 and the expectation is that by 2025 wind will supply 20% of Canada's electrical consumption.

Solar Energy

Solar energy is the most direct form of energy collection we currently have available to us. The source of all of our energy (aside from nuclear) is ultimately the sun and when the heat from the sun causes air to move and rain to fall, energy is transformed into flows of wind and water which we can harness at some point.

When rays from the sun pass through a window in a passive energy home, the irradiated surfaces in the home reradiate the absorbed energy as heat within the envelope of the house. Roof mounted solar hot water and air heaters are also used to pump heat into the house or commercial building. When the photons from the sun

Fig. 2.19 House roof optimized for solar energy collection (author)

strike a photovoltaic (solar) panel they knock electrons free from their atoms and these are transformed into electrical current.

Solar panels actually comprise many smaller cells which are linked together and output a DC current to inverters. The inverters convert it into alternating current to be either used within the house as 120 V or 220 V household current or shunted into the electrical grid for use locally or hundreds of kilometers away.

The potential for solar energy varies across Canada and is lower in coastal areas, due to increased cloud coverage and humidity, and higher in the prairies. And then there is winter, snow cover, and northerly solar irradiation levels to take into account. However, a great deal of Canada's residential electricity requirements could be met through the installation of solar panels on the roofs of residential buildings. This would require significant investments in the panels and associated inverters and cabling and even more in the retrofitting and reorienting of roofs originally designed for esthetics rather than energy collection (Fig. 2.19).

House with roof 100% oriented near due south currently with solar hot air panels installed with room for 40,300 W PV panels for a 12 kW capacity solar system (Fig. 2.20).

The rooves in the picture above will be difficult to retrofit for solar panel installation and will require what will amount to a rebuild of the top story, possibly quadrupling the cost of the solar system itself.

Canada's use of solar energy has increased in recent years, although total market penetration remains relatively small. But installed capacity has seen annual compound growth rates of 13.8% since 2004. In 2014, installed PV capacity reached 1843 MW (Natural Resources Canada 2019).

In Central Ontario, a 10 kW solar system would produce 87,600 kWh of energy if the sun shone 24 h a day, 365 days a year (24 × 365 × 10 = 87,600). The real world experience of systems in this region, which includes night time, clouds,

Fig. 2.20 House rooves with difficult upgrade path to solar energy collection (author)

snow cover, and seasonal variation in the strength of the sun, indicate that they are likely to produce 12,000 kWh a year. This yields a capacity factor of 13.7% (12,000/87,600 = 13.7%).

For comparison, one of the best sites on the planet for solar panel installations is Atacama, Chile. It is considered to be the driest place on Earth and is 2400 m above sea level. A good installation here would produce 22,000 kWh annually giving it a capacity factor of 25%. In contrast, Inuvik, several degrees above the Arctic Circle, is almost everything Atacama is not and solar systems there could expect to see a capacity factor of well under 10% and possibly as low as 6%. With solar energy, more than any other source of power, latitude counts.

Solar energy can also be harvested in the form of hot air and hot water. Passive solar homes store the energy coming in through their southern windows in their thick concrete pads and foundations, thereby reducing power consumption. These home designs feature extremely high levels of insulation and can therefore make the most of the energy they harvest (Fig. 2.21).

The locations suitable for productive hydroelectric dams are relatively few and far between. The same holds for good tidal and geothermal sites. Although wind blows over every square centimeter of the surface of this planet, viable sites for wind generators or large wind farms are far less numerous than many would think. Solar panels appear to be the one technology that can be installed virtually anywhere and on anything with a suitable orientation. This was once the universal hope of the green energy movement but photovoltaic panels are not the ultimate savior of a clean energy society and certainly not at northern latitudes. They are certainly on

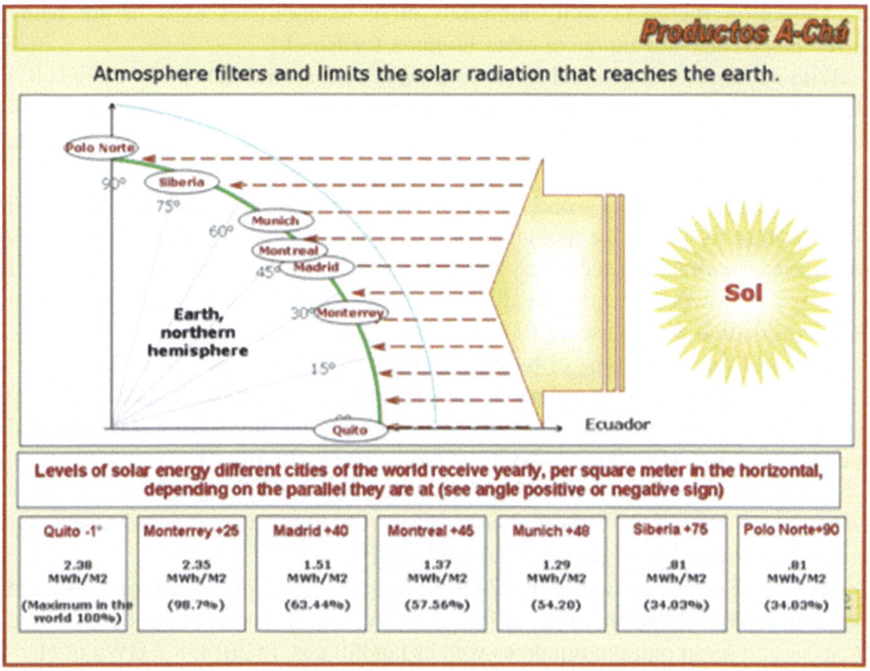

Fig. 2.21 Solar irradiance by latitude (Ayala 2005)

the team but they aren't the franchise player so many, including this author, assumed they would be.

Biofuels

Bioenergy comprises different forms of usable energy obtained from materials referred to as biomass. A biomass is a biological material in solid, liquid, or gaseous form that has stored sunlight in the form of chemical energy. Excluded from this definition is organic material that has been transformed over long periods of time by geological processes into substances such as coal or petroleum. From peat to hard maple, all organic material can be considered to be potential biofuel.

Several types of biomass can be used, with the proper technology and equipment, to produce energy. The most commonly used type of biomass is wood, either round wood or wood waste from industrial activities. Wood and wood waste can be combusted to produce heat used for industrial purposes, for space and water heating, or to produce steam for electricity generation. Through anaerobic digestion, methane can be produced from solid landfill waste or other biomass materials such as sewage, manure, and agricultural waste. Sugars can be extracted from agricultural crops and, through distillation, alcohols can be produced for use as transportation fuels. As well, numerous other technologies exist or are being developed to take advantage of other biomass feedstock.

With its large landmass and active forest and agricultural industries, Canada has access to large and diversified biomass resources that can be used for energy production. Currently, bioenergy is the second most important form of renewable energy in Canada.

Historically, the use of wood has been very important in Canada for space and water heating, as well as for cooking. It is still important today, as 4.6% of households use wood as a primary or secondary source for space heating. Every year, over 100 PJ of energy from wood are consumed in the residential sector, representing more than 7% of residential energy use. (1 PJ = 280 million kWh or 160,000 barrels of oil).

The most important type of biomass in Canada is industrial wood waste, especially waste from the pulp and paper industry, which is used to produce electricity and steam. Every year, more than 400 PJ of bioenergy are used in the industrial sector. The pulp and paper industry is by far the largest industrial user of bioenergy, which accounts for more than half of the energy used in this industry. The Atikokan Generating Station is the largest plant in North America to run on wood waste having been converted from coal in 2014 (OPG 2019).

At the end of 2014, Canada had 70 bioenergy power plants with a total installed capacity of 2043 MW, and most of this capacity was built around the use of wood biomass and spent pulping liquor, as well as landfill gas. In 2014, 8.7 GWh of electricity were generated using wood refuse, spent pulping liquor, landfill gas, and organic municipal solid waste. Most of the biomass-fired capacity was found in provinces with significant forestry activities: British Columbia, Ontario, Quebec, Alberta, and New Brunswick (Natural Resources Canada 2019).

The key point that needs to be made clear with respect to biofuels is that, given their very low EROIs at Canadian latitudes, biofuels can never be more than a marginal source of energy. But biofuels can be distinctly positive sources of energy if the sources are both local and a secondary result of a previous process like sawdust and bark from a lumber mill. On the other hand, liquid biofuels from crops (ethanol) have EROIs around the 1:1 level and may be actually negative in the northern latitudes (Murphy 2010). They are not viable sources of net energy and they should not be considered as an energy source in any national energy strategy.

Firewood obtained locally is clearly a good source of renewable energy. If harvested from one's own property, it is hard to do better than the five or six liters (50–60 kWh of energy) of gasoline involved in cutting and splitting a cord of hardwood with an energy potential of 3000 kWh. Burned in a 60% efficient woodstove, a cord of seasoned hardwood would produce 1800 kWh of heat posting a healthy EROI of around 30:1.

The energetic costs involved with food, tools, woodstove and miscellaneous inputs, not withstanding, firewood, for most of the people who use it, is an excellent energy bargain. Conversely, ethanol from crops is not and only survives on the basis of dollar-based tax incentives which obscure its energetic bankruptcy. The EROI of crop based ethanol production will be discussed in a later chapter.

Animal Power

Historically, horses and oxen provided a large amount of the mechanical power on farms and in logging operations as well as supplying transport in both rural and urban settings. Although very important to the operation of the country, the amount of animal power is not taken into account in this analysis as the energy content of the work the beasts of burden did was hard to quantify. Overall, like gunpowder, the energy content may not have been large but the work done was critical. Integrating animals into an urban environment however was not without its problems. Horses produced an average of 10 kg of manure each day and also deposited 1 L of urine on the streets of towns and cities.

Geothermal

Once the hope of many, including Canada's Green Party, as the future of abundant clean energy, it is now clear that potential geothermal sites are both few in number and relatively small in size in this country. Unlike the geothermal super star Iceland, few countries are sitting atop large bodies of flowing magma relatively close to the surface. Even then, Iceland's Hellisheiði Power Station, which supplies both 130 MW of hot water and 300 MW of electricity to the capital Reykjavik, must be careful not to draw too much heat from its magma bed lest it cool and alter its flow.

The University of Victoria (2017) identified 11 prospective sites in with a capacity delivering of 8% of annual electricity demand. However, the costs of exploiting or even exploring this resource are high. Test holes mean that an investment of up to 50% of the full cost of a working installation must be invested before it is possible to determine if the site is likely to be productive. In other terms, the cost of determining the viability of a project is up to 20 times higher for geothermal than for wind.

Heat below 90 °C might be considered low grade since it can't travel very far or be used to produce electricity which is a high grade form of energy. Here is a description of the requirements for the production of electricity by means of geothermal heat from the Planete Energie group.

> Geothermal water at temperatures of 90 to 160°C can be used in liquid form to generate power; this is called medium-temperature geothermal energy. This technology involves power plants that harness groundwater via geothermal wells. This type of power plant is built near aquifers located at depths of 2,000 to 4,000 meters. In volcanic areas ("hotspots"), where the subsurface holds more heat, the water used by the power plants is sometimes found closer to the surface, at depths of less than 1,000 meters.
>
> In these plants, water that has been pressurized to stop it boiling circulates through a heat exchanger. This equipment contains pipes filled with geothermal water that are in contact with pipes filled with another fluid, generally a hydrocarbon. When it comes into contact with the water-filled pipes, the fluid heats up, boils and vaporizes. The steam obtained drives a turbine that generates power. In the process, the steam cools, returning to its liquid state before being reused in another production cycle.

> If the geothermal water is hotter than 160°C, it can be used directly in the form of steam to drive turbines and generate power. This is called high-temperature geothermal energy. This principle was applied as long ago as 1913 in the world's very first geothermal power plant, in Larderello, Italy. This type of power plant uses water from water tables in volcanic regions, at depths of 1,500 to 3,000 meters. (Planete Energies 2015).

But the earth has a constant temperature within 5 m of the surface which would allow low grade geothermal heat to be pulled up from the ground and used in homes and businesses via heat pumps. But the relative low temperature levels, in the range of 5–10 °C for the southern areas most Canadians inhabit, cannot be used to create higher grade energy such as liquid fuel or electricity. In this application, geothermal energy acts as more of an energy conserver than as a source of raw energy.

As is discussed in subsequent chapters, geothermal may be more of a savior from a storage point of view than an energy generation point of view. There may be great potential for geothermal storage using the mass and insulative properties of the earth to store the surplus heat of summer underground for use months later.

Tidal

Strong, predictable water flows are music to the ears of any renewable energy developer and Canada has some of the largest tidal flows in the world. The Bay of Fundy features the world's highest tides and anyone standing on a rocky point at its mouth is certain to be impressed by the power of the sea as it surges past at speeds of up to 17 kph. For comparison, a tidal generator with a diameter of 10 m can generate 700 kW of power with a flow speed of 11 km/h. A wind turbine capable of generating 600 kW would require a 45 m diameter rotor and wind speeds exceeding 50 km/h (UniLeipzig 2019).

It is early days in tidal power development. A test turbine lost its rotors to Bay of Fundy currents and there are dozens of experimental designs being tested around the world. The corrosive effects of salt water and the force of the water and the wildlife and debris which it contains present unique challenges to developers. Although it is too early to estimate the EROIs or impacts on marine life, there are many projects underway. And they will continue, given estimates which put the theoretical power generation at over 2000 MW from the Bay of Fundy alone. With that kind of potential and the ability to schedule power delivery as reliably as the tides, development at some level is assured (Energy BC 2006).

Hydrogen

Hydrogen is not a source of energy, it is a means of energy storage or an energy "carrier." It is mentioned here only to clarify that common point of confusion. Hydrogen storage of electricity is a very energy inefficient process but it does have some critical strengths and will be discussed in a later chapter dealing with storage and full renewable energy systems.

Energy Transfer: Pipelines

Energy is rarely found exactly where it is needed and therefore must be transported to satisfy demand. There are very few "Niagara Falls" on the planet. The idea of large, continuous output, renewable energy sources located at the center of the transportation, agricultural and industrial hub of a nation and next door to the largest cities and towns, is a rare happenstance. Since power sources and people and their activities are usually located at some distance from each other, pipelines, canals, power lines, and trucks and railcars are needed to transport the energy from where it is produced to where it will be used.

How dense the energy is and the losses inherent in a given means of transport are what determine the longest practical length of energy transmission system. Hot air can be moved relatively efficiently around a house but, given its low density, and the high losses of ducting, distances of over 15 m are impractical. Water is 800 times more dense than air, can be heated (under pressure) to hundreds of degrees Celsius and can be pumped at over 10 km/h through a large pipeline. Icelandic engineers calculate it might be practical to build a pipeline to supply the 700 km distant Faroe Islands with heat from their volcanically driven geothermal hot water systems.

Natural gas and oil can be transferred across continents with low energy. Similarly, electricity moving through high voltage lines say 500,000 V AC (alternating current) lines loses 2% per 1000 km while 3 million volt DC (direct current) line losses would be 30–40% less.

In the USA, total transmission and distribution losses are estimated at 6.5%. Given Canada's more widely distributed population, ours are more likely somewhat higher even though our colder temperatures would improve the efficiency of the copper and aluminum conductors.

There are 40,000 km of oil pipeline in Canada and 77,000 km of natural gas pipelines. The majority of the oil pipeline network transports the oil from the wellhead (or upgrader) to refineries. Pipelines are not used to deliver oil to end users. In contrast, natural gas is delivered to millions of Canadian homes and is available in virtually every town of any size in the country (Fig. 2.22).

The Electrical Grid

All but the most remote villages in Canada's north are connected to the country's electrical grid. In those cases, typically diesel generators, possibly augmented by hydro, wind, and solar sources supply power to a community-sized grid (Dialogue Canada 2007) (Fig. 2.23).

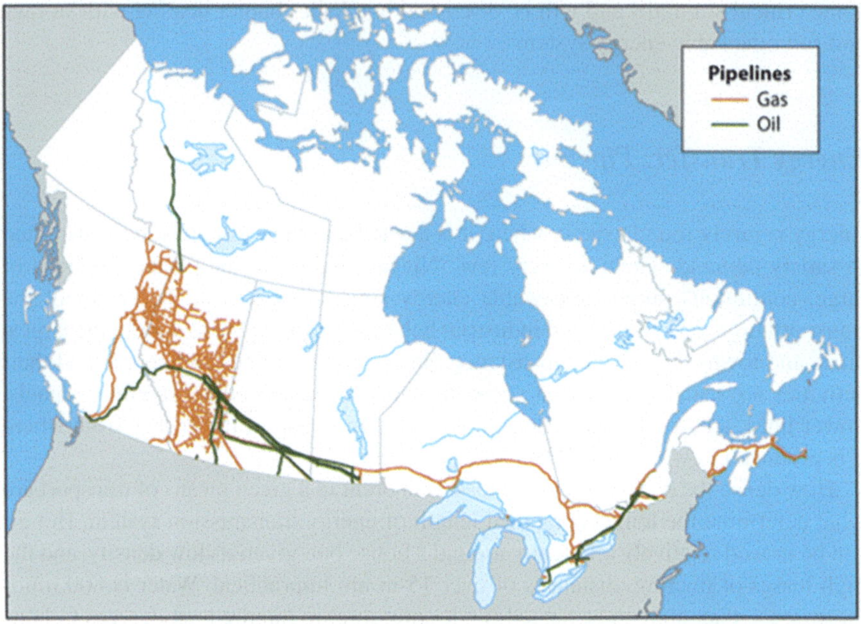

Fig. 2.22 Natural gas and oil pipelines (Oil Sands Magazine 2018)

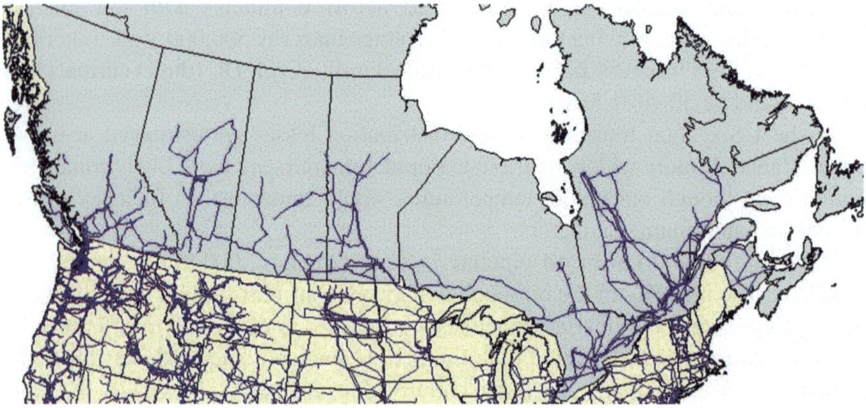

Fig. 2.23 The Electrical Grid Canada and Northern USA (GENI 2019)

Given the very high energy density of fossil fuels, energy can be delivered to any point in the country and, in fact, on the globe. However, cost then becomes a significant factor and it becomes progressively more expensive to move energy once sources and points of consumption become impractical for pipelines and the electrical grid to service.

The Human/Energy Relationship

With access to abundant, cheap energy, the heavy exploitation of natural resources, from minerals to forests to fish, and the creation of a sophisticated society across the nation, was no longer a matter of energy, it was a matter of time and will.

The energy lever expanded human activity by applying the potential of all fuel sources to ubiquitous use in every imaginable application. So dependent are we on raw energy, that the health of our society cannot now be separated from the availability of abundant cheap energy.

References

Alberta History (2019). https://www.alberta.ca/alberta-energy-history-up-to-1999.aspx

Ayala G (2005). https://www.altenergymag.com/article/2005/08/solar-energy-potential-at-different-latitudes/120/

BP (2018). https://www.bp.com/content/dam/bp/en/corporate/pdf/energy-economics/statistical-review/bp-stats-review-2018-full-report.pdf

Calgary Herald Mining and Ranch Advocate and General Advertiser (1884) www.petroleumhistory.ca›archivesnews›phs1995-11

Canada Hydro (2008). https://canadahydro.ca/wp-content/uploads/2015/09/2008-hydropower-past-present-future-en.pdf

CAPP (2019) Canadian Association of Petroleum Producers. https://www.capp.ca/canadian-oil-and-natural-gas/oil-sands

Carmody R (2016) Genetic evidence of human adaptation to a cooked diet. Genome Biol Evol 8(4):1091–1103

CoalScience (2018) International Journal of Coal Science & Technology

Degreedays (2019). www.degreedays.net

Dialogue Canada (2007). http://dialogue-can.ca/points-de-mire-our-focus/sujets-dinteret-issues/pan-canadian-power-grid/

Dodge D (2014) The history of windmills and wind turbines. Canadian Geographic. http://www.canadiangeographic.com/atlas/themes.aspx?id=windenergy&sub=windenergy_basics_history&lang=En

EIA (2019) US Energy Information Agency

Energy BC (2006). http://www.energybc.ca/tidal.html

Euro Nuclear (2016). https://www.euronuclear.org/info/encyclopedia/n/nuclear-power-plant-world-wide.htm

Fairbank CO, Cochrane RO (2000) Oil heritage tour of Lampton county: the birthplace of the Canadian oil industry. Oil Petroleum Institute, London ON

Fame (2019) Canadian petroleum Hall of Fame

Garland W (2019) private communication

GENI (2019). https://www.geni.org/globalenergy/library/national_energy_grid/canada/canadian-nationalelectricitygrid.shtml

Hall (2019) Charles Hall, private communication

Heritage Newfoundland and Labrador (2019). https://www.heritage.nf.ca/articles/economy/oil-economy.php

Hydro Power (2018). https://www.hydropower.org/country-profiles/canada

Loster M (2006). http://www.ez2c.de/ml/solar_land_area/

Masnadi MS, Brandt AR (2017) Energetic productivity dynamics of global super-giant oilfields

Murphy D (2010) Year in review-EROI or energy return on (energy) invested
Natural Resources Canada (2019) Oil sands facts. https://www.nrcan.gc.ca/energy-facts/crude-oil-facts/20064
NEB (2019) National Energy Board https://www.cer-rec.gc.ca/nrg/ntgrtd/mrkt/nrgsstmprfls/cda-eng.html
Niagara Falls Information (2019). https://www.niagarafallsinfo.com/niagara-falls-history/niagara-falls-geology/niagara-geological-areas/drianage-of-the-niagara-river/
Nolan DP (2019) Handbook of fire and explosion protection engineering principles for oil, gas, chemical, and related facilities, 4th ed
NREL (2019). https://pvwatts.nrel.gov/pvwatts.php
O'Connor C (2015) photographer resident of Fort McMurray
Oil Sands Magazine (2018). https://www.oilsandsmagazine.com/projects/crude-oil-liquids-pipelines
OPG (2019). https://www.opg.com/generating-power/thermal/stations/atikokan-station/pages/atikokan-station-biomass-conversion-project.aspx
Planete Energies (2015). https://www.planete-energies.com/en/medias/close/high-temperature-geothermal-energy-power
Powermag (2017). https://www.powermag.com/history-of-power-the-evolution-of-the-electric-generation-industry/
Reguly E (2019) The globe and mail. https://www.theglobeandmail.com/business/commentary/article-is-peak-oil-closer-than-we-think-saudi-aramco-reveals-a-troubling/
Schmitt DR (2013) Heavy and bituminous oils: can Alberta save the world?
StatCan (2017) Statistics Canada Archives Series Q19-25
The Canadian Encyclopedia (2015). https://www.thecanadianencyclopedia.ca/en/article/canadas-first-railway-feature
UNENE (2019). http://www.nuceng.ca/candu/
Unger RW, Thistle J (2013) Energy consumption in Canada in 19th and 20th centuries. Consiglio Nazionale Delle Ricerche, Italy
UniLeipzig (2019). https://home.uni-leipzig.de/energy/ef/15.htm
University of Victoria (2017). http://onlineacademiccommunity.uvic.ca/2060project/2017/06/29/why-arent-we-using-geothermal-energy-for-electricity-in-canada/
Wateroffice (2019) Environment Canada website https://wateroffice.ec.gc.ca
Wikipedia (2019a) History of coal
Wikipedia (2019b). https://en.wikipedia.org/wiki/List_of_generating_stations_in_Ontario
Winchester J (2005) Concept aircraft prototypes, X-planes and experimental aircraft. Grange, Kent
World Nuclear (2019). http://www.world-nuclear.org/information-library/current-and-future-generation/plans-for-new-reactors-worldwide.aspx
Zohuri B (2018) Small modular reactors as renewable energy sources. Springer

Chapter 3
Energy Budgets for People and Nations

Abstract The energy required to prosper in northern regions is vastly higher than the amount necessary to merely survive. When the Europeans arrived in North America, they brought their sophisticated and energy-intensive tools and lifestyles, developed over centuries in an environment which had been heavily exploited and, in many instances, depleted. The environment they encountered in Canada was more severe than anything they had experienced in their homelands, but they had the tools to harvest the resources and energy required from the virgin richness of the New World. Amerindians obtained just enough from the food they ate and branch wood they burned to subsist. Colonists could cut wood and rock efficiently with their steel tools and, in order to maintain their European lifestyles, consumed vastly greater amounts of energy than the native peoples they displaced. Energy evolved into a separate commodity and was finally able to be produced, counted, consumed, and accounted for like any other material good.

The Need for Budgets and Quantitative Analysis

Citizens in every country both produce and consume energy out of necessity. Given the ambitious (or avaricious) nature of humans, typically we will consume as much as we produce, and we will always try to produce a surplus to support luxuries or investments for further growth. Most discussions of energy revolve around how much we produce and rarely include consideration of how much we actually need.

People who use a lot of energy are deemed to be wasteful, while those who use little or appear to use little, are deemed to be frugal. But a budget discussion should involve a set of more complex questions: how much energy is produced, how much effort and energy is required to produce the consumed energy, how much energy is needed for a standard of living deemed to be basic, and how much energy remains which can be termed "disposable" or discretionary.

In the restaurant business, the maxim is "Location, Location, Location". In the energy and survival business "Latitude, Latitude, Latitude" is a more apt summation of the critical factors which determine the relative wealth renewable energy is able to produce.

J. E. Meyer, *The Renewable Energy Transition*, Lecture Notes in Energy 71,
https://doi.org/10.1007/978-3-030-29115-0_3

The Progression of Energy Budgets

Inuit Precontact

The energy inputs for the Inuit before European contact were very simple. The energy in their system came from animals or fish they could harvest and convert to food. Storage was in the animals themselves. There was very little in the way of resources to convert to heat, light, or mechanical energy. In this sparse environment, the Inuit were eking out survival in what amounted to a resource desert. Yet this desert demanded extremely high levels of energy just to survive.

The Inuit energy budget therefore required a great deal of effort to harvest every kWh of energy and the harsh environment required a great deal of energy merely to stay alive. Over the 10,000 years the Arctic was inhabited, there was no surplus left over to build permanent housing, any kind of social infrastructure or to develop tools and learning which were not directly related to short-term survival.

Forest Amerindians

Compared to the Inuit, Amerindians living in more southern regions of Canada faced less extreme climate and had access to vastly richer resource bases. In some regions, agriculture was possible and on the west coast it was even possible to go through an occasional winter without snow or ice. But all areas featured an abundant supply of wood which allowed a much higher level of heat and light to be included in their lifestyles.

These lifestyles were as varied as the environments in which they took place. Plains' Indians and those in the northern boreal forest faced grimly bitter winters and no possibility of building up supplies of food through agricultural production. They, like the Inuit, were forced to move seasonally in pursuit of game and fresh supplies of firewood.

Probably the most environmentally rich and least demanding environment was located on the coast of British Columbia where the natives, blessed with abundant forest and ocean resources, were able to develop many different types of technologies and art forms.

Their huge, permanent log buildings, wide range of clothing and consumer goods, as well as exquisitely finished totem poles, displayed the creative abilities of a people with time to devote to pursuits beyond those necessary for bare survival.

In the most southern central regions of Canada, from Southwestern Ontario to the Eastern Townships in Quebec, agriculture was possible, as were semipermanent structures and villages. These were probably moved every few decades as the soil depleted.

It is also possible that these areas also spawned "forest agriculture" of the type that was practiced by natives in the northeastern USA. The practice of growing stands of nut-producing trees such as the American chestnut, walnut, beech, hazelnut, and butternut may have occurred in the regions of the mildest climate, but records of this practice are hard to come by for Canada.

Plantations of these trees would have provided a huge improvement in the local diet, due to the protein and fat rich nuts which could be stored for months. In the Appalachians, prior to the arrival of an Asian bark fungus in the late 1800s which decimated the American Chestnut, it was possible for people to lay away a winter's supply of rich food with a mere week's worth of gathering work in the fall (Bolgiano 2007).

Such plantations would have added greatly to the energy budget of any band of natives lucky enough to inhabit the favored regions. And, as European settlers moved into those regions, they most likely failed to understand this resource and cleared the stands of the very productive nut bearing trees to make room for fields of the vastly nutritionally inferior grains of wheat, corn, and oats.

Early Settlers Circa 1700

The early settlers arrived in North America with the same iron tools they had used in Europe. But the reason for fleeing Europe was to escape the declining resource bases of that region and establish a more promising life for themselves and their children. When the Europeans arrived, they applied their superior tool kit and technology to what was to them, a rich, huge, and limitless resource base.

There were endless stands of old growth trees the likes of which had been erased from Europe many centuries before. There were many areas of rich soils and the lakes and seas teemed with fish while abundant wild game roamed the forests and prairies. All these resources were ready for exploitation by a desperate, well-equipped pulse of colonists. And they did exploit them, allowing their populations to surge, often at a rate of 3% a year, which meant a doubling every 25 years. Over a period of 150 years (6 doublings) a population of 10,000 would become 320,000 and could become 10 million in 250 years (10 doublings) if nature was willing to accommodate such numbers. It wasn't.

In Canada, these settlers had to endure the same harsh winters as the Amerindian natives but they were able to build permanent dwellings equipped with fireplaces and, later on, stoves. Despite the higher efficiency of these homes and heating systems, the energy used to heat them was vastly larger than that used by the Amerindians because the square footage per person was larger and the temperatures inside were kept higher.

The iron implements wielded by the settlers were also more efficient than the stone axes and knives of the natives thus allowing much more fuel to be produced in much less time. It also made inevitable the destruction of the forest resources but in the period of an expanding frontier, this went largely unnoticed. The American Benjamin Franklin was quoted as saying he hoped that people would not destroy the

New World the way they had the Old but his concern was shared by few in the face of what appeared to be, in the late 1700s, the unlimited potential of the Americas.

Not only could the settlers exploit renewable resources more thoroughly than the Amerindians, they brought a completely new dimension of development to North America: mining. They mined minerals with which to make a multiplicity of articles and machines, but they also mine energy, initially in the form of coal. Coal replaced wood for many heating applications just as it had in England in the early 1600s, when their "timber famine" (forest collapse) forced the population to substitute coal for the vanishing stands of timber (Ward 2007). It is likely the onset of the "Little Ice Age," with its extreme cold, accelerated the decline of forests in Europe.

It is hard to overstate the importance of firewood in the development of Canada in the early years. The European colonists who arrived in the 1600s walked into the full teeth of Canadian winter, as this was during the period called "The Little Ice Age" when temperatures across the globe were several degrees colder than normal. Add in the perhaps 15C to 20C lower winter averages in Canada than in southern England and France and the need for heat becomes apparent.

Of the 24 men who sailed with Samuel de Champlain, explorer and cartographer, to set up the first permanent building in "New France" at Quebec, 16 died during that first winter. Small wonder Champlain had reason to complain that his men seemed to spend all of their time cutting wood. On average, Canadians even today consume twice as much energy as do residents of Britain, and three times more than the Portuguese which gives a clear indication of the impact climate has on energy consumption.

These impacts are not limited to heating, but to all manner of activities and work demanded by the severe change of seasons compared to more temperate regions. As one individual who moved from Montreal to Nag's Head, North Carolina, which is over 1000 km further south, wrote to the author "You wouldn't believe all of the work you don't have to do here."

In 1800, the first year for which records are available, firewood was the source of 40 times more heat than coal. By 1868, 1 year after Confederation, firewood still provided 84% of the energy budget for Canadians (Unger and Thistle 2013).

It was not until 1906 that the combined total energy from oil, coal, and natural gas exceeded that of firewood. It is easy to understand the attraction of firewood for rural and small town residents. If not available on their own property, abundant hardwood and birch were almost certainly available on the nearby property of a relative or close friend. It was likely harvested communally with little in the way of cash exchange.

Coal, oil, and natural gas had to be paid for, and they required special stoves or heating equipment to use. Additionally they were not pleasant substances to handle. After the broad acceptance of woodstoves, which increased efficiency by up to ten times over that of open hearths, firewood became a very refined source of heat used and enjoyed by millions of Canadians to this day.

The energy budget of the early colonists quickly became larger and more complex than that of the Amerindians. Their tools held the embedded energy of their manufacture in Europe and they could exploit the renewable resources much more efficiently and access nonrenewable stored energy in the form of coal. Energy consumption and production in the Americas had begun to surge. The per capita energy budget of North Americans was expanding rapidly.

Canada at Confederation 1867

By the time the Nation of Canada was formally created in 1867, overall the population had increased to 3.5 million, despite the near collapse of the Amerindians, and the infrastructure of the energy economy was firmly in place. Coal had driven the expansion of the railroad network and crude oil was just beginning to replace whale oil for lighting and some other purposes.

The use of coal-driven steam engines and coal-fired forges had spawned a broad manufacturing base and coal was widely used to heat commercial and residential buildings. Wood heat was still the dominant source of residential heat but, with the convenience, consistency, and low cost of fossil fuels, the stage was set for it to fade into the background as a major energy source.

The transition from renewable energy to nonrenewables was well under way. By the late 1800s oil began to appear on the scene and it opened the floodgates of mechanization/energization of the both the economy and lifestyles.

1945: End of the Coal Age, Beginning of the Oil Age

By the end of World War II, all of the basic fossil fuel energy infrastructure for Canada and most western nations was in place. It would continue to grow in capacity and complexity but the essentials were there for the generation and distribution of energy from coal, oil, natural gas. These three fuelled transport, manufacturing, mining, agriculture, forestry, fisheries, and a growing leisure and travel industry which was, in many ways, the ultimate expression of our new found wealth of disposable energy.

The automobile made daily travel possible for a large percentage of the population and cities began to spread suburbs outward onto the surrounding farmland.

Hydro power, coal, oil, and natural gas all served to generate the ultimate means of power transmission—electricity. Society was swimming in and resting upon controlled energy flows. And the supply of available energy was increasing while costs were falling. This allowed new resources to come within reach, not only in Canada but around the world and these were aggressively exploited. These new resources were incorporated into the production of finished goods which swelled through new processes and ever increasing economies of scale. During this period, limitations were rapidly disappearing into the distance.

2019

In current day Canada, all of the energy sources which were available to the previous generations have been joined by technically more sophisticated processes. Fossil fuel energy systems are heat based and must be burned to release their energy. Technically, nuclear power is also heat based as the uranium fuel is used to boil water and create steam pressure to drive turbines.

Today, in addition to heat-based systems, raw energy can be derived from spinning wind generators and stationary solar panels.

All of these energy systems have combined to give Canadians a very large daily budget of energy. Currently, the average net consumption of energy per day amounts to 250 kWh per person. This equates to 25 L of gasoline or 1/6 of a barrel of oil. That is the amount of energy it takes to support the modern lifestyle of Canadians in our northern climate.

Contrasted to the precontact Inuit and forest Amerindians, the average Canadian now uses over 50 times more energy. Unfortunately, the majority, 70%, of that energy is mined in the form of fossil fuels or uranium and, as is the case with all mined resources, the supply is finite. The amount of energy we obtain from renewable sources, such as hydroelectricity, wood heat, solar, and wind power and, of course, food from crops is much smaller. If we were to convert to renewables tomorrow, our energy budget would shrink dramatically (Fig. 3.1).

Although 30% renewable may sound like a very low proportion, it is much higher than the world average of 11% because few nations have the huge river systems which underwrite our hydroelectric capacity. Our Rock Star energy source, Niagara Falls, is only the eighth largest hydro producer in Canada and contributes approximately 7000 GWh annually or 2% of our total energy consumption (Mining and Energy Canada 2015).

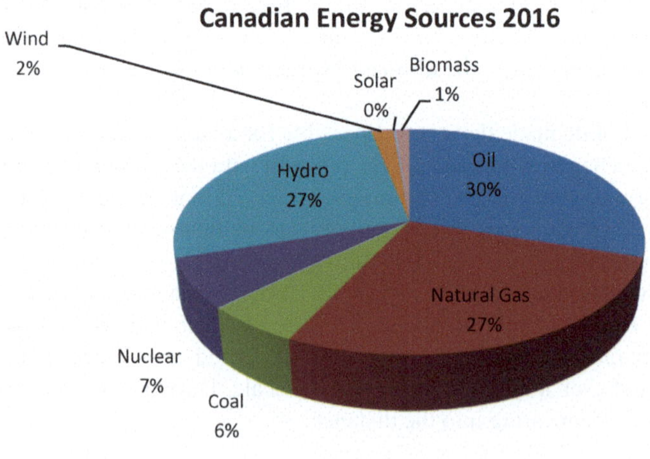

Fig. 3.1 Primary energy sources in Canada (Hughes 2018)

In terms of our energy consumption patterns, we spend 30% of our energy in producing energy, a great deal of which we export. Industrial energy consumption, which stands at 28% of our current energy budget, would probably be 10–15% higher if we produced all of the finished goods we consume, since we import more manufactured products than we produce. These contain a huge amount of embedded energy which must be taken into account in any examination of energy budgets (Fig. 3.2).

The types of energy we consume outside of the energy production sector are led by natural gas and oil based fuels. Here it can be seen that well over 70% of our final energy consumption is supplied by fossil fuels (Fig. 3.3).

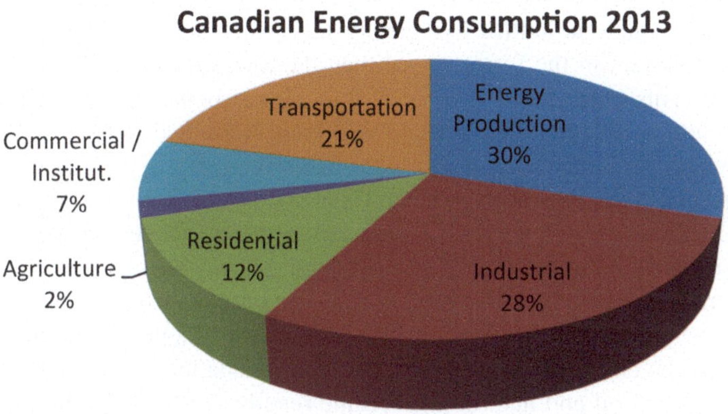

Fig. 3.2 Final energy consumption in Canada (Hughes 2018)

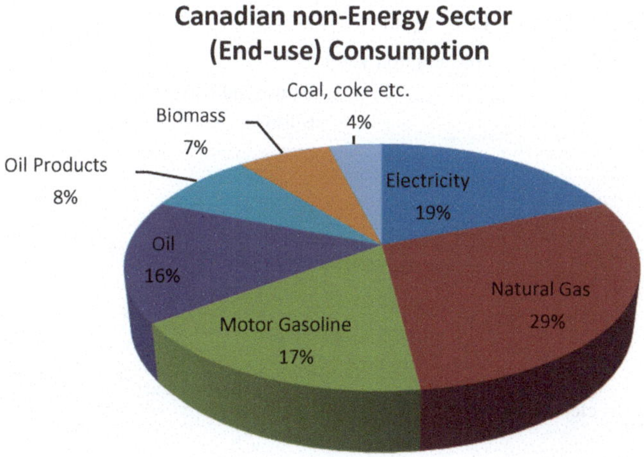

Fig. 3.3 Energy consumption by end use (Hughes 2018)

Summary of Canada's Energy Resources

For the last 100 years our economy hasn't just been using raw energy for some specific tasks like firing projectiles or providing dim, flickering light in the dead of a winter's night, it has incorporated large amounts of energy into every process our society has developed.

The biomass/renewable energy potential of Canada is still very large despite the paving over of huge swaths of prime farm and forest land since the arrival or Europeans. But what have inevitably declined are the stocks of fossil fuels. Despite being used heavily for only 150 years in the case of coal, and 70 years in the case of oil, there is significantly less stored energy remaining underground than that existed 200 years ago.

As David MacKay points out in his elegantly articulate YouTube TED Talk, "A Reality Check on Renewables," originally there was more carbon energy lying under Great Britain in the form of coal than there was under Saudi Arabia in the form of oil. This once great resource is now almost depleted after 350 years of heavy extraction. This reminds us that fossil fuels are mined resources, and mines eventually play out as the now energy-poor British can attest.

The original stores of energy beneath Canada took the form of coal, oil, natural gas, and uranium. These huge reserves are still largely intact despite a considerable portion having been drawn down. The chart below shows the total energy consumption of Canada over the past 150 years. If firewood figures from the early 1600s to 1800 were available, the total consumed energy would be somewhat higher.

However, the bulk of energy consumption in Canada has taken place in the last six decades when oil and natural gas became readily available. In that period, the population increased rapidly as did material consumption levels (Table 3.1).

Table 3.1 Cumulative energy consumption in Canada

Cumulative Canadian energy consumption 1800–2017		
	Equivalent barrels of oil	
	Billions	Percentage (%)
Coal	13.5	11
Oil	39.8	32
Natural gas	37.8	30
Total fossil fuels	91.1	73
Hydro electric	23.5	19
Nuclear	5	4
Total electric	28.5	23
Firewood	5.6	4
Total fossil fuels, electric, and wood	125.2	100

Why Kilowatt-Hours?

Joules are the energy unit commonly used by scientists, researchers, and many engineers. However, the aim of this book is to discuss the ability of societies to change, not only the energy they use but the way they use it. This boils down to the relationship between people and energy and increasingly, our energy will be made and delivered in the form of electricity. Individual consumers, whether homeowners, business managers, or public servants, monitor their electricity use and pay for it in kilowatt-hours. We measure the energy we use to run our houses and charge our cars in kilowatt-hours. Solar photovoltaic and wind farms typically measure their output in megawatt-hours. Watt hours is the metric of the electrical grid.

One liter of gasoline contains almost 10 kWh of energy. One horsepower is equal to 746 W. Run a 1 horsepower engine for an hour and it will produce 746 Wh or 0.75 kWh. A typical wall plug in North America runs at 120 V and uses a 15 A fuse or breaker. Voltage times current equals power so the maximum power that can safely flow from a typical wall plug is 120 × 15 = 1800 W. Use it at this rate for 1 h and you will have consumed 1800 Wh of energy or 1.8 kWh. Europe uses 240 V for most domestic and commercial circuits.

One kilowatt-hour is equal in power to 3.6 million Joules (3.6 MJ). As it is, there will be some large numbers used in the book as the energy quantities addressed span daily individual budgets to annual national consumption. Measurements in kWh keep this somewhat manageable whereas the use of Joules would require the constant use of scientific notation. Below is a table laying out the terms of magnitude for the Watts mentioned throughout this book (Table 3.2).

The amount of energy available to a person each day has increased dramatically since Europeans first arrived and the sources of energy have changed from renewable to nonrenewable.

Over its history, Canada, from 1800 to 2107, has used the equivalent of 125.2 billion barrels of oil. In 2017 it was using energy at the rate of the equivalent of 2.5 billion barrels annually and growing.

Table 3.2 kWh scale

Scale	Common term	Number of zeros
Single		–
Kilo	Thousand	3
Mega	Million	6 or a thousand thousands
Giga	Billion	9 or a thousand millions
Tera	Trillion	12 or a thousand billions
Peta	Quadrillion	15 or a million billions

There are still more terms covering numbers with up to 30 zeros but figures of this magnitude aren't necessary for national energy policy purposes

Energy Budgets Through History

Energy Budgets Through History per Person in kilowatt-hours (Table 3.3 and Fig. 3.4).

In the figures above there is no inclusion of power produced by work animals which was substantial, and may have been in the range of three or four times the energetic content of the values listed for human food. Although both solar and wind are beginning to make contributions, in terms of 150 years of cumulative energy production, these amount to very little. As well, the energy from waterwheels and small farm windmills was not estimated.

A basically flat per capita energy budget from the early days of colonization to the middle of the twentieth century, is not something most people would expect to see.

Table 3.3 Energy budgets through history per person per day in kilowatt-hours

Time period	Food	Wood	Coal	Oil	Natural gas	Hydro	Nuclear	Total
Inuit precontact	3.4	1	0	1	0	0	0	5.4
Forest Amerindians	3.2	6	0	0	0	0	0	9.2
Early Settlers circa 1700s	3.2	78	3	0	0	0	0	84.2
Canada 1868	3	77	14.5	0	0	0	0	94.5
1906	3	38	38	1	0	0	0	80
1948 end of coal, beginning of oil	3	16	32	32	3	10	0	96
1968 Cheap Oil Era	3	4	26	86	56	47	1	223
2017	3	2	16	112	122	76	17	348

Note the 1968 and 2017 figures include large amounts of coal and natural gas and lesser amounts of oil used in electricity and oil sands production

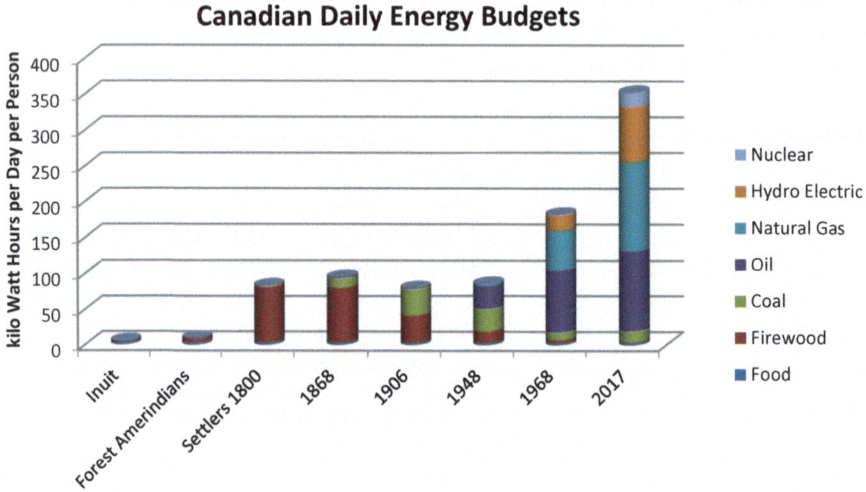

Fig. 3.4 Daily energy budgets through history (Unger and Thistle 2013)

In fact, one could make the case that the energy budget of Canadians remained stable over three centuries from the mid-1600s to the mid-1900s. As counterintuitive as that might seem, the shift in energy sources along with leaps increases in efficiency have enabled a great deal more work to be done than that was possible 300 years earlier.

From the time the first settlers landed to the time their descendants began to replace firewood with coal, natural gas, oil and electricity, immense quantities of firewood were burned very inefficiently to heat all manner of sheds, barns, homes, and commercial buildings. The settlers had no intention of living off the land; they were transforming it to fit the image and lifestyle of their homeland.

Amerindians did not make use of the abundant firewood resources available most likely because they lacked steel axes and saws. The process of taking down a large tree with a stone axe could easily take days of labor. Once the tree was on the ground, it was not yet firewood; it was a horizontal tree which still had to be cut up into smaller pieces, repeating the tedious process of its downing. Regularly turning large trees into cut-to-size firewood would have been impossible with stone implements. Steel saws and axes were required for large scale firewood-production to be at all practical. The technological advances of stone hearths and later the much more efficient iron stoves would only be useful if firewood was cut to the proper size.

Hence, the Amerindians' used of deadfall and small branch wood in very inefficient open fires in their lodges. This was the most practical use they could make of the resources at hand with the tools available to them.

Settlement in Canada took place in the regions covered with forests of maple, oak, beech, ash, birch, and elm, and farms and towns were carved out of this standing stock of Grade A firewood. It was only after the large scale urbanization of the Canadian population began in earnest that firewood became too inconvenient and uneconomic for most Canadians to use.

Also in that period, clear-cutting was often the standard method of harvesting timber. This led to declining soil health and consequently smaller trees with longer maturation times. If the cutting was done on the deep rich soils near the St. Lawrence and Southern Ontario, the ill effects might be smaller and take many more decades to manifest themselves, but clear-cutting on the rocky Canadian Shield could well result in a large loss of soil richness and forest productivity lasting generations.

As woodstoves replaced open hearths and as more technically advanced and efficient coal, oil, and natural gas furnaces became widespread, the total energy consumed may have stayed at the same level but the useful heat generated would have been much higher and applied to many more tasks than simply heating buildings.

Technology's Contribution: Efficiency

Efficiency increases allow more productive work to be done with less effort. One liter of gasoline contains the equivalent of around 10 kWh of energy per liter (actually ~ 9.1 kWh) which is released by burning it. A typical internal combustion automotive engine might be 30% efficient which means that, of the energy potential in the fuel it burns, 30% is converted to mechanical energy pushing the car and all of its ancillary systems down the road while the remaining 70% is lost as heat.

Early steam engines burned coal or wood in bulk to heat water in a boiler to create pressurized steam that drove pistons up and down in cylinders. While they worked and showed potential, and obviously were able to displace human and animal labor, the first steam engines had incredibly low efficiency levels. Early gas and diesel engines were better because they burned a liquid fuel, in aerosol form, in the cylinders themselves so that more of the heat from detonation (rapid burning) was trapped in the cylinder and could be applied to pushing the piston.

Electrical motors are even far more efficient because they do not burn fuel; they use energy to create two magnetic fields which are arranged in such a way that they alternately attract and repel each other causing the one of the field generators (the rotor) to spin. Usually, one of the fields is created by permanent magnets and the other by alternating current in an electrical circuit. Heat is a by-product of this process but it is not the means by which the process is energized. The energy comes from a source at the other end of a short or possibly very long wire. The source can be a nearby battery or a wind, solar panel, or hydro dam or coal plant hundreds or even thousands of kilometers away.

In terms of efficiency gains through history, there has been a great deal of progress. Burning a log in an open campfire under the stars would warm the people very near the fire with radiant heat but the vast majority of the heated air would rise straight up without transferring any warmth to those on the ground. In an Amerindian longhouse or teepee, in addition to the somewhat contained radiant heat, the hot air would be partially trapped by the walls and roof of the structures and impart some warmth to everything within them before escaping through a hole in the top. It must be noted that along with the heat from the fire, all of the ash, soot, and combustion gases would also be partially entrapped in the structures and saturate the air that the occupants breathed.

The stone hearths of the early settlers would have been at least as efficient as the lodge fires of the Amerindians and they channeled the nasty pollutants which wood fires created. Possibly some of the lovely hardwood smells lingered. Despite absorbing a lot of energy in their stone mass and re-radiating it into the room, hearths burned firewood in an enclosure open on two sides and sent large volumes of hot air up the chimney. This hot airflow was expelled with its heat content plus a great many unburned wood particles into the atmosphere.

The leap forward to the modern woodstove sees possibly a tenfold increase in efficiency. Hot gases from combustion are completely contained and only released upward into the chimney when the majority of their heat has been radiated into the room. Temperatures within the stove are much elevated over the more open methods of burning wood resulting in much more complete combustion of all wood fiber thus exploiting the heat potential of the material being burned to its fullest.

Once completely enclosed fireboxes were developed, it became practical to use coal in the same manner as wood. Coal required less space and was a more consistent fuel than wood making it practical for distribution and use in the larger urban areas. A fully enclosed stove would allow coals' major drawback, its incredibly distasteful fumes, to be exhausted from the building. However, there was no hiding the fumes once clear of the chimneys and the air quality of even medium-sized towns whose major source of heat was coal became distinctly unhealthy during periods of low winds. In larger centers, "killer fogs" with very high levels of particulates and exhaust gases were a danger when thermal inversions occurred. The coating of soot which covered every surface in London, England even into the late 1950s, was an indication of the toll taken on the lungs of its residents.

An open fire might have been 5% efficient, a hearth maybe as high as 10%, and early woodstoves were certainly no more than 40% where as a modern woodstove can approach 60% efficiency. A consistent fuel with digital monitoring and finely controlled combustion, like a modern natural gas boiler can turn over 95% of the heat potential of its fuel into heat for the home.

Similar efficiency gains were realized in the machines which converted heat to mechanical motion. The basis of the improvements followed the pattern of development in wood burning, namely, more completely burning the fuel and conserving and containing its heat to where it was doing work.

- Electrical engines, of course, started out impressively efficient 130 years ago and have become incredibly efficient today. Below is a list of the efficiencies of various engines through their development lifespan.
- Steam engines first applied to do real work in the early 1700s had an efficiency of 1% rising to 10% in the 1820s and plateauing at 25% by the late 1800s.
- Gasoline internal combustion engines were running by the 1860s at 5% efficiency, and rapidly improved to 20% by 1900 plateauing in the low 30% region by 1950.
- The diesel engine started out with an impressive 25% efficiency around 1900 and increased quickly to over 40% in a few years, finally plateauing at just over 50% by 2000.
- The gas turbine (jet) engine was given its first practical application during the Second World War with an efficiency of 15% rising to the start of its plateau of 45% by 2000.

The electric motor of 2019 is in the neighborhood of 95% efficient as it takes the pure energy of electricity and converts it directly to mechanical energy without the middlemen of combustion and heat. Heat losses come from mechanical friction and the elevated temperatures of the copper conductors as electrical current flows through them at high rates. If in the future, science produces a material with next to zero resistance, and replaces copper, which has the lowest resistance of any workable metal currently available, then the efficiency of an electric motor could exceed 99%.

What was the overall efficiency improvement across North American societies over time? Vaclav Smil estimates that in the largely wood fueled American economy of the 1860s, people were able to harvest 10% of the potential energy their burned fuel contained. By the early twenty-first century, this had increased to 40%, which resulted in an immense increase in the work that could be done by the same amount of fuel (Smil 2017).

The Miracle Savior That Never Existed

Devices can always be made more efficient but sometimes our expectations for dramatic improvements are misplaced. Late in the twentieth century, there was an urban myth speculating the 200 miles per gallon carburetor had been invented but that big oil companies had bought the rights to it, mothballed it, and silenced the inventors. Certainly big oil has been guilty of some heinous crimes against society ranging from the dismantling of public transit systems in the mid-twentieth century to launching black campaigns to undercut climate change science.

But killing the 200 mpg carburetor is something they are neither guilty of nor needed to do. The best internal combustion engines at mid-century may have been 30% efficient. Given the tires and aerodynamics of cars of the day, plus their weight, it may have taken one-third of a kilowatt-hour of energy to shove a large American car 1 km down the road at 100 km/h.

Keep in mind that 1 L of gas contains the equivalent energy of about 9 kWh. If 0.33 kWh were required to move our aged classic 1 km, then 1 L of gas would push it 27 km if the engine was 100% efficient. An Imperial gallon contains 4.5 L of gas.

This means that even with the perfect efficiency of the motive power, an ungainly 1960s era American sedan could be driven 4.5 × 27 or only 120 km (75 miles) on the energy potential of 1 gallon of gas. The 200 MPG Carburetor would have to find a way to deliver 2½ times the energy contained in the gasoline. In fact, it would have to deliver several times of that because it would be fueling an internal combustion engine with scores of moving parts, high levels of friction, and a process of combustion which involved compressing its air–fuel mixture by a factor of 9 or 10 to 1 before the spark plug detonated the now very dense mixture.

In other words, the internal combustion engine suffers from very high mechanical and pumping losses aside from the heat losses of the combustion process and no miracle of fuel atomization leading to perfect combustion and release of 100% of its energy potential is going to negate those losses.

It is possible to achieve a 200 mpg rating on a vehicle. Simply bolt a small carbureted 5 horsepower engine onto a very aerodynamically slick go-cart with very narrow and hard tires and the vehicle could certainly surpass the 200 mpg mark. But it would hardly provide the solution to a society's transportation needs.

Efficiency's Siamese Twin: Conservation

While efficiency had gained the full attention of engineers anxious to squeeze out the last bit of "horsepower" from every drop of gasoline, the issue of conservation received less attention. This is unsurprising given the increasing availability and declining cost of energy over the 300 year period ending in perhaps the 1970s. In that energy-rich environment, it made more sense to simply produce more energy than even thinking about saving it. It was "demand driven" thinking which still resides in the models of many commercial economists.

Conservation comes in two flavors. The first emphasizes saving energy in the way one does things. Insulate the house, drive a more efficient car, and buy LED light bulbs. By installing upgrades and adopting new technology, it is possible to maintain the current lifestyle and activity patterns while simply using less energy.

The second type of conservation takes the form of modification of our hopes, dreams, and lifestyles. This involves living in a smaller house which loses less heat and keeping the temperatures in a wider comfort zone by wearing more or less clothing, thus requiring less energy for heating and cooling. If an automobile is even needed, drive a smaller more efficient car and use it less by substituting walking, bicycling, and public transit for the fewer and shorter trips you will make.

Keeping Up with the Mesopotamians

What we feel we need is strongly influenced by our need for prestige and acceptance. This is nothing new as human consumption aspirations have existed as far back as recorded history extends. One of the first applications of writing—cuneiform—was applied to business record keeping and trade. Personal letters also were transmitted on these clay tablets and one recovered letter was from a wife to her husband on a trading mission in another town. The gist of the letter was that another trader in their town was beginning construction of a second house and she asked her husband in her letter, "when can we do the same?"

Bigger, grander, more opulent, more comfortable, faster, quieter goods and more visually arresting devices are at the core of consumer markets. In many cases, these goods can be classed as "toys" rather than needs but the need to have what others have is a fundamental human and perhaps, a mammalian, characteristic. Perhaps the "greed society" is substantially composed of the "emotional need society."

Building self-confidence and a wide range of personal abilities might go some distance to subduing these counterproductive aspirations which are the feedstock of much of our current consumer good industry and its advertizing campaigns.

The amount of energy available or the supply side of energy budgets is an important part in determining how well we live. But tempering our aspirations and shaping our activities to match the realistic supplies of energy while making the machines and processes we use as efficient as possible are just as important as the supply. In any financial budget, tailoring the spending to match the income is a well-accepted principle. We need to apply that standard approach to energy.

What Did We Have to Start With?

Canadians have cut down a lot of trees, mined a great deal of coal, and pumped large quantities of oil. The forests of Canada no longer feature vast stands of hardwood in the southern regions of the country as they've been displaced by farms and

Table 3.4 Original reserves, current reserves, and how long they will last

Resource	Remaining reserves	Units	Barrel of oil equivalent (BOE)	Barrels of oil (billions)
Coal	6582	Millions of tons	4.78 barrels per ton of coal	32
Conventional oil	4.5	Billions of barrels	1	4.5
Oil Sands	165	Billions of barrels	1	165
Natural gas	1200	Trillions of cubic feet	5800 ft^3 per barrel of oil	206
Total BOE remaining				408
Cumulative fossil fuel energy consumed in Canada's history				91
Original resources of fossil fuels in Canada precontact				499
Percentage of original reserves consumed to date				18%
Years of fossil fuel lifespan at the 2017 rate of 2.4 billion BOE[a]			Domestic consumption only	170 years
Years of fossil fuel lifespan at the 2017 rate of 3.7 billion BOE[a]			Total including exports	110 years

[a]This assumes stable population of 37 million and no export increase

pavement but they do exist in patches and could readily eventually return to the places humans vacate.

Fossil fuels don't renew themselves but Canada is far from exhausting its reserves. There were and still are large reserves of fossil fuel buried beneath the Canadian landscape. These reserves are estimated to be as follows by the National Energy Board. The reserves are then converted to barrels of oil equivalent to yield a picture of the complete Canadian fossil fuel inventory that exists now and existed originally.

The original endowment of all fossil fuels in Canada was the equivalent of 499 billion barrels of oil. There is still the equivalent 408 billion barrels remaining (Table 3.4).

The calculated lifespans above assume that Canada makes no effort to reduce its greenhouse gas emissions and maintains its current level of exports. Increasing the population and exports will reduce the lifespan of the reserves accordingly while reducing fossil fuel consumption will extend them.

High Built-in Energy Use

Today, Canadians are among the highest users of energy in the world. This is due to the distances we travel, the resource intensive nature of our current economic structure and the severity of the climate in which we live (Fig. 3.5).

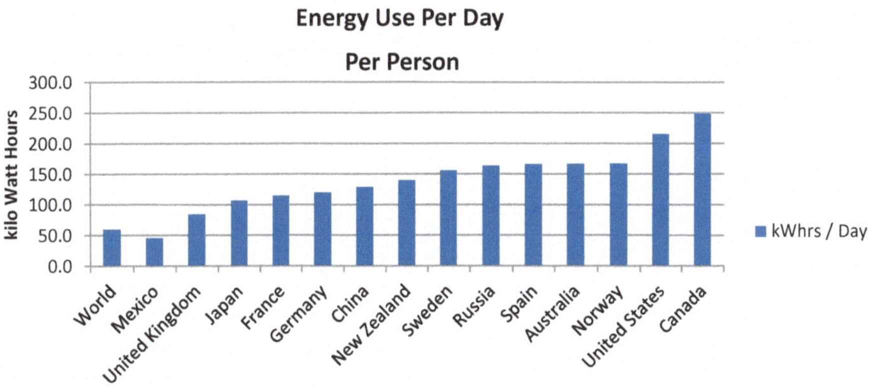

Fig. 3.5 Energy use per day by country (BP 2018)

As the chart below indicates, the energy demand of necessities varies greatly by latitude. In fact, at the most extreme northerly points, heating is required year round, with buildings in Igloolik using more energy for heat in July than ones in Wilmington, North Carolina or Athens, Greece do in January. How much energy is a kilowatt-hour? A large finned electric heater plugged into a 120 V outlet and set to its lowest setting will produce heat at the rate of about 1 kW. If left on for an hour, it would produce 1 kWh. At its maximum setting, it would produce approximately 1600 W or 1.6 kW.

According to Natural Resources Canada data, homes heated in Hay River in the south end of the Northwest Territories at 61° latitude, use 45,000 kWh of electrical energy annually for an incredible daily consumption of 123 kWh per day. Homes in Paulatuk, 7° further north, use the equivalent of 173 kWh daily in fuel oil. Along the southern coast of British Columbia, Canada's most temperate region, BG Hydro estimates the average family home would use an average of 30 kWh per day if using a high-efficiency heat pump. The consumption in every other region of Canada would be higher.

The 100,000 residents of Canada's north spread over 3.5 million square miles use almost twice as much energy per person compared to the average Canadian. And the Canadian average is 25% higher than that of our southern neighbors, the Americans, and almost five times the world average (Fig. 3.6 and Table 3.5).

The consumption of large amounts of energy does not necessarily result in a high material standard of living. In the case of northern locations, a large amount of energy is required simply to survive. Energy must first be spent on housing, transport, and other infrastructure. The amount of energy remaining after supplying the necessities is the amount that can be applied to transforming other available resources into the goods and services we consume on a discretionary basis. The level of "discretionary energy" determines the potential "standard of living."

The accessibility and richness of the resource base determines how much energy will be required to process those resources into the commodities which can then be transformed into the final end products we consume.

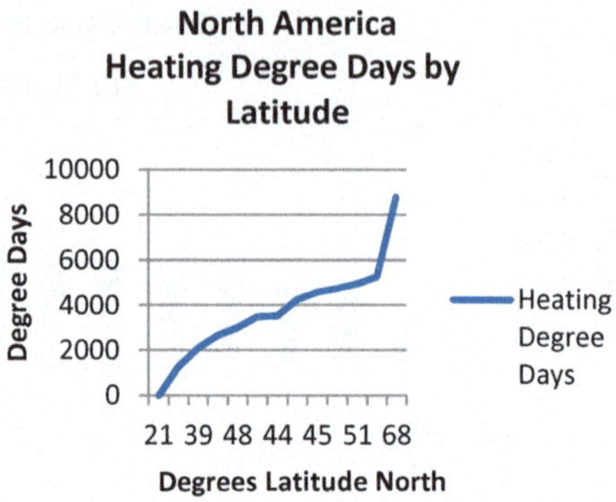

Fig. 3.6 Degree heating days by (BizEEE 2019)

Table 3.5 Degree heating days by latitude (BizEEE 2019)

	Degree heating days	Latitude North
Gaudalajara	0	21
Wilmington, NC	1291	34
Washington, DC	2117	39
Harrisburg, PA	2684	40
Victoria	3017	48
Windsor	3500	42
Toronto	3545	44
Halifax	4250	45
Ottawa, ON	4600	45
Parry Sound	4765	45
Calgary	4975	51
Winnipeg	5285	50
Inuvik	8821	68
Arctic Circle		66

Embodied Energy in Trade

Canada exports a great deal of raw energy and the energy required to produce those exports can be said to be embodied in them. But we import much more in the way of manufactured goods than we export, and the energy used to create those products in other countries, constitutes energy we consume. Net imports of energy must be incorporated into our daily per capita budget.

In Canada, the "net" figure is not a large figure, as we do export a lot of commodities with high energy content. In countries like the UK and Switzerland however, the net energy import figures can be very large. In the UK, the addition of net imported energy to the raw energy they consume would increase total energy consumption by 17% (Kejun 2008). The UK is after all, a net importer of virtually every commodity one can list, from oil to food to steel to fertilizer.

Switzerland is an even higher resource and energy deficit nation which imports, in the goods it consumes, over double the amount of energy it produces. Although not directly part of energy equation, it should be noted that most developed countries are net importers of energy via goods manufactured in developing countries. In those countries, particularly China and Korea, coal is the largest primary fuel and consequently the greenhouse gas emissions embedded in these goods is much higher than if those goods had been made in countries with "cleaner" energy mixes.

The Amount of Energy Produced Is Important But When It Is Available Is Almost Equally as Important

Fossil fuels have provided immense amounts of energy. Almost as importantly, fossil fuels are self-storing which means that they can produce their energy any time we ask for it. Not only has humanity gotten used to the idea of using huge amounts of energy, we have built our societies as well on the expectation that energy will be available consistently. But renewables, with the exception of firewood, are by definition, as inconsistent as clouds, snow, and wind.

The historical human relationship with renewable energy has always conformed to the ebbs and flows of energy as it presented itself for use. In the era of fossil fuels, energy became a constant companion and effectively a slave to be summoned at will. As we return to renewable energy for a larger and larger portion of our energy supply, the issue of variability will play an increasing part.

Instead of expecting to always being able to flick a switch or press a throttle to access whatever level of power we require at the moment, the variability of renewable sources will force us to integrate our consumption patterns more closely with the limitations of natural energy flows.

A critical element of renewable energy systems is the flexibility or quick reaction time of a base load to fill in during the periods in which solar and wind power are operating below capacity and below the level of demand (Fig. 3.7).

Hydro energy is ideally suited to the smooth integration with the variability of solar and wind power. This will be dealt with in a later chapter, but Canada has a clear advantage in this area over most other nations. While our northern location imposes large and highly variable heating demands on the energy system, the consistency of hydro sources will be a huge asset in integrating solar and wind power with grid demands.

Still we will have to learn to modify our demand for energy to fit the times when it is available. The ability to "make hay while the sun shines" is a skill that must be re-learned by Canadians. That skill has been a key component of the toolkit of every sustainable society. Storage of energy and the reaction time of the base load are important parts of the energy equation, but flexibility of demand, and of the people making their demands is still very much a necessity.

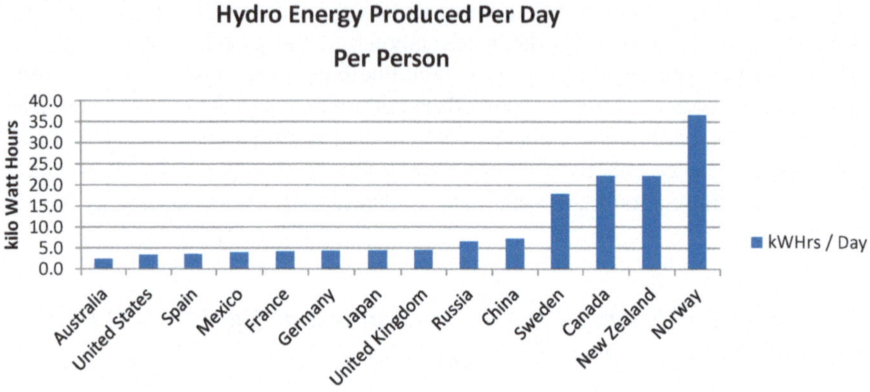

Fig. 3.7 Hydroelectric energy produced per person per day by country (BP 2018)

Necessity or Luxury?

Andrew Nikiforuk's energy slaves analogy (Nikiforuk) conjures up images of the economy effortlessly churning out goods and services like an overflowing spring of consumption. However, once the thin veneer of discretionary spending is peeled off this illusion, the reality of the work necessary to fulfill the essential work in this country becomes more clear. We may well regard ourselves as cosseted consumers with a huge buffer of energy between us and the primitive life we left behind. But this "to the manor born" feeling is purely illusionary. The difference between what we consume and what we need to survive with even a moderately sophisticated society at Canadian latitudes is much thinner than most would imagine.

If fossil fuels were removed from the energy mix, what sort of life would we live? Is the base energy that renewables—over 85% hydro currently—deliver sufficient energy to support even a marginally sophisticated society? One must also keep in mind that our current infrastructure of roads, copper wires, dams, and buildings were made largely with fossil fuel energy. This question will be dealt with in a later chapter.

Energy Availability vs Energy Demand

From a renewable energy point of view, matching seasonal need with availability becomes more difficult as one goes further north. Solar panels are only productive if there is a considerable amount of sunlight available. Wind turbines need air moving at sufficiently high speeds to push their blades and rotate their generators.

Fortunately, while solar energy falls off to effectively zero across much of northern Canada for the winter months, wind speeds increase. But that is not enough to overcome the severe demands that norther climates make on energy systems.

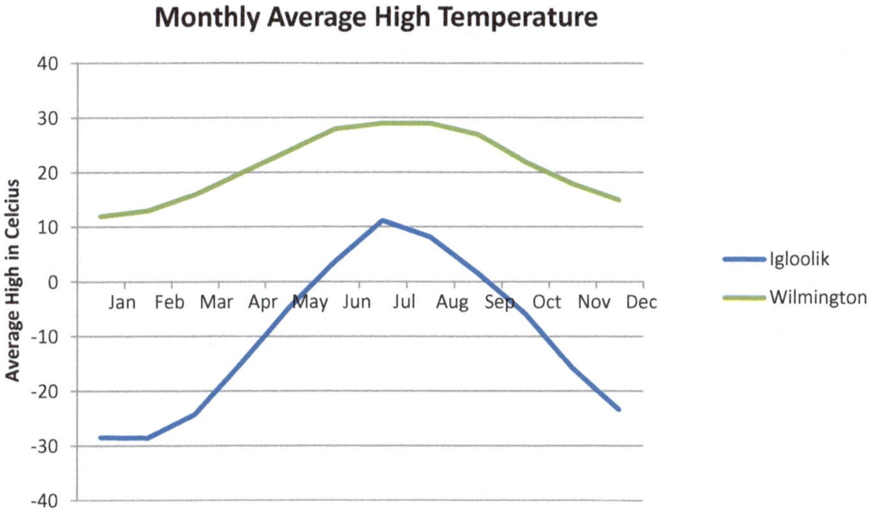

Fig. 3.8 Monthly average high temperature Igloolik, Wilmington (Degreedays 2019)

A residence in Igloolik, above the Arctic Circle, can require 60–80 kWh per day in heating during the winter, when the average daily high temperature does not get above −15 °C between November and April. For reference, the average daily high in July is 11 °C. Average nightly lows are below freezing for 10 months a year.

In Wilmington, North Carolina, average nightly lows don't drop below 5 °C at any time while their solar radiation levels remain strong enough to produce useful heat and electrical energy all winter. The potential for producing solar power in Wilmington in its worst month of December is higher as that of Igloolik 7 months a year (Fig. 3.8).

Looking at the capacity factor for solar energy collection systems, the tremendous variation in potential output based on latitude becomes obvious. Capacity factor is the percentage of total potential output achievable under perfect conditions (24 h a day strong sunshine) vs what the system actually produces in the conditions in which it is installed (Fig. 3.9).

At temperate latitudes, significant solar energy for hot air, hot water, and photovoltaic systems is available year round. Wilmington solar systems see energy levels which never drop below 60% of their summer peak. Not so at more northerly locations where solar energy drops effectively to zero for over 4 months a year.

The timing mismatch between available energy and required energy can only be solved by energy storage. Fossil fuels "easily" fulfilled the storage requirement because they are themselves storing energy simply waiting to be used (burned). Renewable energy mostly produces electricity which is much harder and expensive to store, especially for long periods. The issue of storage will be dealt with in a later chapter, but it is a critical consideration in the transition to renewable energy.

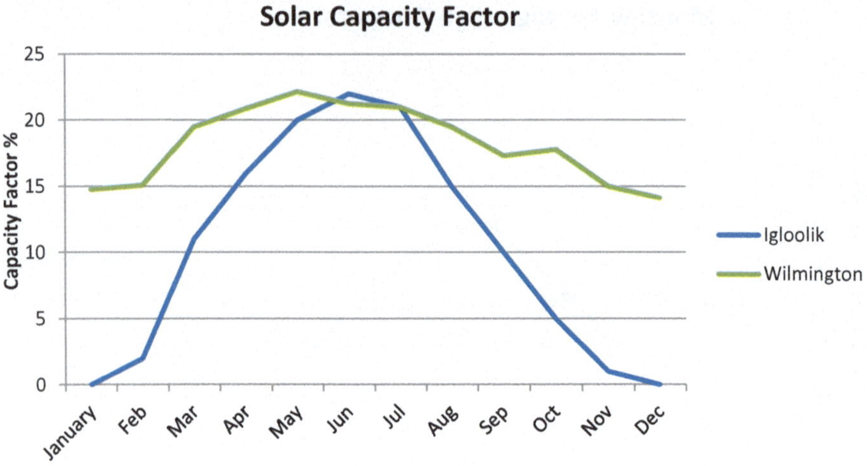

Fig. 3.9 Solar capacity factor Igloolik, Wilmington (NREL 2019)

National policies tend to be based on aggregated measurements, but energy policy needs to be different. The variation in energy generation and energy demand for different parts of the country requires regional analysis, which is much more than fine tuning. Solutions from Southern California cannot be dragged and dropped into more northern climates, as franchise business models and house designs were in the 1970s. Those examples had poor outcomes and the mistake of applying assumptions from unrepresentative regions to all of Canada will have disastrous outcomes.

Richness of the Resource Base

Huge amounts of energy are of little use unless there is a viable resource base they can be used to exploit. Once the basic needs of shelter and food have been assured, whatever energy is left over can then be turned into goods and services, which take a culture beyond mere subsistence.

Energy may be the uber-commodity without which no other resources can be developed, but the richness of those other resources determine how productive humans can be with their disposable energy. And since Canada was essentially a virgin country for mining, hydroelectric development, and agriculture, energy could be put to very good use.

But resources can be depleted. Renewable resources like soil can be overused and allowed to erode and decline. Water resources can be utilized to their maximum while forests can be overharvested and made useless for generations. Fisheries can crash. It is over 30 years since the Canadian Atlantic cod fishery collapsed yet the 2017 landings were still just 6% of the 1990 level.

Nonrenewable resources can be mined until the grade (richness of the "paydirt") of the remaining deposits becomes so low that they are no longer economically or energetically feasible to mine.

As the resource grades declines they require more energy to work and therefore the energy intensity per unit of output increases. Technological advances can mitigate this trend but, as can be seen from the chart below on the decline of copper ore grades in Canada, increasingly larger amounts of rock must be mined and processed to remove increasingly smaller amounts of metal.

Where once the average Canadian copper mine worked seams of 4% copper, (Mudd) the same amount of effort now yields barely one-tenth of this critical metal than it did 125 years ago circa 1886. This is a far cry from the virgin resource found thousands of years ago in the early Bronze Age when humans lit fires under rich rock outcroppings and the metals simply melted onto the ground. It is virtually impossible to find an example of a resource which is yielding more today for a given input of energy than it was 150 years ago.

This trend holds true all over the world. Potosí was founded as a mining town in 1546 Bolivia. Over the next 200 years it produced 40,000 tons of silver—1.4 billion ounces—but by 1825, the silver had largely run out.

Resource decline has been masked and offset by the availability of cheap energy. As we move to renewable energy, the issue of declining resource productivity will become more critical. "Infinite substitution" is an economic term that implies there will never be limiting shortages of important materials because the ability, in theory, to replace and substitute for every material is deemed to be "infinite." This applies to monetary constructs of the economic system but does not apply to real world commodities or the availability of energy. In fact, it makes the unsupportable and dangerous assumption of the availability of infinite energy (Fig. 3.10).

During its brief history, Canada has been so rich in buried energy and easily accessible resources, that we have easily been able to overcome the very high

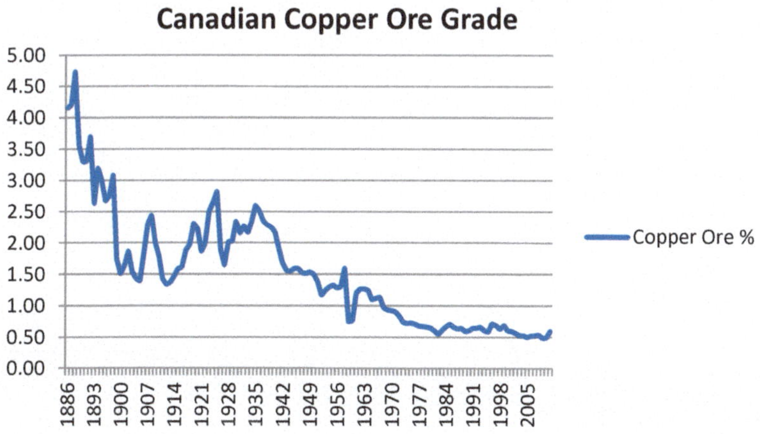

Fig. 3.10 History of copper ore grade in Canada (Mudd 2017)

demands made by our severe climate to support a very high standard of living, even with a growing population. The existence of our country was made possible first by huge stores of biomass energy sustaining early settlements, and then by coal driving railway development, and finally the ascent to high levels of prosperity built on the abundance of oil, natural gas, and hydroelectricity.

To the point of 2019, the Canadian relationship with energy has been one of ever expanding possibilities. Why would we ever seek to move away from those fuels, which amount to the lifeblood of our society?

References

BizEEE (2019) degreedays.net

Bolgiano C (2007) Mighty giants—an American Chestnut Anthology, images from the past, Bennington

BP (2018) Statistical Review of World Energy. https://www.bp.com/en/global/corporate/energy-economics/statistical-review-of-world-energy.html

Degreedays (2019) degreedays.net. https://www.degreedays.net/

Hughes JD (2018) Canada's energy outlook

Kejun J (2008) Embodied carbon

Mining & Energy Canada (2015). https://www.miningandenergy.ca/energy/article/top_10_hydroelectric_dams_in_canada/

Mudd G (2017) Resources and supply-demand over the very long term

NREL (2019) National Renewable Energy Laboratory. https://pvwatts.nrel.gov/

Smil V (2017) Energy transitions. Praeger, Santa Barbara

Unger RW, Thistle J (2013) Energy consumption in Canada in the 19th and 20th centuries. Consiglio Nazionale Delle Ricerche, Italy

Ward P (2007) Energy consumption in England and Wales 1560 to 2000

Chapter 4
Abundance Abounds, Why Change?

Abstract The hallmark of good business sense is the ability to recognize the inevitable structural decline in a particular business model and to move resources into models which have better potential. In resource models such as mining or fishing, it is relatively easy to see the writing on the wall for resource depletion as the effort to maintain production goes up as richness of the resource declines. It is difficult to turn away from something that has been so beneficial for so long but those with the best instincts for survival are able to manage it. Businesses, no matter how complex, pale into simplicity compared to nations. Nations can't simply fold up and move to a better location or change what they produce in a matter of several years. They contain huge amounts of infrastructure, both physical and human. Attachments to ways of life are strong. The most wealthy and powerful interests in the country became wealthy and powerful through the current business model. For a nation to successfully transition to renewables, the clear need to change must outweigh the resistance to change by a very large margin.

Fossil fuels perform vital tasks in most countries and, in northern ones, they are critical to the continued survival of the societies themselves. In Canada, it would be impossible to provide food, transport, or heat at all but the most minimal level for even small numbers of people without fossil fuels. The easy availability of huge amounts of energy, combined with their inherent storage characteristics makes these uber-commodities the near ideal fuels. With our present level of technology, supporting our current level of population, in our current lifestyle fossil fuels are vital and they will be very difficult to replace.

There are only two problems with carbon-based fuels. The first is they are non-renewable and, like any other mined commodity, will become more expensive and eventually run out. This is not to imply there will ever be a time when oil, coal, or natural gas will not exist on our planet, it is simply whatever remains will require so much work and energy to extract that to do so will be both economically and energetically unfeasible.

J. E. Meyer, *The Renewable Energy Transition*, Lecture Notes in Energy 71,
https://doi.org/10.1007/978-3-030-29115-0_4

The second problem is that burning carbon fuels produces carbon dioxide which, upon entering the atmosphere, acts as an insulator and tips the delicate balance of the planet's climate mechanism toward retaining more of the sun's heat. Carbon fuels, buried underground, represent the cumulative stored energy of millions of years of solar-driven plant growth. Releasing all of this energy and its chemistry as we have over a short 300 year time span—not even a glint of light on the geologic calendar—has inevitably produced changes in the composition of the atmosphere.

Although climate has changed dramatically and continuously throughout history, the global temperature have never risen as rapidly as it is now and the rate of increase appears to be accelerating. Even more importantly, this increase in temperature looks certain to end the uniquely stable climate humans have thrived in for the past 10,000 years as the average global temperature is now warmer than it has been in 11,000 years (Climate Analytics 2015).

Prior to 10,000 years ago, earth's climate was vastly more volatile. In these unstable conditions, humans were not able to develop agriculture on any scale that would allow permanent settlements and the increased population densities necessary for the launch of social and technical learning. But about 10 millennia ago, it became possible for bands of hunter-gatherers to begin to grow crops and remain in one place over many generations.

They were able to spend more time producing goods and structures which would last rather than striving to keep their inventory of tools and goods to a minimum as they moved across a varying landscape with the seasons. And the crops they produced, although inferior to the varied and protein rich diet of hunter-gatherers, offered more calories for less work. Additionally, the grains these new farmers grew lent themselves to storage which underwrote a stable food supply all year long. Human numbers grew and communication and learning began in earnest (Figs. 4.1, 4.2, and 4.3).

Compare the broken stone tools humans used 2½ million years ago to the state-of-the-art Clovis points of 12,000 years ago. From those spearpoints, 2 million years in the development, it only took us 12,000 years to walk on the moon. That technological leap off the planet was made in 1/200th of the time it took to progress from broken stone to crafted stone.

Climate determines the health and richness of the environmental systems humans depend on. It determines how much of our time will be devoted to producing food and, consequently, how much time we will have left to improve our learning, our society, our technology, and our infrastructure.

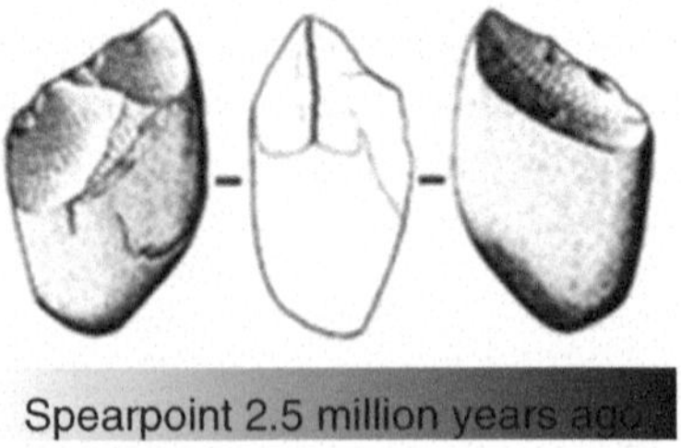

Fig. 4.1 Spearpoint from 2.5 million years ago (unknown)

Fig. 4.2 Spearpoint from 12,000 years ago (unknown)

Fig. 4.3 Space Shuttle launch (unknown)

Humanity's Relationship with Climate

The fairly minor fluctuations in global temperatures over the past 10,000 years have produced pivotal events in history. A 300 year drought circa 1200 BC spelled the end of many societies in the Mediterranean. For decades, these collapses were believed to be the result of raids by an unknown "Sea People" but researcher and author Eric Cline clearly brings into the mix the element of social disintegration concurrent with a centuries-long drought (Cline 2014). At the start of the Middle Ages warming, Vikings struck out across the Atlantic in very favorably warm conditions, while the Anasazi—The Pueblo People—in the already arid American Southwest, experienced drought and collapse.

Humanity clearly has a strong relationship with both climate and energy. The term "relationship" may be misleading since neither the climate nor energy resources need humans. "Critical Dependency" better defines our need for both abundant energy and a stable, moderate climate. Without both of these, human civilization and possibly the human species as we know it will disappear into the sedimentary layers of history.

For tens of thousands of years, humans survived and sporadically prospered using nothing more than the renewable energy sources of biomass, sun and wind but now we are dependent on using many more times the energy than the world can provide on an on-going basis. Since the early 1700s, the energy which has launched human populations and societies into an exponential trajectory of seemingly never ending growth has all been mined from finite stores from below the surface of the earth.

Climate Matters

Climate matters to human society. No matter how severe climate change has been or will be, it is likely there will be an environment somewhere on the planet able to support some humans until the next massive asteroid strike. It is necessary to differentiate between the survival of the human species and the survival of enough of the species to maintain human society and learning.

The famous Black Death and the less famous decline in the 1600s where perhaps 35% of the world's population perished and revolts, wars, and famines were widespread. It was even possible to walk across the Bosporus from Europe to Asia during the very worst winter (Parker 2013). These climatic fluctuations damaged many societies and killed millions of people but they did not derail the march of progress. In fact, in Europe, the decades of social chaos caused fundamental religious and political beliefs to be challenged, leading to the Age of Enlightenment.

Next to the oxygen in the atmosphere, climate is the most important factor in human life since it determines food availability, clean water, and favorable temperatures upon which all bio-systems depend. Climate mandates what proportion of the planet will support human life on a consistent basis.

Climate Hasn't Always Been This Kind

It is only in the past 10,000 years that humans have become sophisticated enough to become a progressively creative species. Human civilization was not created by an improvement of our physiology, it has resulted from the climate given opportunity to slowly accumulate knowledge. A relatively stable climate allowed our ancestors to develop agriculture and towns to grow. The result has been higher forms of learning and more elaborate social structures.

The graph below shows the exceptional stability of the climate we have enjoyed in the past 10,000 years compared to the prior 40,000 years. This plateau or tightrope represents a range of temperatures within a window of 1.5–2.0 °C. Human history has largely been written in this period of relative stability, but even then, small changes in climate have laid waste to numerous civilizations. Even if we had unlimited oil, would any moderately sophisticated level of civilization be possible in the conditions of those 40,000 years (and millions of years prior to that) given the 10–15 °C range of temperatures? (Fig. 4.4).

The magnitude of the fluctuations in the Holocene relative to the previous periods are stark and illustrate the severity of conditions faced by early humans. Those gross climate fluctuations would have meant far less food availability, short lifespans, and much more frequent nomadic wandering to seek necessities. Under these conditions, it would have been almost impossible to accumulate material goods, tools or any higher level of learning necessary to allow a continuous surge of social and technical development.

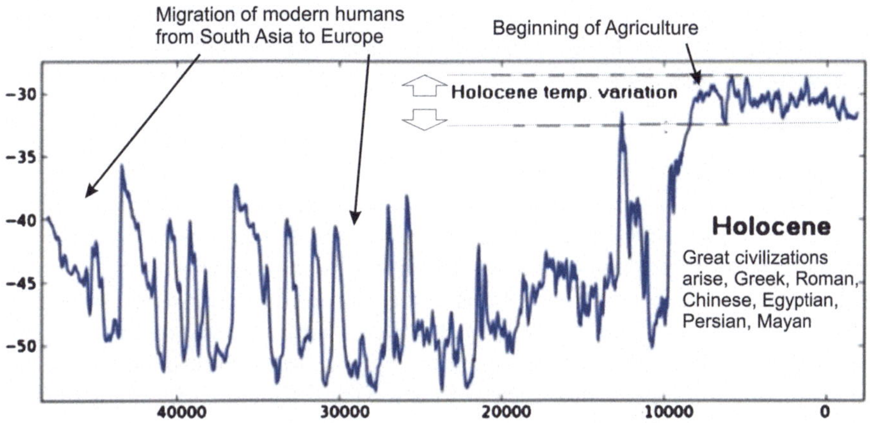

Fig. 4.4 Climate Stability Tightrope of the past 10,000 years shown using Greenland ice core to determine the surface temperature of the ice (−30 °C to −55 °C) (Hansen 2019)

The warming of the planet and alteration of the weather patterns modern humans are used to is underway, and it is likely to approach the scale of the changes our ancestors experienced over the previous hundreds of thousands of years ago. This degree of climate change will present immense challenges for human society. Outcomes, particularly for any specific region, are impossible to predict with precision but change and increased variability will likely be the universal experience.

Some Possible Consequences of Climate Change

- The well-known sea level rise due to glacial melt and the expansion of warming sea water will make all low lying areas more prone to flooding and storm damage. Coastal cities and agricultural lands are vulnerable as is any infrastructure within the expanded inundation zones.
- Larger more powerful storms and more severe rain events. This would apply to most populated regions of the globe.
- Longer and more severe droughts.
- Increased frequency and severity of forest fires.
- Invasive species, in effect "species change" as species live or die depending on the environment they are adapted to.
- Unusual weather, more ice storms, localized flooding.
- Planting season and harvest time disruption and unpredictability.
- Travel and economic disruptions.

No matter how moderate or severe these impacts become, they represent a change to the environment our complex society is currently based on. We have engineered our modern infrastructure based on essentially the past 150 years of our experience with weather patterns. As well, our food system is tuned to the current water cycle and planting and harvest times while our trade networks implicitly assume the continued flow of large amounts of material and energy around the globe.

This period of benign environment we have become used to is the most favorable in recorded human history and the overall impact of large changes in the dynamics of our climate is unlikely to be positive.

These changes, no matter how big or how small, cumulatively add up to higher costs (both dollar and energy) of doing business and higher costs of living, meaning, there will be less discretionary income. The increases of basic systemic costs will be felt most by lower income groups because they spend a greater proportion of their income on the basic living costs of food, shelter, and transportation. As our relatively benign climate goes dynamic, costs, ranging from more money spent on infrastructure to property insurance to food to snow clearance will put additional stress on already tight budgets for segments of society.

Taking our carbon foot off the climate change accelerator is the first step in dealing with the changes that are already underway and locked in. While we are working to reduce the damage we are doing, we will also have to work at minimizing the damage to which our social and economic systems are susceptible by the dynamics we have already set in motion.

One Potential Impact Climate Change: If the Gulf Stream Stops Streaming

The phrase "global warming" was coined by the American climate scientist Prof. Wallace Broecker in his 1975 paper (Broecker 1975) tying rising carbon dioxide levels in the atmosphere to warming of the biosphere. "Global Warming" is a simplistic term for the complex changes which will take place in the earth's flows of water, air, and heat. But "Destabilization of Weather Patterns Characteristic of the Past 10 Millennia with a General Trend Toward Warming," although more descriptive, would have been far less likely to become a popular catchphrase. "Global warming" can be thought of as a proxy term for the complex changes in the planet's weather systems which will occur as the biosphere warms.

Broecker studied the Ocean Conveyor Belt, the circulation pattern of global ocean currents. In the Gulf Stream portion of this system, cold, salty water in the North Atlantic sinks, drawing an ocean current from the warm Caribbean toward Europe which keeps Europe's climate much milder than its latitude would suggest.

Without the Gulf Stream, Broecker claimed Europe would suffer winter temperatures lower by about 11° with London feeling perhaps more like Calgary, Alberta, since both of them sit at 51° north latitude. The average winter low temperature for Calgary is −13 °C in Dec–Jan, while in London it is +4 °C, – a 17 °C difference.

An additional comparison could be made between Fort Smith, Alberta and the capitals of Norway, Sweden, and Finland. Like Fort Smith, Oslo, Stockholm, and Helsinki sit at 60° latitude but moderated by the Gulf Stream, the Northern European capitals enjoy much more livable temperatures. The average January high temperature in Oslo is 0 °C and in Fort Smith it is −19 °C.

Broecker described the Gulf Stream as the "Achilles heel of the climate system" which could change rapidly given only a slight rise in temperatures to keep water from sinking in the North Atlantic. A slowing or cessation of the stream of warm water eastward across the Atlantic would have dramatic impacts directly on Europe as well as weather systems across the globe. Several recent studies have cast doubt on the Gulf Stream waters as the critical means of heat transfer from the southwestern Atlantic to Europe and have identified air currents above it as a principal conveyance. Whatever the mechanism, a reduction in flow of heat to Europe will have dramatic impacts.

Prof. Broecker, who passed away in early 2019, described this process as "a tipping point" leading to abrupt climate changes rich with unwelcome "surprises." How climate change plays out will only be known as it happens but the predictions of larger, more powerful, and longer lasting weather systems have certainly proved to be accurate.

The relatively benign climate that nurtured many generations of increasingly sophisticated human civilizations is now reverting to something more dynamic. This new dynamism will demand significant changes to our economic and social structures presenting large challenges to governments, businesses, and individuals. The challenges presented by "global warming" will be less about heat and more about the physical impacts of increased flooding, forest fires, droughts, and hurricanes that will exceed the coping abilities built into our infrastructure which was designed using the past 150 years of climate data.

Canada's Record on Climate Change: Cool Words, Warm Actions

In 1997, Canada signed on to the Kyoto Protocol which called for the signatories to commit to lowering their greenhouse gas emissions below the base year of 1990 by specified amounts by the year 2012. Canada agreed to cut its emissions to 6% below its 1990 levels by 2012. In fact, Canada had no strategy whatsoever to achieve its targets and simply continued with its pursuit of economic growth.

By 2011, it had become obvious to everyone that Canada had made next to no effort to achieve its targets, and the government withdrew from the agreement, the only country in the world to do so. By 2012, Canada's greenhouse gas (greenhouse gas) emissions had gone up by 20% equaling a level 26% above our target of 560 mega tons. The only other nations to post worse results were Saudi Arabia and Australia.

What happened? The two drivers of Canada's increased greenhouse gas emissions were the expansion of the oil sands, responsible for almost 40% of the increase and the six million person increase in population driven largely by immigration. This contributed almost 60% of the increase as the average Canadian emits the equivalent of 20 tons of carbon dioxide annually and increasing the population by six million, in this very energy intensive country, produced an increase of 120 mega tons in emissions.

Added to the 60 mega tons contributed by the oil sands, emissions would have risen 180 mt above the target but the rise was moderated by the Green Energy program in Ontario. By heavily subsidizing wind and solar energy and by converting coal generating plants to natural gas, levels were reduced by over 30 mt, keeping the net increase to "just" 150 mt above the target.

The Paris Agreement uses consumption levels in the year 2005 as the base. Going forward, if we expect to meet our future Paris Agreement targets for 2030 and 2050, our fossil fuel consumption budget would look like this in "Barrel of Oil Equivalent"—BoE:

- Canada's current 2017 fossil fuel consumption = 54 BoE per person
- Canada's Paris Accord greenhouse gas emission targets
  - 2030—30% reduction in greenhouse gas emissions over 2005

    38 BoE
  - 2050—80% emission reduction over 2005

    11 BoE

The Paris targets are national targets, which mean that if our population increases, the allowable fossil fuel energy consumption per person will have to drop. If the unlimited growth policies of past governments remain as national policy and Canada's population triples by 2120, the fossil fuel energy diet would be cut to just 4 barrels of oil equivalent from our current level of almost 48 barrels now.

Nationally, Canada consumes oil, natural gas, and coal on the level equivalent to about 2 billion barrels of oil annually. To meet our commitments, in 2030 that would lower our consumption to 1.4 billion barrels and in 2050, the level would be 400 million barrels annually. These are both massive declines.

How the prospect of these reductions in fossil fuel use will be greeted by a wide variety of parties with vested interests is hard to guess. But change has to come to the pursuit of endless growth. Canada has embraced growth in the commercial economy as its prime social and national goal for the past five decades. Even ignoring the negative impacts on individual welfare and the land, the possibility of meeting future emission reduction targets are next to zero given the energy intensity of increased resource extraction and population growth.

Humanity's Relationship with Resources

Biological systems are resilient. Like humans, most species of animals and plants reproduce quickly when given the opportunity. Their numbers can increase at a compounded rate for as long as a favorable environment can sustain their numbers.

Renewable systems, whether plant, fish, mammal, or insect can snap back in a matter of five or ten generations once the pressure of over-exploitation has been removed. Nonrenewable stocks of minerals however will not grow back once we stop mining them. The reality of mineral availability is that the grade of resources we will be working with in the future will always be lower than it is now.

Advances in technology can reverse this trend for periods of time and allow us to extract minerals more easily from progressively lower grade ore bodies. But over the long term, most of the ores we have been mining have been demanding higher inputs of energy. In the energy boom days, this was not a problem. Stepping away from abundant and cheap fossil fuels however, will bring the scarcity problem into sharp relief as less abundant and harder to store renewable energy will struggle to subsidize a depleting resource base.

Our access to resource bases and our ability to process them may well decline.

Minerals

Following are examples of some of the more critical commodities demonstrating the increased effort and reduced yields as we mine increasingly less rich deposits of our mineral resources (Fig. 4.5).

Depletion rate is the term used to describe the long term annual decline in output from oil fields. Conventional oil fields very slowly decline with depletion rates between 2 and 6% and can have useful lifespans on many decades. Tight oil fields, with their far lower densities of oil spread through seams over large areas, have much higher rates of decline.

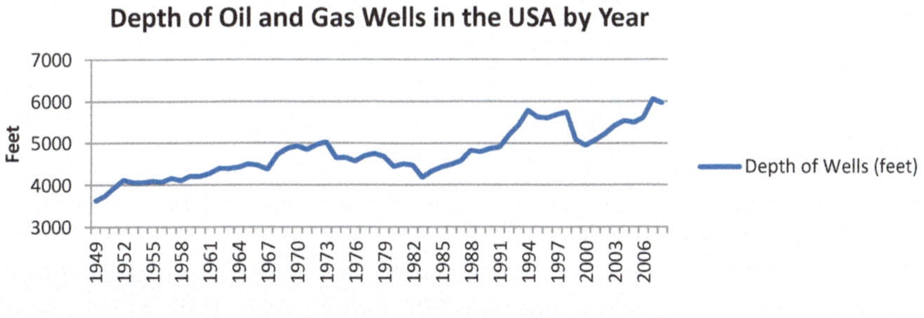

Fig. 4.5 Average depth of oil and gas wells in the USA by year (EIA 2019). In 2019, to exploit oil and gas, we have to drill deeper and more often to produce less energy

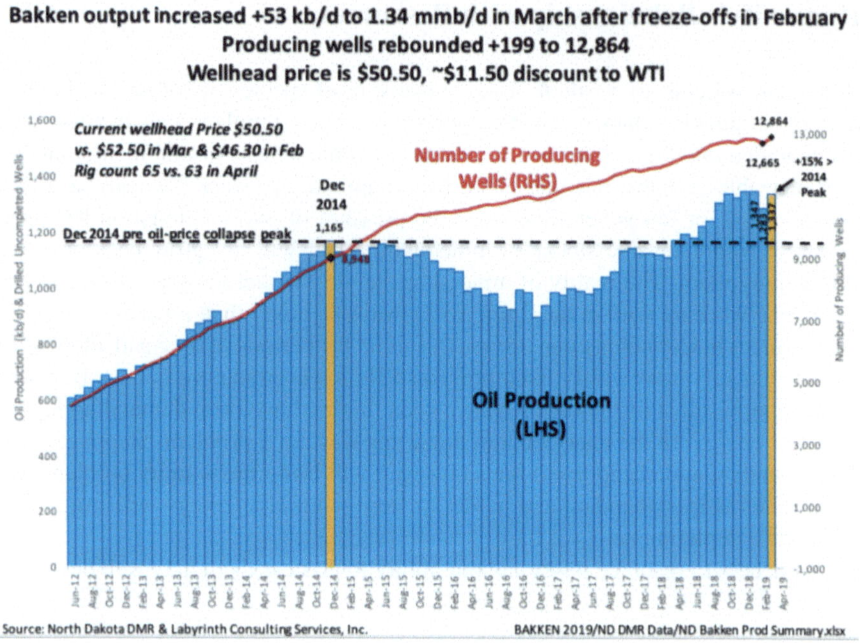

Fig. 4.6 Bakken fracked oil field, lower output per well (Berman 2019)

Tight oil has depletion rates of close to 70% annually and so output can only be maintained by drilling an increasing number of wells until the field eventually "runs dry" (output is so low that more energy is expended in getting oil out than is contained in the oil produced) (Fig. 4.6).

In December of 2014, there were 8948 producing wells in the Bakken filed and output was 1.165 million barrels per day for an average of 130 barrels of oil per day per well. In March of 2019, there were 12,864 wells producing 1,337,000 barrels per day for an average of 103 barrels of oil per day, a reduction of 21% per well. This is likely the result of an increasing proportion of old wells and of the exhaustion of "sweet spots" as we exploit progressively less rich plays.

Gold

This is a good example of diseconomies of scale whereby the more that is extracted, the higher the unit costs climb (Fig. 4.7).

The yields and material processed by the operations of Barrick Gold show why the price of gold has increased so much over the past 20 years. Humans have been mining gold for scores of centuries and the largest and richest fields have been discovered and developed. As the richest seams play out, attention is turned to smaller,

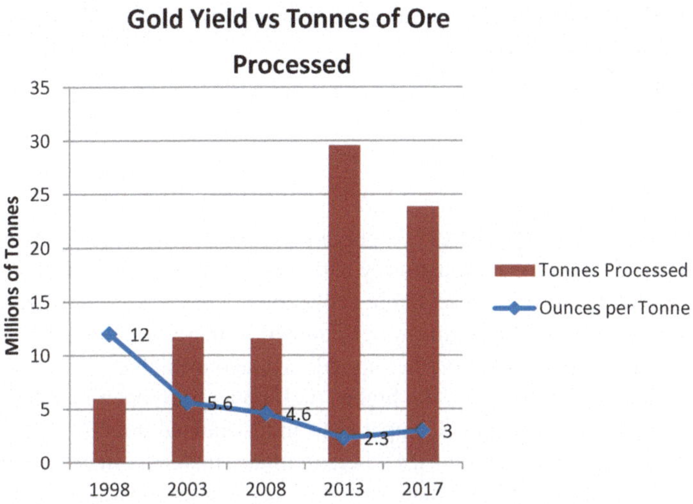

Fig. 4.7 Tons of ore per ounce of gold by year (Barrick Gold 2018)

less rich ore bodies which require more work to reach and more material to be mined and processed.

In Canada as well as in the USA, "gold rushes" have occurred in the past 150 years. These occurred when gold savvy Europeans discovered virgin deposits of gold. These deposits may have been in the form of nuggets and flakes in rivers and gravel formations or in rich seams of the mineral that had never been subjected to any extraction effort. The flood of would-be miners into the California goldfields or the Klondike constituted a "rush."

In Canada's Klondike, Dawson City was erected to accommodate the trickle of men into the area as it turned into a flood after the 1891 strike, swelling the population in the region to 25,000. The easy gold was quickly found and accessible sources depleted leaving far fewer opportunities requiring much higher cost extraction methods. A decade later, the population had declined to a few hundred individuals. Any gold produced was done so by sophisticated miners with advanced, energy intensive equipment working low grade deposits in a methodical process.

In the mining process, sweat, picks, and shovels gave way to steam shovels and gunpowder and these were in turn replaced by a wide variety of diesel-engined tractors, drill rigs, processors, and TNT. This machinery, advanced technology and a great deal of energy were required to cut through and process the scores of tons or rock that held small traces of gold.

Typically, large, rich resource bodies, whether fossil fuel, mineral, or renewable, decline gradually over a long period of time. The types of resource plays we are developing now are typically smaller and less rich. The result is not only higher unit costs but shorter production lifespans and more rapid depletion and abandonment. It may also be true that the legacy costs associated with smaller, lower grade mineral developments are inherently higher as more mine tailings are produced and there is less profit to remediate them.

Food

Agriculture in developed societies is highly mechanized with large inputs of energy for fuel to make the complex machinery and keep it running through the cycles of plowing, fertilizing, seeding, insecticide application, and harvest. Fertilizer must be mined or chemically produced and water pumped in increasing amounts as we try and push yields per hectare up on the limited farmland available.

From the late 1900s to 2008, the higher fuel inputs have been matched by proportionally higher food output. Compared to industrialized countries, the story is different for less developed countries where dramatic increases in crop output have come at the price of even greater increases in energy and fertilizer inputs in the same period. How much of this is a direct result of climate change has yet to be determined, but as more marginal land in both LDCs and developed countries is pressed into service, the sensitivity to climate dynamics will inevitably show up in a demand for higher energy inputs or lower crop outputs.

Data from 2008 onwards is being processed by researcher Prof. Dr. Uwe A. Schneider of the University of Hamburg but clearly the energy–food relationship is an area that deserves very focused attention. Can technology increase or even maintain crop yields in the face of changing weather patterns? In which regions are there significant changes? What role is climate change playing? These are questions on which society must have clear answers to avoid critical supply shocks in the future.

As with mineral and energy resources, we need to describe the productivity of the production process with a metric which represents the energetic costs per unit of output. Failure to do so will cloud our fundamental understanding of the physics of the production process and perhaps result in decisions which lead to poor and hard to interpret outcomes.

Fisheries

Scarcity in fisheries involves the trawlers traveling further to land smaller catches of lower quality. Fish farms are a substitute but they require processed, energy intensive food and they produce fish which, by most accounts, are inferior to their wild cousins. Replacing natural bio-systems with fossil fuel-based systems can increase the density of food production greatly but the energy input per unit of output invariably rises as well. Sun-based energy systems are free and eternal while systems based on mined energy require constant upstream inputs.

Pollution: The Almost Universal By-Product

Higher levels of pollution are the direct result of processing larger amounts of material to produce smaller yields. Since the early 1960s, we have been very successful in reducing the output of many harmful pollutants but there are material, time, and energy costs in pollution reduction. We can make progress in specific areas like CCFs and the ozone layer and acid rain but large problems like plastic in the environment and CO_2 emissions seem to defy easy resolution.

Taking "clean coal" as an example, the processes of carbon emission removal from coal production and the storage of this carbon underground has not been made practical in any of the many large-scale experimental projects in the world.

The theory is that the carbon emissions from the burning of coal (carbon dioxide) can be removed and then this gas can be compressed and stored underground in suitable geological formations. Even if a large enough supply of ideal geological storage formations existed, the process of scrubbing the coal emissions and compressing and pumping the carbon gasses underground requires considerable energy. In fact, it requires enough energy that it makes coal, the cheapest energy source available, more expensive than solar or, critically natural gas.

Larger effort for less output inevitably leads to higher costs of legacy cleanups relative to the amount produced. Small amounts of pollution in a large healthy environment can mean relatively low cost cleanups if any are actually necessary at all. Conversely, in a highly stressed environment with high levels of pollution, the mitigation costs per unit of pollutant can be vastly higher as the ability of the eco systems to self-remediate are lower, requiring that human efforts and energy inputs must be proportionately greater.

Scarcity: The Energy/Resource Relationship

Compounded scarcity occurs when energy begins to be more difficult to obtain. In an era of increasingly abundant energy, it is easy to overcome the declining fertility of the soil or lower fish landings by adding fertilizer to enhance crop output and perhaps "sailing" further to chase richer fish stocks.

Similarly, abundant and even growing supplies of energy allow the declining grades of ore bodies to be processed with little or no increase in cost. In the times of abundant energy, there is very little apparent food or mineral scarcity. But when energy begins to require more effort to produce, the availability of all other resources declines because energy is required for the production of every commodity.

We are working harder and expending more energy to get the energy we are using. EROI is dropping and net disposable energy is dropping as a percentage of gross energy produced. Energy is the prime commodity. The availability of every other commodity is based on the availability of energy to find, extract and produce it. Our relationship with commodities and our monitoring of them should be based on the amount of energy it takes to produce them. Long-term data on the energy input/material output ratio should be a critical consideration of national policy in order to determine future options.

In our energy rich world, we have assumed that we could go and get any resource we needed. Given the obvious decline in resource richness, it has been speculated that in the future, we will mine humanity's garbage piles or lasso mineral or water-rich asteroids and comets and bring them back to earth. The energy required to lift a rocket loaded with mining apparatus into space, retrieve a distant space rock and return with it, and then lower it slowly into the atmosphere wildly exceeds the energy required to harvest the same amount of mineral from even badly depleted fields on earth.

And in terms of leaving earth for livable planets, we have no exit strategy, no planes, boats, ships, or spacecraft capable of carrying even one human to interstellar greener fields in a voyage of less than 1000 lifetimes.

So we will have to make do with what we have, where we have it. The earth is not going to change so we must adapt our expectations to what it can provide sustainably rather than dwelling on the dream of "if only …." And certainly, making do with what we have should not be seen as a punishment because the verdant planet we live on is by far the best planet humans have discovered out of the hundreds of billions that have come within range of our furthest seeing radio telescopes.

The Energy Scarcity * Resource Scarcity Multiplier

The specific mineral we choose to mine and ultimately transform into useful goods is one part of the resource availability question. The other part is energy availability. If the EROI of all energy sources in a society was over 50:1 as in the glory days of the 1950s, then even very poor ore grades could be mined economically. There would be almost no shortages of any useful substance because the energy used to produce it would be virtually free.

If, on the other hand, society had to try to exist with EROIs of 6 or 7:1, still significantly higher than those of the oil sands, energy would not be abundant and low cost. It would be a struggle to provide basic health and education services. Any resource body, whether renewable or not, would have to be either absolutely critical or very rich to justify investing the energy for its exploitation.

What would this metric look like? EROI × Yield percentage? What would it be called? Net Effective Resource Yield Index? (NERYI—"nearie") It hardly rolls off the tongue. But here are two examples to establish the concept.

Assume the year is 1955 and the EROI of oil is, for a fleeting moment, 100:1. Assume the yield of gold is 0.75 ounce per ton for a rating of, say 0.75. The mineral rating is 0.75 and the energy rating is 100. Then the NERYI would be 100 × 0.75 or 75.

Next, assume the year is 2019 and the worldwide EROI of oil is 17:1. Assume the yields of gold averages 0.48 ounces per ton. NERYI would be 8.2.

Looking at the drop in the overall availability index from 75 to 8.2 over that 64 year period, one could assume that the real cost of the gold has risen. Maybe not by a factor of almost 10, because we will have learned a great deal in those 64 years and built much more efficient processing infrastructure, but a significant cost increase would be indicated.

Illustrating and quantifying the changes in resource availability will be a key factor in understanding the human situation as the world changes around us.

Timing: Why Change Now?

Whether driven by the desire to avoid catastrophic climate instability or by the reality that fossil fuels are no longer readily available, the transition from fossil fuels to renewables will require a great deal of time. Not only will the sources of energy have to change but so will the way energy is distributed and used.

The BP Statistical Review gives the world's oil reserves a 53 year lifespan but is careful to note that such forecasts are impossible to make accurately. However, the BP estimate does establish a rough timeframe to the oil era. It is certain that some oil will be available after 50 years and it is highly probable that affordable oil supplies will have disappeared decades before.

When finally confronted with the rapidly approaching scarcity of oil, governments will be forced to prioritize use with agriculture, mining and emergency air transport, research rocket launches, etc. in the top echelon. Unnecessary uses like space tourism, all manner of nonessential travel and consumption, plus any process that can be readily electrified can expect to be very quickly cutoff from access to strategic energy supplies.

It appears the world has at least 50 years of natural gas available at current consumption levels while the reserves of Canada could last 200 years given no increase in population or exports. However, it should be noted that for many countries, Canada included, any large-scale shift to electrified transport and renewable energy will make natural gas fired electric plants a very attractive option in many grid systems. In countries embracing these strategies, natural gas consumption will increase dramatically.

Economic patterns and material flows will also be rearranged as energy becomes a very limited commodity and a critical strategic good. Jeff Rubin dealt with these ramifications in his book "Your World is About to Get a Whole Lot Smaller." Jeff called the trend, but based his timing estimates on conventional oil without the inclusion of unconventional sources. He was early but, as far as the transition away from fossil fuels is concerned, early is better.

Government policy will have to adjust its policy making process to enable consistent progress over decades toward the day when fossil fuels form only a tiny portion of our energy budget. And eventually form no part at all. No matter whether you are concerned with your home, business, or public building, in the distant future, any energy you require that you do not produce on site will come to you, not in a pipeline or tank but down a wire.

If we ignore the impact of greenhouse gas emissions on climate and begin to be concerned about the renewable energy transition only when oil supplies begin to decline dramatically, we will have fewer policy options and a much more compressed time frame for implementing them. Or, if begun in earnest now to head off the most severe climate consequences, the transition to renewable energy, although still very painful, can be accomplished while maintaining a workable social order. Planning a coherent staged withdrawal from fossil fuel sources rather than reacting to worsening fuel shortages will give society time to adjust to changes and fix mistakes.

Reacting to rapidly declining fuel availability in a comparatively short time will force major changes upon an unprepared populace and business community. The transition will then take place in an era of climate surprises and a continuously changing environment making it much more difficult to manage and evaluate results.

The graph below illustrates the fairly minor changes in the time frames between acting now, in accordance with the Paris targets or waiting until market forces, in the form of regular and worsening energy shortages, plus fluctuating and skyrocketing energy prices, deliver a strong response.

The conditions existing over the 30 year gap between the planned vs reactive responses will be very different. A response now will take place in the presence of few energy shortages or cost instability during the onset of climate change and will have a multi-decade planning time line. A response to fuel shortages will have no coherent plan and will be determined by spot prices and futures pricing markets to allocate real physical energy investments determined not by energy metrics but by dollar metrics.

National policy will be driven by energy measurements whereas commercial reactions can only respond to flows of fiat currency.

Monetary metrics do not represent biophysical assets or flows with any reliability nor were they ever designed to do so.

Here It Comes Ready or Not

Waiting to implement the transition will also force the transition to be made in a shorter time span. The depletion rates of conventional oil fields range from 2 to 6% annually—a very gradual and predictable rate of decline. But by well before mid-century, most conventional oil fields will be largely depleted and oil production will be coming mainly from unconventional sources such as fracking and the oil sands.

Oil sands depletion rates are quite predictable and gradual but those of fracking oil fields are not. They have depletion rates on the order of 70% annually and require very aggressive and continuously growing new well drilling to just maintain current output. As less rich fields are exploited—humans always pick the highest hanging, lowest quality fruit last—the depletion rates and drilling rates will both increase until the resources are exhausted in a final very steep decline.

Conventional oil peaked gracefully and as it declines relatively smoothly, unconventional oil has been both filling the gap and increasing overall production. Unconventional oil will peak differently given their low EROIs and, in the case of fracking, extremely rapid depletion rates. The transition from unconventional oil to renewable energy involves a great many more variables and investments than switching from a ready source of oil to distinctly more difficult sources of oil.

In the end, even governments with sound and flexible energy policies will be confronted with many surprises and failures during the transition and will have to make significant adjustments. If government policy is driven by short-term energy price signals, individuals, businesses, and governments will likely be repeatedly blindsided by unpredictable and sudden social and economic shocks which will be very hard to recover from.

Paris Accord vs BP Oil Depletion Forecast

If the world puts off fossil fuel reduction, all it gains is 30 years of business-as-usual and we will then have to deal with a vastly worse climate scenario, and deal with it in a much shorter time frame. As well, the business-as-usual path leaves virtually no viable reserves of oil available whereas the aggressive fossil fuel reduction path mandated in the Paris Agreement would preserve substantial oil assets for use by future generations (Fig. 4.8).

Human Responses to Crisis: Pick an Analogy

Perhaps the best known and well-documented disaster in history is the sinking of the RMS Titanic. From the causes of the disaster to the decisions made during it, illuminating records and eyewitness statements give a very complete picture of the

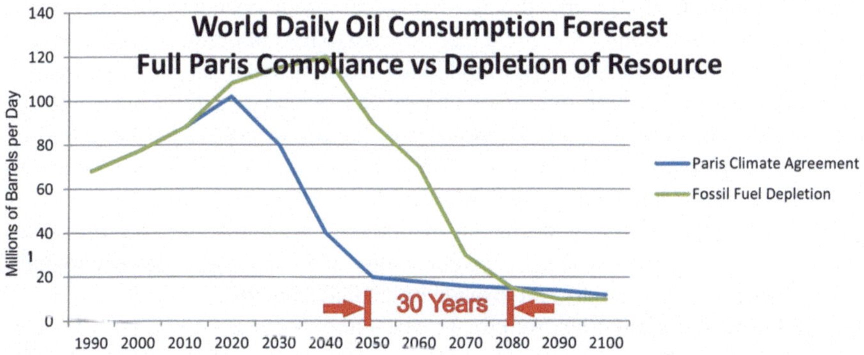

Fig. 4.8 Fossil fuel consumption reduction time difference by choice or by scarcity (BP 2018)

sequence of events. The "unsinkable" Titanic was 4 days into her maiden voyage from Southampton to New York when she struck an iceberg late on the evening of April 14, 1912 and sank in the early hours of April 15.

Throughout history, ships have sunk in great numbers with great regularity and any ship of any significant size, even to this day, is required to have water pumps and lifeboats as part of its basic design. Despite being hailed as unsinkable, the Titanic was fitted with 20 lifeboats and large pumps. She had been designed for 48 lifeboats but 28 had been removed for esthetic reasons. They made the ship look "cluttered". The capacity of the 20 lifeboats was 1178 while, on that maiden voyage, there were 2208 passengers and crew aboard, well short of the ships 3547 capacity.

The Titanic's builders met the letter of the law in force at that time, since the specification for the number of lifeboats was based on the faulty metric of the tonnage of the ship rather than on the number of people who might be on board.

The cause of the collision with the iceberg was the high rate of speed the ship was travelling in what were known to be dangerous waters—seven iceberg warnings had been received on the day of the 14th. The reason for the speed was believed to be the desire on the part of the ship's owners to set a record for the shortest travel time to New York. As it transpired, they may have set a record for the shortest voyage but not for the shortest complete voyage.

The iceberg was struck at 11:40 at night and the ship very quickly took on huge amounts of water. The ship's designer was on board and was able to give an expert, and, as the event transpired, a highly accurate estimate of the time remaining and the manner in which the ship would sink. The crew had 2 h and 40 min to make decisions and act.

The situation the crew faced on the Titanic was vastly less complex than the dual challenges humanity faces with climate change and the post fossil fuel era. Additionally, they had the advantage of unchallenged authoritative advice and there were likely no interest groups on board which favored doing nothing.

In the context of the planet's current situation, having the ships designer present to evaluate the situation would have given a degree of clarity akin to having Mother Nature standing at a lectern giving a PowerPoint presentation on how global events will unfold. She might open up with "Humanity has thrown itself off the tightrope of climate stability. Consequently, this is what is going to happen and this will be the timetable for those events." She could then proceed to list our options. Such a presentation from an unchallenged deity would make decision-making a great deal easier.

But humanity's reaction to climate change and resource scarcity plays out like the "frog in warm water" myth. The story was that if a frog were placed in boiling water, it would jump out, but when put in warm water which is slowly heated to boiling, the frog wouldn't move and thus perish. In reality, frogs jump out of water when it gets uncomfortably warm, no matter how gradually the warming takes place. But the story is a parable of incrementality and human ambivalence to making any small change that doesn't provide immediate benefits.

The Titanic crew had cries from the lookouts and the massive impact of the collision with the iceberg to spur them into immediate action. It is unlikely that the ship's chief engineer waited until he got an estimate of the time-to-sink before he turned on the pumps. In contrast, humanity is experiencing increasingly novel climate regimes and gradually degrading resources but no singular worldview-altering event has yet occurred. We may have become more wary about our climate future but, in the absence of absolute clarity, we have accepted this as the new normal without taking decisive action to deal with the root causes of climate disruption and resource depletion.

We have expert opinion documenting the changes the planet is undergoing, many of which can be observed by most people. The forecasts of what will happen are even accepted by the great majority of both populace and leadership of countries around the world. But we currently lack the iceberg impact impetus or "Pearl Harbour"-like event to galvanize us into effective action. From the beginning of the first frantic actions to start the pumps and disembark the passengers, to the ship's final moments, the band continued to play to maintain some semblance of normalcy and calmness. This was certainly a high point of social cohesiveness in a sophisticated society if ever there was one.

In 2019, the band is certainly playing, as it was on the Titanic, but very little else is happening.

The final element of Titanic's journey was a 10 min long, 12,500 ft accelerating plunge to the seafloor which she impacted at a velocity of about 56 km/h. The earth, however, is not about to spin into the sun or otherwise disintegrate. The issues relate more to how human society and the biosphere around us will fare over decades if we do not reverse the declining health of the planet.

Titanic's crew and passengers had 2 h and 40 min to deal with how best to survive in a new hostile environment—the North Atlantic—before their life support system disappeared from beneath their feet. But their actions immediately affected their own fates. Forward-looking actions to deal with climate change will impact mostly those under 60 and those yet to be born rather than the people who are actually making the decisions.

Humans have a time window in which to act. By all expert accounts, the sooner we begin to act, the more effective our actions will be and the cheaper it will be in the end. But timing a peak or inflection point is necessarily imprecise in a complex and variable environment. Picking a perfect time to begin with exactly the right level of aggressiveness is an impossible task. But trying to achieve the impossibility of perfection should not handcuff decision-makers into inaction. It is much easier to identify trends and mechanisms and those should be the focus of our efforts to deal with both climate change and the declining availability of high EROI energy supplies. The best timing, of course, is now.

Are We Improving?

Several days after hurricane Florence had finished inundating North and South Carolina with record rains, the CNN weather specialist Chad Myers discussed the accuracy of the hurricane track predictions from the various weather models. He noted the predictions showed unusual consistency as well as exceptional accuracy. The eye of Florence came over the coast only 2 km away from the point that had been predicted a full 4 days before.

Certainly, there were circumstances which gave the guiding weather systems exceptional stability and allowed the predictions a higher than usual level of certainty. But overall, the accuracy of hurricane forecasting, in terms of strength, path, and timing have improved dramatically in the past three decades.

When hurricanes were first given human names in 1953, female names were used. This was because the males who were involved at all levels of the weather monitoring process had little idea of what the hurricanes would do, found them almost completely unpredictable and therefore named them after women who they felt shared those same characteristics. Male names were not introduced until 1979 but the unpredictability level still remained very high.

Many people in the potential paths of the storms used to ignore warnings to prepare or evacuate because the forecasts had been so inaccurate in the past. Now, approaching the third decade of the twenty-first century, the vast majority of people do heed the warnings. They spend great sums of money preparing their properties to withstand the predicted winds and rainfall and flooding and they also pick up and move hundreds of miles out of the storms anticipated path. Why?

TV networks flood the "airwaves" with coverage of incoming hurricanes and the real time impacts of every type of unfolding disaster. Everyone who might be subject to hurricane impacts is vividly aware of what those impacts might be. And people are also aware that the warnings they receive are now highly accurate. There is now a general belief that the hurricane warnings they receive are accurate enough to spend money on and disrupt their lives for. Clearly people now believe and act on the warnings they are given whereas decades ago, they may have had a dismissive attitude and were more inclined to "ride the hurricane out."

Riding Out Climate Change

Inaction on climate change can be interpreted as stemming from a dismissive or disbelieving attitude. But this is unlikely to be the case for any national government. Although the USA pulled out of the Paris Agreement on climate change, in doing so, during his 30 min speech, President Trump never once questioned the science. If Trump had any genuine misgivings about the scientific validity of climate change, his lengthy speech, broadcast around the world, would have been the ideal opportunity to voice them. But he didn't.

Speaking to his base on the campaign trail and in 3 am tweets, Trump has railed against any measure to fight climate change and even called it "a Chinese plot." But standing on the lawn of the Whitehouse and effectively speaking directly to other world leaders, Trump complained about the fairness of the Paris Accord rather than about the validity of the reasons for its creation. He left the door open to rejoining the agreement.

Technology to the Rescue

In moments of crisis and profound change, humans tend to develop a deep seated longing for salvation by a higher power. When no immediate solution is apparent and threats are imminent, the wish for a game changing event to occur very naturally takes center stage in our thoughts and hopes. In days past, these hopes and dreams may have taken the form of the heavens opening and a series of divinely inspired events taking place or the last minute arrival of the cavalry appearing from over the hill.

In the modern world, we look to, and many expect, science and technology to deliver the next big breakthrough that will sweep away problems and bottlenecks. Many people not only assume technological advances will eliminate major problems, they fully expect these advances to improve the human circumstance on a continuous, if uneven, path of endless progress. "The next big thing," "the next long wave" are all expressions of confidence that the human race has risen to the level of, not only self-salvation, but of limitless expansion of possibilities, both on earth and into the rest of the universe.

In terms of energy and the environment, the manifestations of these expectations come in the form of some sort of black box which will produce copious amounts of electricity, silently, cleanly, and cheaply. This is a natural extension of the nuclear dream which, in many ways, has delivered on its promises. Many of the miracle solutions are based on more advanced versions of the nuclear paradigm. Fusion or thorium reactors come to mind as examples of container sized black boxes which could be installed easily in countless locations producing abundant and virtually limitless energy for all.

Small versions of this dream feature beer cooler sized boxes producing unlimited energy for cars, trucks, and airplanes as well as any moving vehicle the human mind is capable of creating. Often these dreams take the form of solutions to specific problems. They could be affordable solar panels with efficiency ratings of 90% rather than their current 15–18% or batteries with five times the energy density of gasoline as opposed to the 1/40th of the power density they offer now. Perhaps technology will deliver the means to drill down through the earth's mantle and allow us to extract "limitless" (there's that word again) heat from the molten core.

None of these things seem beyond the realm of possibility given the progress we have made to date. Perhaps we will stumble on and develop a new form of energy that has been staring us in the face throughout our existence. Electricity can be thought of as a "new" form of energy even though it has existed since the universe began. Other forms of energy may well exist, just as other forms of mass, we are not able to see or feel, have been shown to exist and with which we are just now becoming familiar.

But none of these dreams ever seem to involve fossil fuels or the discovery that the earth is really a large tootsie-roll pop with oil, instead of magma, as its core. No one, it seems, expects fossil fuels to do more than keep the wheels turning for more than a few decades until we can switch to other forms of energy.

Conveniently Coincident Crises

Dealing with the large and complex challenges of climate change and fossil fuel depletion at the same time would appear to be a daunting prospect. It is. But fortunately, although they are very different in their causes and impacts, the solutions to them may cover a lot of the same ground. That overlapping footprint consists of the measures of conservation and efficiency.

Very quickly reducing our consumption of fossil fuels may well save us from the most extreme consequences of climate change. It will also preserve stores of precious and accessible (read lower cost) fossil fuels for future generations to use should the situation absolutely demand it. Additionally the moderation of the rate of decline in fossil fuel availability will go a long way in preventing the extreme price spikes and supply shocks that mark the increasing scarcity of a vital resource.

In a period of rapid change, any effort helping to minimize instability has to be seen as a critical contributor to the success of the transitions we will be undergoing throughout this century. An early start and consistent follow-through on transition programs will be critical to their ultimate success.

Change can take the form of be proactive strategies or desperate reactions. Accelerating climate change is happening as is resource depletion on our finite and delicately balanced planet. Certainly these two forces will be extremely difficult to deal with effectively and doing so will challenge the political and power structures in many countries. But difficult adaptation is ultimately a more effective course of action than denial and attempts to stay the course. Most countries have accepted the need for deep changes on paper, and it is past time for this acceptance to manifest itself as concrete actions. When words do turn to action, the learning process will encompass a blend of the technical, social, and natural disciplines.

References

Barrick Gold (2018) Barrick Gold annual reports

Berman A (2019) Labyrinth Consulting Services monthly report

BP (2018) Statistical Review of World Energy. https://www.bp.com/en/global/corporate/energy-economics/statistical-review-of-world-energy.html

Broecker W (1975) Climatic change: are we on the brink of a pronounced global warming? Science 189(4201):460–463

Climate Analytics (2015) Global warming reaches 1°C above preindustrial, warmest in more than 11,000 years. https://climateanalytics.org/briefings/global-warming-reaches-1c-above-preindustrial-warmest-in-more-than-11000-years/

Cline EH (2014) 1177 B.C.: the year civilization collapsed. Princeton University Press, Princeton

EIA (2019) U.S. average depth of crude oil and natural gas wells

Hansen B (2019) History of Earth's climate 7. http://www.dandebat.dk/eng-klima7.htm

Parker G (2013) Global crisis—war, climate change & catastrophy in the 17th century. Yale University, New Haven

Chapter 5
Renewable Energy Learning Curve

Abstract Direct comparisons among nations to determine their relative success or failure in transitioning to renewables are fraught with difficulty yet given world standards now in place clearly identifiable targets can serve as a baseline to focus the conversation. Renewable energy is diffuse and more energy is required to support life the further north one goes. Hence, more land area will be required to support life in the north and population density will be lower than in more moderate regions. Energy systems will be more complex in the regions which require more energy and greater storage capacity. Understanding the difference in characteristics of fossil fuels and renewable energy and electricity is key to understanding how our current energy production and consumption habits will change as the transition evolves.

We Have Skills

The transition from fossil fuels to renewable energy will be difficult for the very reasons that made the transition from scarce renewable energy to abundant fossil fuels so easy. Renewables will never offer the ease of use, storage capacity, energy density, or the sheer abundance fossil fuels currently do, but they do have some large positive attributes of their own. When we walk off the plank of the end of fossil fuel era we can be very glad that renewables will be there to break the fall. How sudden a fall it is depends on how quickly we start on the transition to renewable energy and on how aggressively we commit to it. Thankfully, the 250-year period during which the use of fossil fuels exploded, was one of rapid scientific advances certain to be immensely useful in enabling a successful transition.

The early Canadian settlers used almost as much firewood-based energy as mid–twentieth century Canadians did from all sources. The improvements in lifestyle and productivity achieved over that 300 -year period were due to technological improvements in energy efficiency and the types of products and applications which simply required far less energy input than heating large, drafty buildings in the dead of winter with very inefficient fireplaces. Have we crossed the technological threshold that will allow us to prosper without diminishing earth's resources or environment

J. E. Meyer, *The Renewable Energy Transition*, Lecture Notes in Energy 71,
https://doi.org/10.1007/978-3-030-29115-0_5

any further? Do we have the skills and the tools to do more with less? Or, realistically, can we live with the concept of doing slightly less with a lot less? In some regions of the world, the answer is very definitely yes. The temperate areas where substantial solar and wind energy are available all year long and where there is very little demand for heat at any time, are well positioned to make a successful transition to renewable energy.

Unfortunately, Canada and most northern nations will find that the basic level of energy required to maintain a functioning society at higher latitudes will be a challenge to supply from green sources all year round. Averages don't matter in this situation. "All year round" is the concept that needs to be emphasized because, in northern climates, a week of downtime for the energy supply system will have impacts far beyond significant financial pain for many individuals and businesses. A week without heat during winter in the far north could be compared to a minute without oxygen in its impact on society (Fig. 5.1).

Beyond the prospect of life threatening cold to residents spending 7 days in minus 30 temperatures, the damage to infrastructure from a deep, long lasting freeze would be massive. Water systems would sustain severe damage and buildings would ice up from the inside out. Food distribution would be disrupted. The effects on social stability would vary from community to community but recovery would be lengthy.

Compare this to the experience of the Great Northeastern Blackout of 2003. Electrical power was lost for 55 million people in Ontario and eight US states with the outages lasting from late afternoon to midnight of the following day during mid-August. One hundred people died from various causes and obviously most business ground to a halt as did most urban traffic. But once power came back on after the 36 -h outage, recovery was rapid and there were very few, if any, lasting effects.

Power dropouts are a big inconvenience in modern societies even in benign weather but, in severe cold conditions, long outages would amount to social calamities. If the variability of energy supply resulted only in temporary economic dislocation, then transitioning away from energy systems based on fuels whose output is completely constant to ones which get the majority of their energy from sources with highly fluctuating output would be straightforward. But northern societies were built on the constancy of energy availability.

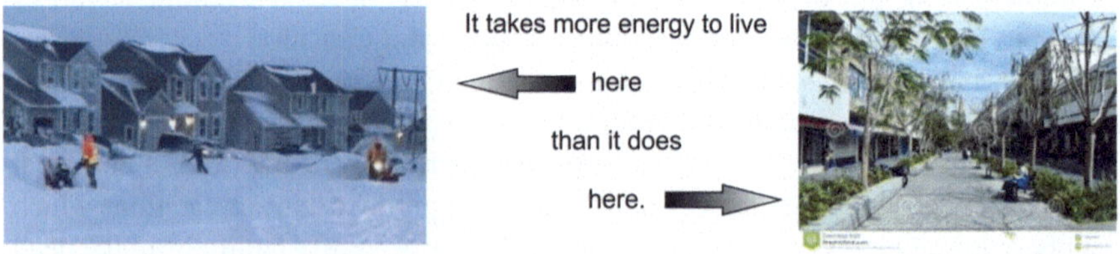

Fig. 5.1 Energy intensity of life in northern regions vs southern regions. The streets of Burlington, Ontario on a winter evening compared to the streets of Guadalajara, Mexico on an average day. Burlington is within 150 km of the most southerly point in Canada

In the more temperate parts of the world, despite the clear challenges in transitioning to renewable energy, the potential to do so successfully is very high. In northern regions, the difficulties are compounded by the inversely related variables of high winter energy demand and low winter energy production. This makes massive energy storage a necessity rather than a convenience. Currently, in 2019, outside of fossil fuels, there does not appear to be a storage option that is either economically or energetically workable for the transition of northern communities to renewable energy.

Northern Is Different: Canada Is Northern

"Northern" can be defined as any area which requires a great deal of heat in the winter and whose access to real time energy supplies is constrained during that period forcing reliance on energy storage. By that definition, virtually all of Canada, northern Europe and Russia can be considered northern. Although Paris, France is just as far north as Thunder Bay, Ontario at 48° latitude, the average January high temperature is 7 °C in Paris and −9 °C in Thunder Bay. Clearly, Thunder Bay qualifies as a "northern town" but does Paris?

We need to create a simple metric using the variables of energy demand, energy available, storage capacity, and duration of time during which the supply of energy falls below the level demanded. Going forward, sophisticated mathematical models will be able to identify the challenges of transition on a region to region basis, but, at this point, it is only necessary to identify the problems and give some broad indication of their scale.

Once the most fundamental personal and social energy needs have been met, the question of how much energy will remain must be dealt with. Discretionary energy availability is a key consideration as it determines the level of sophistication of a society can be maintained. And how well that energy and its end products are distributed among its citizens will determine how progressive and stable the society can become.

How Well Prepared Are We?

What does the human energy toolkit look like compared to the stone and iron axes of several hundred years ago?

Five hundred years ago the human toolkit was not that dissimilar to the toolkits which existed 4000 years ago during the early Bronze Age. Transport was handled by sailing ships, carts, and wagons pulled by horses and oxen. People walked a lot. Stone axes were still in use in many parts of the globe but the vastly more efficient iron axes, saws, and ploughs had completely taken over in the more advanced countries.

Energy was supplied in real time by wind and water while coal and wood provided heat whenever it was required. Gunpowder had been in use in firearms since the mid-1300s but that remained the most sophisticated use of energy until the invention of the first truly practical steam engine by James Watt in 1776.

Watt was far from the first to work with steam which transformed heat energy into mechanical energy. A first century A.D. Greek inventor named Hero of Alexandria built a steam turbine, but apparently made no attempt to connect its extremely low power output to any mechanical device. Spaniard Jerónimo de Ayanz used steam power to remove water from mines in 1606 and Thomas Savery built the first real steam engine in 1705.

Despite the extreme inefficiency of these early devices they established the ground-breaking principle that stored energy converted to heat could be used to replace the energy supplied by wind, water, humans, and beasts of burden.

And even more importantly, the steam engine established the connection between heat and mechanical work. Humans have been using heat from stored energy (firewood) for half a million years for warmth and cooking and later to break rocks for mining. But it was only 400 years ago that we learned to use heat to do mechanical work.

Heat Performs Work, Humans Leap Forward

This immense breakthrough opened the door to replacing human and animal power with a wide variety of mechanical devices driven by heat. These devices quickly became very large and able to do things humans and animals simply could not. As time went on, we were also able to build smaller machines to fit every conceivable application.

But the key concept here is the ability to do work with heat, specifically a heat differential. If everything had the same temperature, there would be no movement. Using stored energy to create a heat differential produces force and, properly controlled, that force can be focused to do work.

Until electricity was discovered, the ability to make heat do work could be viewed as the most important discovery in human history. Electrical energy is superior to heat energy in a number of ways, the foremost being the extremely high efficiency with which it can be used. However, our environment is not made up of electrical currents, whereas heat differentials are abundant. Dare we use the term "limitless"? Actually, it is probably best to leave that term on the scrapheap of human fantasies.

Nevertheless, heat potential can be found throughout our planet and atmosphere, although most of that potential is of too low a grade to enable us to harvest it. That is, too low given our current technology, but perhaps, in the future, we will be able to harvest energy, maybe in electrical form, from the earth's heat flows.

But energy has a very useful characteristic and that is it can't be destroyed, once it has been used, it just carries on in some other form or place with diminished intensity. Heat, being a form of energy, can be used many times. We don't always have to blow heat out of our exhaust pipe and watch it disperse (not disappear) into the atmosphere. German farmers used to build their houses over their cattle stalls. The heat from the animals, probably amounting to more than 500 W per good-sized cow, would radiate and convect up into the house and add considerably to the energy budget of the farm family during the winter.

Similarly, communities which burn what garbage they can't recycle will generally set up a district heating network that distributes the heat from the burning refuse to homes and business during the winter. In these projects, perhaps 10% of the energy potential in the refuse is turned into high-grade electrical energy and the rest is captured as heat which (almost certainly in the form of hot water) is piped short distances to local consumers. The heat then dissipates out of the buildings and into the atmosphere.

Mike Nickerson describes this process well in his books, "Planning for 7 Generations" and "Life, Money and Illusion" making the point that we should use something as many times as possible before throwing it away. In the case of heat, it should be used and channeled and "recycled" as many times as possible before its energy has atrophied (entropy) to the point where it can no longer be put to any practical use.

Electricity Flows

Although our environment is not made up of electrical energy flows, they still do exist. Voltage differentials exist in abundance in nature and electrical currents therefore flow. The best and most spectacular example of a natural electrical energy flow is lightning. This is a phenomenon that equalizes the level of electrical charge between the earth and the atmosphere, or different parts of the atmosphere via a stream of electrons.

Again, current levels of technology do not allow us to tap these electrical energy flows efficiently but, we've seen spectacular efficiency gains in the evolution of our heat based machines, from the 1% efficiency of the first steam engines to the 70% efficiency levels of jet turbines. Perhaps natural heat and electrical flows will be able to provide stable and carbon free energy sources for us in the future.

What Have We Learned Since Heat Was First Made to Move Water?

The tiny spark of heat-powered innovation grew into a blizzard of development and learning as we applied power to every conceivable process and even developed that new form of energy, electricity.

Power units became smaller, more powerful, more efficient and applicable to a range of applications limited only by the creators' imagination. Clearly, not all of these new devices and abilities had positive outcomes or even positive intent but, by and large, technology and the increased use of energy, has produced immense benefits for humanity.

These developments have moved in lockstep with the newest form of human creativity, science. The discoveries that science produces can have many practical applications but science itself is the process of learning for learning's sake. Science investigates and tries to understand why the world functions as it does while defining and quantifying the rules of our universe.

Such learning would not have been possible without the devices powered by fossil fuels and electricity. The toolkit of twenty-first century humans is immensely more capable than the one we possessed a mere 15 generations ago. Humans have not changed, however, and it could be argued that neither have our social structures. So the question is, can the immense technical learning and investigative power we possess be effectively applied to the problems presented by the forced transition to renewable energy?

Do we have the social will and intelligence to apply the tools we have to building a society that can function sustainably? Technically, we may already know enough to be successful.

Technical Strengths

Humans have the ability to measure energy and biological systems in extreme detail. We can monitor their performance over time and predict the outcome of changes in many complex systems. We have the ability to represent complex systems with mathematical models which, once "calibrated" to the real unit measurements of the past can make instructive predictions useful for selecting the best policy options. We can communicate and accumulate knowledge from many different sources. Existing renewable energy systems can harvest energy from the wind, water flows, and directly from the sun and store it for future use.

We can do more things, more quickly with less energy than ever before. Our current capabilities may not guarantee success but they do indicate potential. Scaling tools and ideas up into fully operational and resilient infrastructure requires a skill set beyond nuts and bolts and extends to the ability of a society to understand itself and implement difficult changes while maintaining its cohesiveness.

This could be described as the technology of social sustainability. The hard sciences might deliver the tools and systems we need to transition to renewable energy but the success of the implementation will depend on how far we have advanced along the social decision-making path. Government can declare states of emergency or set ambitious targets but if the public will to follow through for the long haul wanes, then even the best plans can falter or even slide backwards.

As the conviction of the planet to address climate change grows more solid, it has to be noted that a number of early initiatives are in decline or have been abandoned. These failures have occurred for different reasons but they have left a number of regions drifting without direction at the very time they should have been leading the way toward a green future. Some regions have implemented extremely successful programs and are positioned for continued progress.

Where Are We on the Learning Curve? Technology Matures

Technological advances pushed solar system efficiency rapidly higher in the early days but the pace of progress has slowed as the technology matures. For example, in terms of the most common type of solar panel used across the world, it would seem that we are approaching a plateau in what that technology is capable of delivering. Other types of solar cell compositions have the potential to achieve much higher efficiency levels and perhaps one of these will prove practical and be cheap enough for broad use. This would allow a leap or a step up in the graph above. It would also deliver higher EROIs which are the main determinant of the value to society of any major energy system. It should be noted that a system efficiency of 15% means that 85% of the sun's power is lost to us and goes into the atmosphere as heat to play its part in the energy flows of the planet. If it were possible to capture some of this heat and apply it usefully, the process would be termed "cogeneration" and would result in higher EROIs (Fig. 5.2).

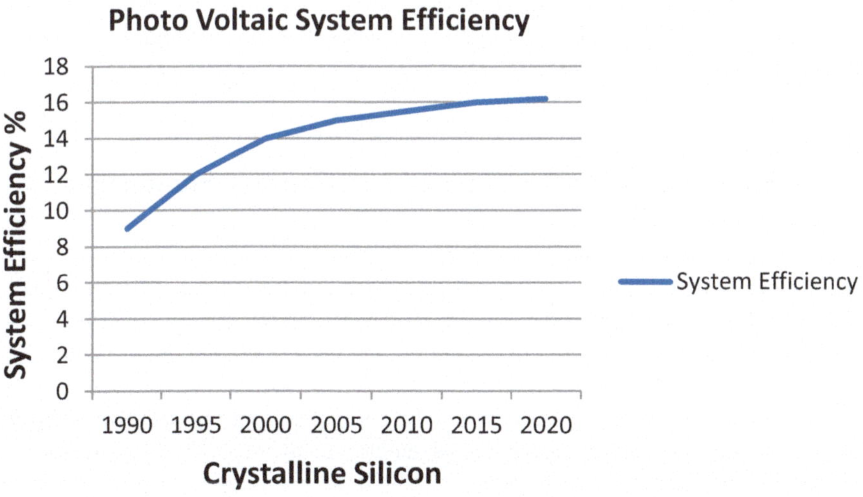

Fig. 5.2 Maturing of one type of solar pv technology (Han 2014)

Through the Threshold of Infinite Learning and Growth

A significant portion of people expect that since humanity has reached such a high level of technological achievement we can deal with any crisis that presents itself. The progress our society has made in the past 500 years has been staggering and most of this is due to the technological breakthroughs which have fueled advances in standards of living, health, education, and even social equality. But these gains have come at the cost of burning considerable natural capital.

On a biophysical level, the health of the earth's biosystems and the richness of its remaining nonrenewable resources, problems are becoming visible. In the past, humans have developed breakthroughs in technology which have spurred rapid progress along broad fronts. But most of these breakthroughs have resulted in increases in consumption of all raw materials including, most critically, energy. Many people expect the looming problems we face to be mitigated by a leap of science that will allow continuation of business-as-usual in addition to the further enhancement of our lifestyles.

Perhaps though, the next breakthrough will come in reaching a new level of understanding of how this planet operates and how we can maintain the progress we've made while restoring and preserving the earth's vital systems. Can we learn to live sustainably on earth? Many cultures before us have done it. They did not have our levels of consumption or technology but they built a pattern of understanding, respect, and even reverence for the life and energy flows that surrounded them.

In the modern age, the many groups which have made the effort to take the first steps toward sustainability have experienced an array of different outcomes. There have been many successes and many failures. But in all cases, these efforts have increased society's stock of learning.

Humanity has a tremendously sophisticated tool kit of ingenious and powerful devices but the one tool missing from this marvel of a tinkerer's workshop is the knowledge of how to apply our largely consumer-based technologies to the biophysical challenges before us.

Greenhouse Gas Emission Reductions: Country Summaries

Every country faces the same list of challenges and opportunities but the composition of the mix of energy sources and energy demands each has to work with is unique. Furthermore, different regions within each country will face their own specific issues, making the use of "country averages," a sometimes misleading exercise.

Time is also a consideration that has to be put in the proper perspective as "averages over time" are just as misleading as geographic averages. The key measuring criteria of any successful energy policy have to include minimum tolerable levels of energy availability and peak demands and how these consumption requirements can be matched to the production schedule of the various energy sources.

CAT The Climate Action Tracker Group (CAT 2018) measures the real greenhouse gas emissions from many different nations and compares them to the reduction targets committed to by those nations. Finally, the group compares both commitments and actual cuts to the level of cuts needed to prevent the planet's climate from warming beyond 1.5 and 2.0 °C. These are the two benchmarks set by the Paris Agreement with a 1.5 °C increase being the ideal target and 2.0 °C being "the last ditch," beyond which it is feared our climate will spiral out of control.

Climate Action Tracker ratings explained.

- **Effect of current policies on emissions**: The policies a government has implemented or enacted and how these are likely to affect national emission over the time period to 2030, and where possible beyond.
- **Impact of pledges, targets and (I) NDCs** on national emissions over the time period to 2030, and where possible beyond.
- **Fair share and comparability of effort**: Whether a government is doing its "fair share" compared with others toward the global effort to limit warming consistent with the Paris Agreement.

These ratings are based on the concept that if all nations performed at the level the subject nation is, the impact on global temperatures would be as described in the rating.

Critically Insufficient—at a level leading to at least a 4 °C increase in global temperatures by 2100.

Highly Insufficient—at a level leading to between a 3 and 4 °C increase in global temperatures by 2100.

Insufficient—at a level leading to between a 2 and 3 °C increase in global temperatures by 2100.

2 °C compatible—at a level leading to no more than a 2 °C increase in global temperatures by 2100

1.5 °C Paris Compatible—at a level meeting the Paris target of 1.5 °C increase in global temperatures by 2100.

Role Model—at a level substantially under the Paris target of 1.5 °C increase in global temperatures by 2100.

The mechanisms for extreme warming might take the form of the feedback loops of a shrinking of the polar ice caps and glaciers reducing the reflectivity (albedo) of the earth and increasing the absorption of solar energy, to the methane releases from melting permafrost across northern Canada and Russia to the increased burning of forests across the globe resulting in large amounts of carbon dioxide being released into the atmosphere. The forests have been a source of hope in that plans to increase their area, size, and health would have resulted in a large increase in their uptake, or storage, of carbon. If exactly the opposite happens and the biomass of global forests continues to decline (destruction of global forests account for ~20% of CO_2 emissions), then the difficulty of reducing levels of CO_2 emissions or even of limiting their rise will have been dealt a heavy blow (Fig. 5.3).

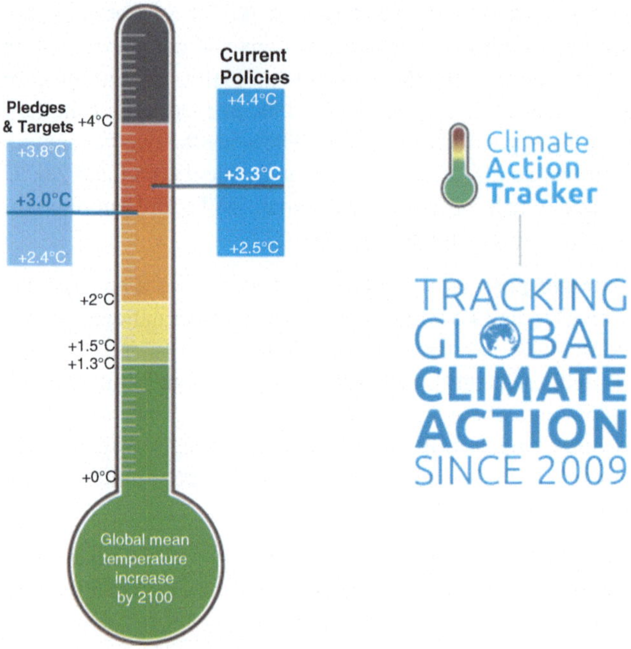

Fig. 5.3 The Climate Action Tracker (CAT 2018) Thermometer 2018. (Copyright © 2018 by Climate Analytics, Ecofys, a Navigant company, and NewClimate Institute)

Greenhouse gas emissions seem to be having very little impact on government planning at the moment. But if and when humanity begins to take climate change seriously, and make mitigating it a priority above the growth of the commercial economy, then those efforts will have a dramatic effect on the timetable of the transition to renewables. Scientists have stated that unless humanity is deeply into a program of fossil fuel reduction by 2030, we will face severe climate impacts. How far and how fast we have to travel on this issue will be determined by the progress we have made to this point. Following is a discussion of successes and failures (Fig. 5.4).

Australia

Much like Canada, Australia is a large country with very strong renewable energy potential. In Canada, that potential largely lies in the areas of hydro and wind, whereas in the sunny, dry continent of Australia, solar PV is the biggest future green energy source. Energy demand in Australia relates to cooling rather than heating.

Australian greenhouse gas emissions growth has even been slightly worse than that of Canada over the past 30 years as it is on the same economic policy track of rapid population growth and expanding exploitation of nonrenewable assets, primarily coal, most of which is exported (Table 5.1).

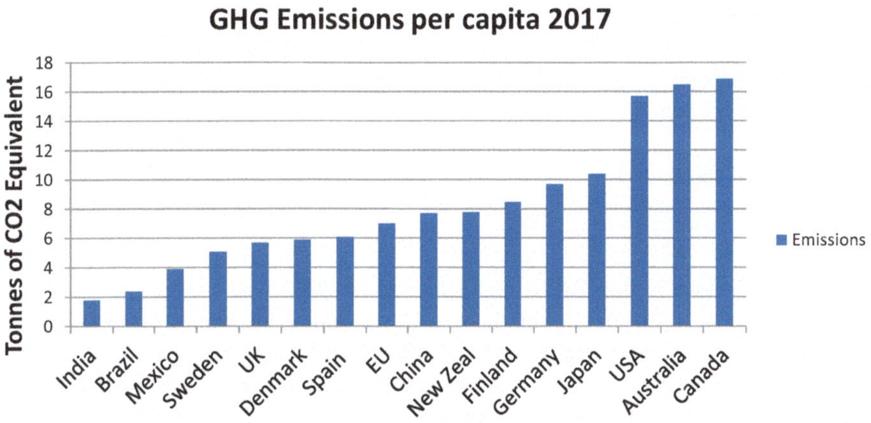

Fig. 5.4 Greenhouse gas emissions per capita by country (CAT 2018)

Table 5.1 Australia emissions record

Australia	Emissions % increase or decrease since 1990	Cuts to current trend to hit 2030 1.5 °C target	Complies with own national plan?	Reasons for change in emissions 1990–2020	Rating
CAT ratings	+29%	+53%	No	Coal production and population growth	Insufficient

Brazil

Brazil is a huge tropical country, with 56% forest coverage, moderate climate and excellent solar and biofuel potential. It also has substantial hydro capacity to supply the needs of a large and growing population. Brazil is one of the few countries on earth that can successfully produce ethanol from crops in a manner which is distinctly energy positive. The EROI of sugarcane ethanol in Brazil is approximately 8:1 and ethanol and biodiesel comprise 35% of Brazil's motor vehicle fuel consumption.

Brazil's new president has stated his opposition to many of Brazil's existing climate policies from reducing greenhouse gas emissions to the maintenance of protected forest areas to cutting fossil fuel subsidies. Per capita emissions are quite low in Brazil and therefore fluctuate greatly with any changes in the emissions or uptake of its extensive forests.

Brazil's progress in forestry emissions mitigation observed since 2005 has stopped due to the budget cuts of 50% to the Environment Ministry and 70% to deforestation monitoring authorities. Land grabbing and removal of land from the protection of "National Forest" designation has been regularized. The new President has proposed eliminating the Ministry of Environment entirely.

Table 5.2 Brazil emissions record

Brazil	Emissions % increase or decrease since 1990	Cuts to current trend to hit 2030 1.5 °C target	Complies with own national plan?	Reasons for change in emissions 1990–2020	Rating
CAT ratings	+99%	+64%	No	Higher fossil fuel use and population growth	Insufficient

The good news is that Brazil had achieved its 2020 land use targets by 2012. However, Brazil has been slipping back with the annual allowance of 3900 km^2 of deforestation being exceeded in 2017 when 6900 km^2 were destroyed.

CAT rates Brazils plan as "Insufficient," implying that its targets are consistent with warming between 2 and 3 °C (Table 5.2).

Canada

Canada is a kaleidoscope of energy sources and geographic regions. Life at any sophisticated level anywhere in Canada is energy intensive. There is a large variety of sources available on the energy palette of this large landscape but replacing fossil fuels will be a great challenge given the high level of the per capita energy demand (Fig. 5.5).

Canadian greenhouse gas per capita are among the highest in the world and since the Kyoto benchmark year of 1990, emissions have increased substantially instead of declining by 6% to conform to the country's commitment. The two largest sources of emissions increase were the increase of population driven by high levels of immigration and the increase in production from the Alberta oil sands.

With balanced levels of immigration and no oil sands production, the many energy conservation and greenhouse gas reduction programs across the country would be having a significant impact in the form of a 20% decrease in emissions from 1990. Immigration policy is on track to triple Canada's population by 2120 and oil sands production is being heavily boosted by both the Alberta and federal governments. Given the strength of the growth lobby, emissions are certain to overwhelm the best efforts of local programs and continue rising in the future.

Canada is a country focused on high growth, both in terms of population, and through increased resource exploitation and consumption. These policies have been even further ramped up by the current Trudeau government with increases to both immigration levels and the development of new pipelines to export increased levels of oil from the Alberta oil sands. Furthermore, the low hanging fruit of vehicle emissions is still being left on the tree. The fuel economy of cars and other light-duty vehicles has slowed down in recent years. After improving by an average of 1.7% per year from more than a decade, fuel economy declined in Canada by 1.5% between 2014 and 2018 (International Energy Agency 2018).

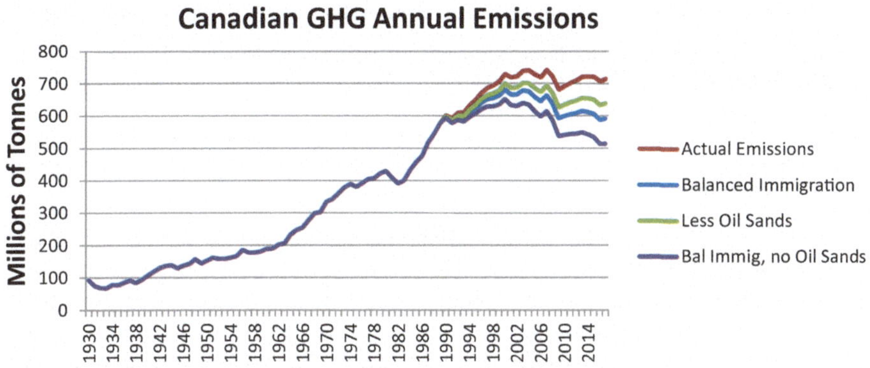

Fig. 5.5 Canadian greenhouse gas emission history with major growth components broken out (WRI 2017; Statistics Canada 2017)

Canadians certainly waste a great deal of energy, but it is also true that large amounts of energy are required here to maintain a sophisticated society and have been since the colonization by Europeans began. But instead of transitioning away from population growth, construction, and resource extraction toward a broader based economy, governments remain locked on an economic structure which requires a huge amount of fossil fuel to sustain.

What is perhaps most damaging to the emission reduction cause in Canada is a complete lack of any coherent strategy. There is no infrastructure model of our energy system in use at the federal or provincial level which would allow a rational discussion of policy options by illuminating the real world impacts of different actions.

The report on Canada from the Climate Action Tracker group finds that even if the government fully implemented their planned policies, although they would be an improvement, they would still be insufficient to meet even the least aggressive Paris targets. In fact, analysis by the Pembina Institute and Energy Innovation indicates that an even larger shortfall than that expected by the government might be underway. Given the economic focus on the energy intensive growth policies, it is difficult to see how such increases could be avoided. CAT finds Canada needs to "significantly enhance both its self-determined target and its proposed level of climate actions to get onto an emissions pathway compatible with the Paris Agreement."

In other words, Canada's action plan is not sufficient to meet its greenhouse gas emission reduction targets and those targets are not aggressive enough meet the most minimal 2 °C standard to begin with. CAT notes that if Canada continues on its current path, and all other countries missed their targets by a similar margin, the world would be pushed into a climate regime of over 4 °C warmer than the one in which human societies have flourished so strongly for the past 10,000 years.

There have been clear efforts to reduce carbon emissions across Canada. Initiatives in Ontario in the early 2000s delivered clear greenhouse gas reductions but suffered from political backlash but the closing of coal plants is almost certain to be permanent..

Canada's current policies can be nicely summed up by the CAT statement below.

> Canada is likely to miss its Paris Agreement (NDC) target to reduce economy-wide GREENHOUSE GAS emissions by 30% below 2005 levels by 2030. It remains unclear whether Canada will rely on carbon sinks in forests, soils, and wetlands to achieve its target. If it does so, and experiences low economic growth, it might achieve its NDC target with current policies. Given the lack of clarity around the potential reliance on carbon sinks, the CAT rates this target "Highly insufficient" (CAT 2018).

The benefits of large and healthy forests were assumed to be a large help in meeting Canada's carbon emissions targets. However, the forest, soils and wetlands sink the government might have been counting on to bail Canada out of its growth-at-any-cost economic policies might be backfiring. As can be seen from the graph below, Canada's landscape is now emitting more carbon than it is absorbing. The amounts are substantial, with the 240 mega tons of 2015 equaling more than double the emissions from the oil sands and 30% of the nation's total.

Up to the year 2000, and certainly during the negotiations of the Kyoto targets in the 1990s, it was easy to assume that good forest management would result in large amounts of "carbon capture" by "fixing" the carbon in the biomass of the forests. But increasing droughts, forest fires, and insect invasions have turned those assumptions upside down in the first two decades of the twenty-first century. Melting permafrost is also a potentially huge emitter of methane which, by weight, has up to 25 times the greenhouse effect of carbon dioxide. Instead of moderating Canada's net emissions, its landmass now seems to exacerbating the impacts of its rapid growth economy (Fig. 5.6).

There is no strategy even on the drawing boards that will deliver the cuts needed to stay within the "safe" 1.5 °C emission limits. Although Canada may be one of the

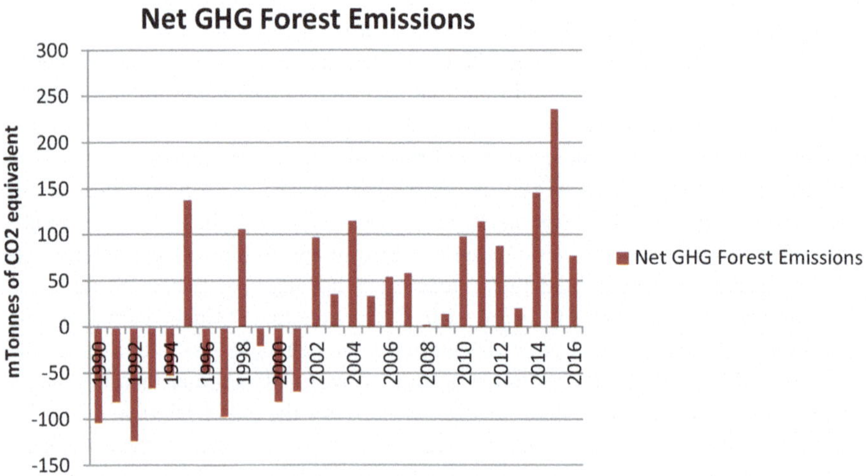

Fig. 5.6 Net Canadian greenhouse gas emissions from forests by year (Natural Resources Canada 2018)

Table 5.3 Canada emissions record

Canada	Emissions % increase or decrease since 1990	Cuts to current trend to hit 2030 1.5 °C target	Complies with own national plan?	Reasons for change in emissions 1990– 2020	Rating
CAT ratings	+19%	+51%	No	Oil sands production and population growth	Highly insufficient

worst offenders in missing their greenhouse gas emissions targets, it is by no means alone in doing so (Table 5.3).

China

China is the world's most populous country and a world wind and solar technology leader in both capacity and rate of growth. China is by far the world's largest consumer of coal which it uses to churn out manufactured goods and export them worldwide. In effect China is, in large part, the world's factory. Despite having large coal reserves, it imports a great deal of energy as it has little in the way of natural gas or oil and chooses to use the reserves of other countries to extend the life of its own. Still, an immense amount of domestic energy is embodied in China's manufactured exports and these more than make up for the energy it imports. The government of China is an authoritarian regime which features top-down decision-making and government policy is integrated with corporate control through clear policies and partial government ownership in many large enterprises. China's large geographic size features very diverse climatic regions. In terms of oil and natural gas, China's small reserves are quickly depleting, which at least partially explains its drive to electrify its transportation network.

If there is one nation with a record of hitting very aggressive targets, it is China, which is positioning itself as a global climate leader, and its actions do have an enormous impact on global greenhouse gas emissions. The record shows China has the world's highest emissions as a country but these are expected to be cut dramatically by 2030. Emissions from the "world's factory" hit a new high of 13 billion tons in 2018 due to a continued rise in coal consumption. This is just over double those of the USA and 15 times that of Canada. On a per capita basis, the Chinese level is one-third of that of Canada and the USA without including the very high levels of energy embedded in Chinese exports which would cause their effective emissions to drop.

While China was not committed to reducing emissions in the 2020 timeframe, it is committed to reducing growth by 2030, with large-scale drops by 2050. Paradoxically, China is simultaneously the largest consumer of coal and the largest solar technology manufacturer, and the choices it makes between the technology of

Table 5.4 China emissions record

China	Emissions % increase or decrease since 1990	Cuts to current trend to hit 2030 1.5 °C target	Complies with own national plan?	Reasons for change in emissions 1990–2020	Rating
CAT ratings	+412%	+43%	Yes	Coal consumption and increase in manufacturing and personal consumption	Highly insufficient

the past versus the future will have a decisive impact on the world's ability to limit warming to 1.5 °C. In the words of the Climate Action Group, "Given that China is on track to achieve or overachieve its climate targets, its next step as a global climate leader could be to set an example by submitting a strengthened NDC to the Paris Agreement by 2020."

China's commitments take the approach of reducing the carbon intensity of its GDP by 60% by 2030 and by increasing the biomass stock of its forests by 4.5 billion cubic meters. Given Canada's experience with carbon fixing in forests, the ability of the Chinese to better avoid forest fires, infestations, and drought remains to be seen.

Even with these measures, CAT rates the Chinese effort as "Highly insufficient."

China is implementing significant policies in multiple sectors to address climate change, and is implementing an emissions trading system, which includes a mandatory renewable energy certificate scheme that sets targets for renewable energy for each province individually. However, the Chinese government abruptly reduced subsidies for solar projects in 2018 and lifted a 2-year ban on new coal-fired power plant construction. How this impacts the achievement of their targets will only be known in several years. Will China now start missing targets?

China leads the world in EV production and now that it has achieved its industrial and per capita income goals, its national strategy is now focused on developing a healthy environment. Chinese national goals are discussed further in Chap. 6 (Table 5.4).

Denmark

Denmark has few energy resources beyond the North Sea winds which blow over its landscape and coastal waters. Consequently, it has become a wind generator powerhouse and exports its large turbines all over the world. Per capita wind power generation is 50% higher than the number two nation, Spain and almost four times higher than both the USA and Canada. It has large cables to Norway so that Norway's hydroelectric potential helps to smooth the intermittent Danish winds.

Table 5.5 European Union emissions record

EU	Emissions % increase or decrease since 1990	Cuts to current trend to hit 2030 1.5 °C target	Complies with own national plan?	Reasons for change in emissions 1990–2020	Rating
CAT ratings	+29%	+78%	Yes	Coal reduction and significant wind and solar investment	Insufficient

Consequently, it has been able to cut its emissions by 28% from 1990 to 2017. There is no CAT rating for Denmark as it is grouped into the EU.

EU

The European Union is currently composed of 28 member states under a unified economic and political structure. Those member states have achieved a very wide range of cuts but the table below only represents their collective numbers (Table 5.5).

Finland

Finland is a progressive Nordic nation that benefits slightly less from the moderating effects of the Gulf Stream than do its more westerly Scandinavian neighbors. It has no oil or natural gas fields but is able to exploit some hydroelectric potential from its river systems. It has more forest coverage than any other European nation at 72% (CIA 2018).

Nuclear, hydro and bioenergy—which alone account for 29% of energy supply—have made Finland's share of fossil fuels in the total energy supply one of the lowest among International Energy Agency member countries. Finland cut its greenhouse gas emissions by 18% from 1990 to 2017.

Finland expects to replace 30% of its transport fuels with renewables by 2030. Further, the IEA reports that "Finland also aims to reduce car ownership by fostering a shift from personal transport towards transport services." (IEA 2018) There is no CAT rating for Finland as it is grouped into the EU.

France

In 2019, France has been beset by Yellow Vest protests since the government announced a plan to increase the price of fuel by a meager 4%. Diesel prices had already risen by 20% since the start of the year. Sorely pressed consumers, the victims of housing inflation, growing inequity and wage stagnation for years, had had enough and the "green tax," as minor as it was, sent people into the streets. Once again, green policies had the misfortune of being bound up in a groundswell of disenchantment against a general social malaise. The fight against climate change involves high prices for energy. Any one-dimensional policy that attempts to simply pass on or impose higher prices as a method of destroying demand for carbon fuels will certainly meet stiff resistance. Unless demand destruction is achieved by both raising the cost of fossil fuels and by making renewable energy more accessible or conservation easier, constant protests will likely continue to delay and derail green initiatives.

France has cut its greenhouse gas emissions by 15% from 1990 to 2017. France derives 75% of its electricity from nuclear reactors and is a large net electrical energy exporter. There is no CAT rating for France as it is grouped into the EU.

Germany

Germany was an energy powerhouse in the early days of coal. It is still an industrial giant with large exports of finished goods but it has become a very large net energy importer as its demands have grown far past what its own coal reserves can support. Germany has relatively small hydro capacity and minimal oil or natural gas reserves. Accordingly, it is a technology leader with very progressive solar and wind policies allowing it to produce over 20% of its total electricity consumption by renewable means.

The population is high for its land mass, and despite being south of the Nordic countries, it still cannot be considered to have a moderate climate. Some heating is required for at least 7 months a year for most of Germany. Although it imports over 60% of its energy, Germany effectively re-exports a substantial amount of energy embedded in the finished goods it ships to other countries.

Germany has developed a clear vision of its power options moving forward as this energy map indicates. The renewable energy systems pictured on the map are the expression of detailed planning but, in Germany's case, an unexpected event interfered with the achievement of its goals (Fig. 5.7).

Germany delivered strong reductions in its greenhouse gas emissions in the 1990s and in the first decade of this century. However, its decision to shutter its nuclear facilities following the Fukushima disaster has forced the increased usage of coal, and lower grade, higher emission lignite coal at that. Eleven countries in Europe have done better than Germany at cutting their coal dependency. Electricity production is not the only problem area for Germany. The number of cars on the road has increased and the trend away from lower emitting diesel engines to gasoline put further upward pressure on emissions for this hybrid-phobic country.

Consequently, Germany's progress on emissions reduction has largely stalled since 2009 and its goal of reducing its greenhouse gas emissions 40% below the

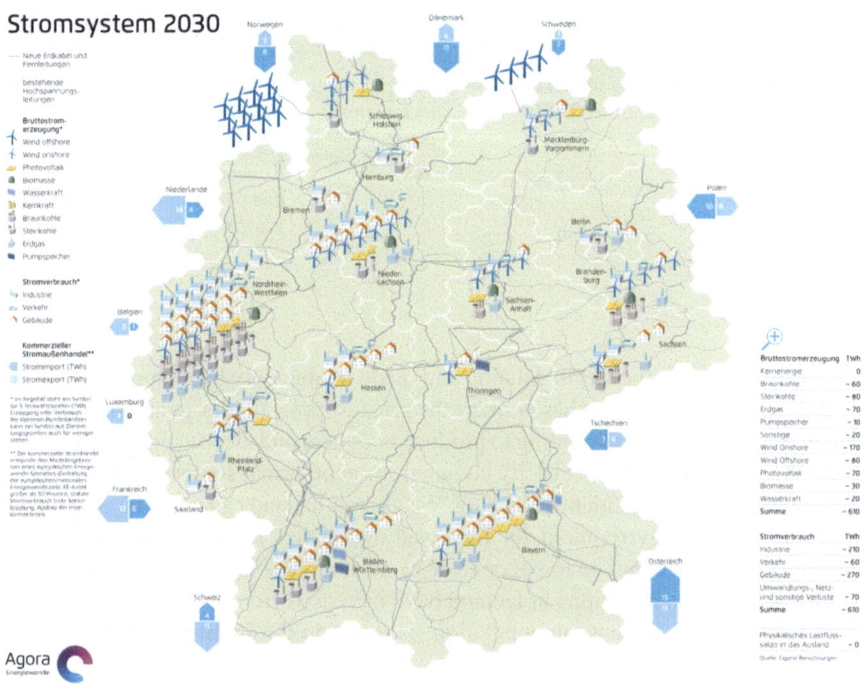

Fig. 5.7 Pictograph of forecast German renewable infrastructure (Agora 2018)

level of 1990 by 2020 is in jeopardy with the government now focusing on its 2030 goal of a 55% cut in greenhouse gas emissions.

One Shining Moment of Renewable Power

> At around 6:00 am on 1 January, 2018, a combination of strong winds and low demand after New Year's Eve celebrations meant that wind power alone produced about 85 percent of Germany's power consumption, according to data provided by the Federal Network Agency. Hydropower and biomass installations covered the rest, as there was no solar power generation before sunrise. Most experts believed it was most likely Germany would cross the threshold on a sunny and windy spring day. Renewables set last year's record of almost 90 percent of total power use on 30 April 2017.

Germany also benefits from a clear strategy in the form of its "*Energiewende*," the country's planned transition to a low-carbon, nuclear-free economy. After a slow start on hybrids and EVs, Volkswagen is expected to produce 27 new electric models by 2022 with sales volume rising to 15 million cars over several years. Germany now appears to be making a strong commitment to EVs.

Germany has cut its greenhouse gas emissions by 27% from 1990 to 2017. There is no CAT rating for Germany as it is grouped into the EU.

Table 5.6 India emissions record

India	Emissions % increase or decrease since 1990	Cuts to current trend to hit 2030 1.5 °C target	Complies with own national plan?	Reasons for change in emissions 1990–2020	Rating
CAT ratings	+310%	+2%	Yes	Coal production and population growth	2 °C compatible

Iceland

Iceland is located in the North Atlantic and made livable only by the presence of the Gulf Stream. As a nation, it has perhaps the greatest potential for geothermal energy which it is aggressively exploiting to produce both heat and electric power. Currently over 25% of the nation's electricity comes from geothermal sources as does almost 90% of its energy for residential heating. Weather in Iceland can never be considered warm and the energy required for the basic necessities is large. But most of the inhabited areas are either on top of or close to substantial sources of geothermal heat.

Greenhouse gas emissions in Iceland have increased from 1990 to 2017 by 28% due to a large 33% increase in population. There is no CAT rating for Iceland.

India

India will have the world's largest population by 2020 since it is still growing rapidly at a high rate of 1.1% annually. It has the fifth largest coal reserves in the world and relatively small oil and natural gas reserves. It has moderate hydroelectric capacity but, being the second most populous nation in the world, on a per capita basis its capacity is quite low. It manages to support such a vast population by virtue of its moderate and warm climate which imposes very little in the way of heat demand. Twenty four percent of India is forested (Table 5.6).

Japan

Japan spans almost 3000 km from north to south. Its regions range from the lush island of Okinawa, home to one of the longest living groups of people on the planet, to northern islands which receive upwards of 5 m of snow annually. Japan has limited energy resources and a large but shrinking population. It is 67% forest covered and having been formed substantially by volcanic eruptions has a high potential for geothermal energy production although they currently produce less than a third of 1% of its electrical energy production (Table 5.7).

Table 5.7 Japan emissions record

Japan	Emissions % increase or decrease since 1990	Cuts to current trend to hit 2030 1.5 °C target	Complies with own national plan?	Reasons for change in emissions 1990–2020	Rating
CAT ratings	−2%	Na	No	Fluctuating due to increase in coal consumption and shutdown of considerable nuclear capacity	Highly insufficient

Table 5.8 Mexico emissions record

Mexico	Emissions % increase or decrease since 1990	Cuts to current trend to hit 2030 1.5 °C target	Complies with own national plan?	Reasons for change in emissions 1990–2020	Rating
CAT ratings	+46%	+53%	No	Industrialization and 52% increase in population	Insufficient

Mexico

Mexico has run through about 85% (BP 2018) of its once huge offshore oil and gas deposits and has now become, once again, a net energy importer. It is the home of ancient sophisticated societies which flourished in its low energy-demand environment. Mexico features 36% forest coverage and although hydroelectric potential is quite low, its solar potential, of which it has developed very little, is very high (Table 5.8).

New Zealand

New Zealand has large amounts of hydroelectric capacity and very limited fossil fuel resources. Its semi-temperate climate keeps heating requirements modest. Medium population density, strong wind potential, and 32% forest coverage allow New Zealand some flexibility in its quest for sustainability.

Beyond being picturesque, New Zealand's mountains provide powerful torrents of water flowing down steep slopes which have been harnessed to produce large amounts of hydroelectric power. New Zealand ranks high in the world for hydroelectric power output per citizen.

Greenhouse gas emissions have been flat since 2001 while per capita emissions have fallen since 1990. Still, they are only 15% lower than those in Canada despite having vastly lower energy requirements for basic needs due to their temperate climate. Where do New Zealand's relatively high demands come from? In 2016, the

Table 5.9 New Zealand emissions record

New Zealand	Emissions % increase or decrease since 1990	Cuts to current trend to hit 2030 1.5 °C target	Complies with own national plan?	Reasons for change in emissions 1990–2020	Rating
CAT ratings	+20%	+44%	No	45% increase in population	Insufficient

agriculture and energy sectors were the two largest contributors to New Zealand's gross emissions, at 49.2% and 39.8%, respectively (Ministry of Environment 2017). New Zealand has an exceptionally high rate of agricultural emissions which is largely composed of methane releases from sheep and cattle.

New Zealand's 2030 target under the Paris Agreement is to reduce greenhouse gas emissions by 30% below 2005 levels by 2030. By 2050, the cuts will be 50% below 1990 levels as it moves toward a zero net carbon economy (Ministry of Environment, New Zealand 2019) (Table 5.9).

Norway

Norway is a northern nation with a great deal of hydroelectric capacity. Its population centers are at higher latitudes than those of Canada but it benefits from the warm Gulf Stream and therefore its winters, although darker, are more moderate than those experienced by most Canadians. Norway produces and exports a great deal of oil and natural gas and has one of the highest standards of living in the world while maintaining comprehensive and progressive social and environmental policies.

Norway has the largest hydroelectric generating capacity per person on the planet. As well, forests cover 31% of Norway, almost as high as the 34% coverage of Canada. Tidal flows at many points along Norway's rugged coast are strong enough to support tidal power. The potential for solar power is low given its latitude, but there are undoubtedly numerous good potential wind farm locations. Their abundant hydro capacity makes other forms of renewable energy somewhat redundant for their own needs but certainly exports might create a demand in the future.

Climate variability does become a factor in their energy mix as the droughts of 2018 forced a reduction in hydroelectric output resulting in higher electricity prices and probably a slight increase in carbon emissions.

The CAT evaluations for Norway again found a large gap between intent and results. They rated Norway's currently implemented policies "Highly insufficient"—emissions are projected to decrease by only 7% in 2030—a far cry from its 2030 NDC target of "at least 40%" below 1990 levels. Norway's currently implemented policies are not consistent with the Paris Agreement, and are instead consistent with warming between 3 and 4 °C if all others followed a similar level of ambition.

But Norway's climate law, which aims to achieve "carbon neutrality" by 2050, has been complemented by an increased budget allocation of $1.22 billion US. Norway places very high import duties on automobiles which effectively almost doubles the cost of cars. These duties are waived in the case of EVs, giving them a very large boost in affordability. Also, EVs get special parking rights in cities and access to express special vehicle lanes. Consequently, now every second new car sold in Norway is electric.

Norway has given users of fuel oil for home and business heating until 2020 to find another source of heat. Fuel oil is the least efficient and highest greenhouse gas per unit of heat means of heating and this law should improve the occasionally very poor air quality in the country's major cities during the winter.

However, Norway pumps about 2 million barrels of oil per day, while its daily output of natural gas stands at around 300 million cubic meters. The 2 million barrel per day output is approximately 40% of Canada's current production of 4.5 million barrels, yet Norwegian emissions from oil and gas production were 15 mega tons per year while Canada's are in excess of ten times that at around 165 mt.

Lower Canadian emissions standards plus the inherently inefficient and energy intensive processing of the oil sands results in a level of emissions per barrel produced five times higher than oil produced from Norwegian offshore fields. Norway is extremely well positioned to be a clean energy leader given its large per capita hydroelectric resources but it has extended beyond these advantages to innovate in many areas. Sale of petro-powered vehicles will be forbidden past 2025 and EVs currently make up 50% of total sales. A project to develop wireless charging of taxis is underway, which could ultimately lead to technology which could apply to the entire EV vehicle fleets of all countries (Table 5.10).

Russian Federation

Russia is by far the largest country in the world and features an extensive array of energy resources including hydro capacity, oil and natural gas while potential for both wind and solar PV is quite high. It is not a progressive state in terms of energy policy and has made little use of its renewable resources. It is a leading exporter of fossil fuels. Russia has a similar mix of energy demand and energy sources compared to Canada and, once fossil fuels begin to decline, will face similar problems

Table 5.10 Norway emissions record

Norway	Emissions % increase or decrease since 1990	Cuts to current trend to hit 2030 1.5 °C target	Complies with own national plan?	Reasons for change in emissions 1990–2020	Rating
CAT ratings	0%	+83%	No	High oil production offset by conservation	Highly insufficient

and opportunities. Forests cover 49% of Russia's surface. Many consider Russia to be an authoritarian regime, and, if so, the priorities of its leadership are clearly different from those of China which is making great strides in building its renewable energy infrastructure.

Nevertheless, in recent years there has been increasing acknowledgement of the need to transition more quickly toward renewable energy sources (RES) at the highest levels of the Russian Government. President Vladimir Putin described RES as "the proper path" for global energy development, while deputy energy minister Andrei Teksler was equally clear last year. "Renewables are no longer referred to as alternative energy, they are traditional," he said, as quoted by the Financial Times. "We are not ignorant of the future."

The road to greater integration of solar and wind energy in Russia will be a long one. Ignoring hydroelectric power, which provides 51.5 GW of the country's approximately 53.5 GW of clean energy generation capacity, renewable energy claims a mere 3.6% share of the country energy mix.

The government's official energy policy document up to 2035 sees this share grow to only 4.9% by 2030, according to the International Renewable Energy Agency (IRENA 2018) (Table 5.11).

Spain

Spain had substantial manufacturing capacity in the middle ages but it declined after it seized possession of immense hoards of gold in the New World. Its manufacturing base withered as it became easier to just pay in gold for whatever it required. The Spanish manufacturing base never made a full recovery but its renewable energy sector has become quite advanced. Climatically, Spain can be considered a fairly moderate country with relatively low energy demands being required to support a moderately sophisticated standard of living. It has good wind and solar potential, and, despite a relatively large population, still remains 37% covered in forest.

Spain has very high solar and wind potential and has employed advanced solar technology, including solar concentrators. With its many solar PV farms, Spain has gained a great deal of experience in integrating renewable energy with the grid.

Table 5.11 Russia emissions record

Russian Federation	Emissions % increase or decrease since 1990	Cuts to current trend to hit 2030 1.5 °C target	Complies with own national plan?	Reasons for change in emissions 1990–2020	Rating
CAT ratings	−29%	+64%	No	Oil and gas production increasing after large drop after breakup of Soviet Union	Critically insufficient

Spain also has significant hydroelectric capacity and is one of the countries in Europe best positioned to successfully transition to renewable energy.

Yet, Spain's greenhouse gas emissions went up over the period of 2014–2017. The vagaries of climate caused a backward step for Spain as a deep drought caused reservoirs to hit near record lows resulting in much reduced hydroelectric power generation. The deficit was covered by coal and gas, the main sources of CO_2 in the electricity sector in Spain. Coal consumption to generate electricity grew 21% in 2017 compared to 2016 and gas 31.8%. This increase was not offset by a decrease in the use of other fuels as oil consumption grew 0.5%.

Spain is aiming for 90% carbon cuts by 2050 under its new climate strategy. This would mandate 100% renewable electricity by 2050, a ban on gas and diesel vehicles by 2040 and the elimination of all fossil fuel subsidies. An interim target would aim for 70% renewable power by 2030. This sun-drenched country intends to install 3000 MW in renewable capacity each year between 2020 and 2030. This would equal either 12 million standard 250 W solar panels or 1500 large, 2 MW wind turbines or a combination of both. Renewables provided 32% of Spain's electrical power in 2017. A 2012 study by Prieto and Hall on the EROI of Spain's existing solar PV installations showed a rather disappointing result of 3.5:1 (Prieto and Hall 2012).

Its greenhouse gas emissions increased by 15% from 1990 to 2017. There is no CAT rating for Spain as it is grouped into the EU.

Sweden

Sweden is blessed with substantial hydroelectric capacity since the same mountains which shed water to the west in Norway, shed water off their easterly slopes through Sweden. Those same mountains block some of the warmth of the Gulf Stream and consequently, Sweden has a slightly higher demand for energy than does Norway. Its 69% forest coverage provides a great deal of residential heat but Sweden has no oil or natural gas resources and did commit to eliminating its nuclear power plants. However, this has proven to be difficult as wind has not yet been determined to be a viable substitute in winter. Sweden still gets 35% of its electricity from nuclear and its phase-out, as in Ontario, may happen by attrition rather than by planning.

Sweden has reduced its greenhouse gas emissions by 26% from 1990 to 2017. There is no CAT rating for Sweden as it is grouped into the EU.

United Kingdom

The UK, like Germany, was an energy and industrial powerhouse in the century and a half after the beginning of the industrial revolution. It sat on enormous stores of coal, most of which are now gone, and used that energy to build an empire which fed raw material back into its then high-tech carbon fueled industrial complex. It exported the finished goods to willing and sometimes unwilling markets. The UK has an enormous population given its small size and a very much depleted resource base. The energy prospects for the UK are very well detailed in the work of Prof. David MacKay, which is covered later in this chapter. Of all the nations on earth, the United Kingdom represents the old metrics of success, whereby one small nation comes to dominate many other larger nations and draws their resources into itself to sustain a level of consumption and social sophistication that would be impossible on its own resource base.

The Empire is now dead, and although the goodwill of the Commonwealth endures, the resource flow model that underwrote Britain's rise is not likely to be possible in a post-oil world of renewable energy and increasing resource scarcity. The UK has decreased its greenhouse gas emissions dramatically by 40% from 1990 to 2017. There is no CAT rating for the UK as it is grouped into the EU.

United States of America

The USA "boasts" one of the highest levels of per capita energy consumption in the world and is seen as the ultimate consumer society. It has large energy capacity from coal to solar and from wind to oil and natural gas. Its regions cover a wide range of extremes from high energy demand northern states and mountain regions to the moderate regions of its east and west coasts to the heat-challenged deserts of the southwest and near-tropical deep South. Its hydro capacity is significant but modest for the size of its population and its forests are extensive, but, like those in virtually every other country, increasingly fragile.

America is a world leader and despite the pronouncements of the current administration about the fallacy of climate change agreements and the need to turn the USA into "an energy (fossil fuel) super power," there is a great deal of progressive change occurring at the state and municipal levels across the country. If the targets of these jurisdictions were fully met, the USA would be within striking distance of their Paris Agreement commitment, resulting in emissions that are 17–24% below 2005 levels in 2025.

Coal plant retirements doubled and emissions per unit of electricity decreased in the first half of 2018 compared to the first half of 2017 (CAT analysis based on EIA). In July, electric vehicle sales broke the 2% mark of all new vehicle sales nationwide, with front runner San Jose, California topping 13% in 2017. These encouraging trends will need to accelerate substantially to avoid the worst impacts of climate change.

The Green New Deal, being now widely discussed, may be impossible to implement by 2030 as its most ardent supporters hope, but it is very much focusing the climate issue and energizing the people and the programs which can make a huge difference in, not only world emissions levels, but in the world's climate conversation.

Table 5.12 USA emissions record

USA	Emissions % increase or decrease since 1990	Cuts to current trend to hit 2030 1.5 °C target	Complies with own national plan?	Reasons for change in emissions 1990–2020	Rating
CAT ratings	+2%	+72%	No	35% coal consumption cuts but 60% oil production increases, 30% increase in population	Insufficient

The USA has abundant potential for solar and wind and has, on average, moderate demand for necessary energy. Being the originator of the consumer society may mean they will experience more withdrawal pain from high consumption levels than most nations, but there are certainly many Americans willing to lead lifestyle changes (Table 5.12).

Lofty Ambitions, Clear Failures

There are strong efforts worldwide to reduce energy consumption and greenhouse gas emissions and increase the production from renewable energy sources. Technological progress has been rapid but there have also been many instances of failure to overcome social resistance to the changes accompanying the transition to renewable energy (Fig. 5.8).

In the 27 years from 1990 to 2017, the greenhouse gas emissions from the largest developing nations, China and India have risen the fastest, followed by the resource based economies of Norway, Canada, and Australia. The most developed nations with largely stable populations have delivered the largest reductions in greenhouse gas emissions. Countries whose national governments do not have exclusive control over resource and climate policies rely on their different regions to develop effective policies. The results have been variable.

Ontario: What Went Wrong?

Ontario was once the "have" province in Canada. It was big, with a large population and a substantial base of cheap hydroelectric power. This attracted manufacturing and there were factories throughout the province. Now, Ontario still has a large and even a rapidly growing population, driven primarily by immigration, but it no longer has the depth or diversity of manufacturing capability it once had and it is very decidedly a net energy importer. The province of Ontario is responsible for 25% of Canada's greenhouse gas emissions and the new Conservative government has

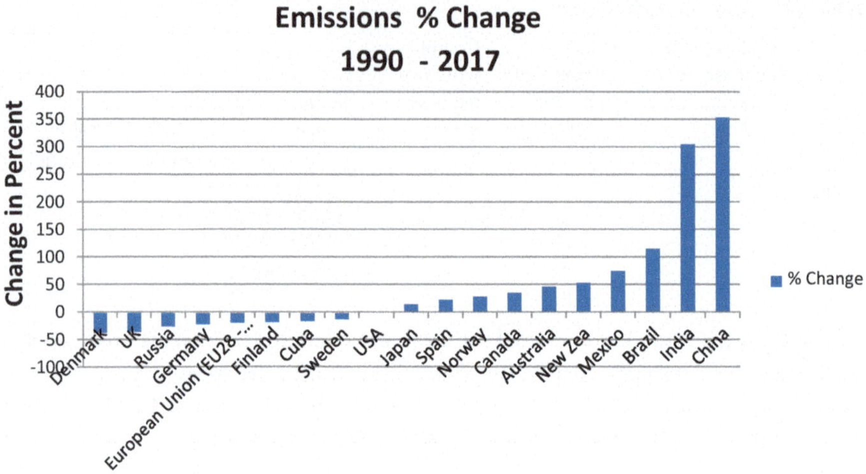

Fig. 5.8 Emission change by country (Global Carbon Project 2018)

canceled the Green Plans of the former Liberal governments, which included incentives for the purchase of EVs, solar and wind electricity production and all energy conservation programs.

Included in the program cuts are any conservation incentives which, according to the Ontario Clean Air Alliance "costs just 1.7 cents per kWh" to enact. Eliminating these programs and the energy savings they generate will just require greater reliance on high-cost nuclear plants, where the cost per kWh will hit 16.5 cents per kWh by 2025. Conservation is the low hanging fruit for any transition to renewable energy strategy, because it costs far less to reduce consumption than it does to add generating capacity in an environment of continuously escalating costs.

Furthermore, the lack of leadership at the highest provincial levels can have a stagnating effect on decision-makers all the way through the energy infrastructure. The province's Independent electrical system operators do not find themselves motivated to take any initiatives on greenhouse gas emissions as indicated by this statement from their Vice President, "The IESO has no mandate to reduce greenhouse gases so DERs are of little interest compared to nuclear." These system operators have no plans to encourage conservation or the connection of small renewable residential sized solar or wind installations to the grid.

The new conservative Ontario government also joined several other provinces in mounting legal challenges to the federal government's imposition of a carbon tax.

Resentment against the greening of Ontario arose when electricity rates increased dramatically over a period of a decade. The whipping boy for these increases came to be solar and wind farms and the high levels of incentives which were key to their creation. At the same time, the cost of the solar and wind generation costs hit the distribution system, so did costs that had been foregone for decades.

Infrastructure: Why Spend the Money?

The Puerto Rico example serves as a model of what can happen when infrastructure is allowed to decline. Money is saved up front but, in the long run, costs can spiral out of control.

Puerto Rico suffered a complete breakdown of its aging and obsolete power system during Hurricane Maria in the fall of 2017. Repairs had been put off and modernization foregone amid charges of corruption and incompetence going back decades. It was an unreliable system teetering on the brink of failure. When Maria struck, failure devolved into absolute collapse. The category 5 hurricane killed over 3000 people and flattened the fragile and dated electricity network. The question of bringing electricity service back to the island quickly turned from one of repair to one of complete rebuild.

Now, 18 months after the disaster struck the island, Puerto Rico has a fragile power grid in place delivering power to nearly 100% of its original customers. But the current system is a patchwork and does not have the resilience necessary to withstand another powerful hurricane strike.

That kind of slow degradation of the grid and its resultant vulnerability to unusual events was the reason behind the decision of the provincial Liberal government of the early 2000s to spend the money to stabilize and modernize Ontario's vital electrical system. The Liberal government in power, confronted with what could eventually become a Puerto Rico-like event and the need to reduce carbon emissions, committed to upgrade the provinces hydroelectric infrastructure. Billions of dollars were spent on upgrading the network of high voltage lines, building natural gas power plants and closing coal plants. The cost overruns of refurbishing the nuclear plants to extend their lives by a few years to decades also ran into the billions.

All the bills from previous neglect came due at the same time the costs and complications of solar and wind renewable energy were coming on stream. Electricity prices effectively doubled over the span of a decade and cries of pain were heard from the smallest consumer to the largest industrial corporation.

Since maintenance and upgrading had not been performed incrementally over time, the Ontario government was forced to do it in a relatively short period, in fact three or four decades worth of capital expenditure in the timeframe of one. The province has spent $70 billion since 2003 amounting to $5000 for every man, woman, and child in the province. Real costs that were never built into the system on a continuous basis now came in a lump bigger than consumers, stoked by various interest groups, were willing to swallow. Blowback was inevitable.

Also, the counterintuitive issue of "curtailment" came to light. Curtailment is a universal problem with solar and wind generation. When the wind blows and the sun shines, these generators produce electricity whether it is needed or not and output has to be curtailed, in effect, wasted. Curtailment rates in China approach 14% while rates in Ontario have varied between 5.3 and 13.9% (Luft 2016). In Ontario, excess electricity was often shunted to the US market and sometimes the Americans were even paid to take it. This process stoked rage among already bruised consumers.

Why Illuminate When You Can Exploit?

Media corporations had a field day. They put people who had their power cut off for failure to pay their bills right in front of the cameras. They featured grandmothers who had to choose between feeding their grandchildren and heating the house. One of the more sensationalist TV media corporations showed a hydro bill from one residential owner that stood months in arrears at several thousand dollars. This residential customer had a hydro bill of $954 for April. Her bill for March was $3662. Surely this should have been a flashing red light to their "Investigative News" team.

Instead of asking why the customer was using so much energy and how they could lower their expenses, the media corporation simply screamed "injustice" to the now howling mob of incensed viewers. The charge of $3662 equates to approximately 17,000 kWh, about 2½ times the amount of electricity a small, efficient home uses in an entire year for all heating, lights and power. The $954 bill for April meant that the household was using approximately 6 kWh of electricity every hour for the entire month, a huge amount at any time but April is a moderately warm month.

Responsible journalism would have pursued the investigation of where all of that energy went and how the excessive consumption could be reduced. Why weren't conservation incentives applied to this consumer's house? Was there a problem with their electrical system? Did they leave windows open on cold nights? But there was no coverage of cause and effect, just exploitation of the consumer's pain.

Politicians were no better. The leader of the opposition published campaign literature showing American electrical prices to be significantly lower than those in Ontario without bothering to convert US prices to Canadian dollars. Doing so would have increased the American costs by 30%, making them roughly equal to those in Ontario in many jurisdictions. The media wasn't interested in revealing such gross distortions.

To be fair, electricity costs in Ontario are higher than many of those in Canada and the USA but lower than ones in Mexico and about the same as China. Quebec has the lowest costs in North America, and possibly the western hemisphere, due to the massive water flows it has been able to tap. Coal is the next cheapest electricity generator followed by natural gas, wind, nuclear, and solar. Electricity costs are a work in progress as different regions are at different point on the path to low CO_2 emissions. Regions which have not yet switched from coal to natural gas and renewables may have low electricity costs now but they have a price hike coming as they phase out coal.

If Ontario had never outgrown the generating capacity of Niagara Falls, it would have the cheapest energy on the planet. But an expanding population and manufacturing base forced the building of more, smaller and more remote dams and then the construction of coal plants and finally nuclear plants. Each time, the incremental costs of every kilowatt-hour went up as more costly energy sources were developed and blended into the system, a near perfect example of "dis-economies of scale."

Another example of media corporations as champions of consumerism came in the news report on a truck owner who complained that his annual licensing fee was over $100 more expensive than the charge for lighter vehicles. This individual was using a 7000 lb truck as his personal transport, not to actually haul any large or heavy freight. The network reporter noted how unfair this was, rather than asking the consumer how using a 3½ ton vehicle to get groceries fit into his perspective on climate change and finite fossil fuels.

With no government strategy in place to reduce consumption and increase renewable production, Ontario is at loose ends in terms of cutting its greenhouse gas emissions. More hydroelectricity will be purchased from Quebec and perhaps the lives of the nuclear plants will be extended, but as Quebec reaches the limit of its capacity, and as the nuclear plants reach the end of their extended lives, Ontario will have few choices other than increasing fossil fuel (natural gas) consumption. It will also have to raise electricity prices in step with the real costs. This would be true even if it stabilized its population. The difference is, with the Green Plan of the previous government, the foundation of a transition to renewable energy was being laid by the higher costs. But with those initiatives now dead in the water, just to keep up with growth, fossil fuel generating capacity will have to be built and later mothballed well short of its useful life as the renewable energy mandate becomes more critical.

The Ontario green initiative collapse is a worst-case scenario for climate and energy policy in Canada but it reflects the difficulties of implementing policies which are not well explained or understood by the people they affect most. No one wants to pay more for what they already have and change is difficult even under stable economic conditions. As inequality increases and people find themselves harder pressed to meet their living expenses, consistent policy implementation can be expected to remain problematic.

Despite the retrograde steps of the current government, Ontario has made very distinct progress in the past 20 years. Even in the face of rapid population growth, its greenhouse gas emissions have declined by 10% from 1990 to 2016 due in large part to the replacement of coal plants by natural gas and wind generation. Solar farms, representing 98% of Canada's current capacity, have also made significant contributions. There have also been strong, if spotty, programs to upgrade windows, insulation and convert to LED lighting and geothermal heat pumps.

There is a moderately extensive system of EV charging stations in the main population centers of the province and this is being rapidly expanded to cover the long sections of empty highway in more remote areas (Fig. 5.9).

Alberta

Alberta is Canada's fossil fuel hub. Despite being, as one professor at the University of Calgary put it to the author, "ground zero" for potential solar electricity production in Canada, this western province leads the country in greenhouse gas

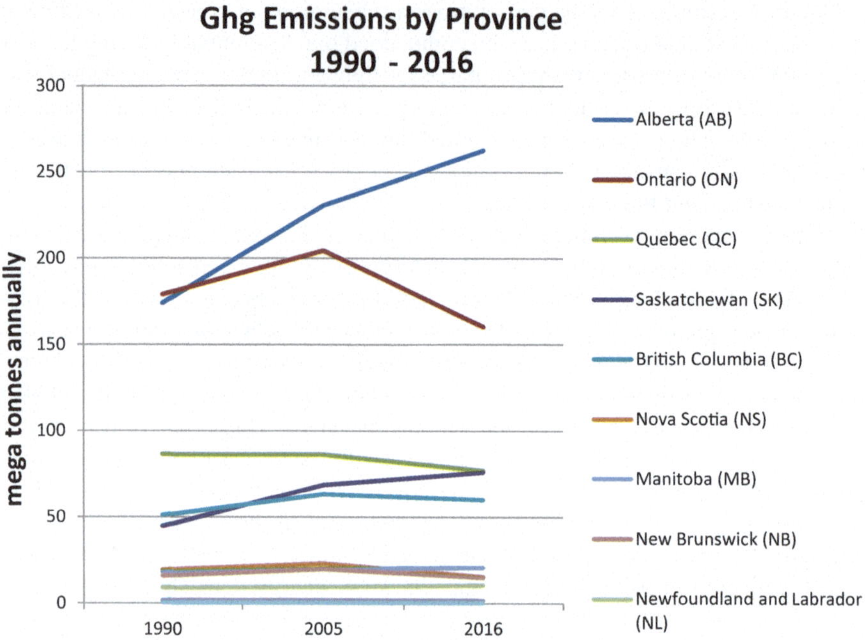

Fig. 5.9 Emission change by Canadian province (Statistics Canada 2017)

emissions. Caps have been established for greenhouse gas emissions from the oil sands but as pressure builds to export more, it is an open question whether these will be respected in the future.

Not only does Alberta have to contend with the emissions from its current oil and gas and coal industries, it must somehow deal with the emissions from 155,000 oil and gas wells which are either abandoned or have no economic potential. The legacy cost of dealing with these as well as 3600 segments of abandoned pipeline has been pegged at $47 billion (Glen 2018). And this is in a province with no clear path to remediating the thousands of square kilometers of oil sands mining sites.

Alberta is not alone in facing huge oil and gas legacy costs. University of Regina geography and environmental studies professor Emily Eaton found that of the 110,000 wells in Saskatchewan, over 24,000 were inactive.

The Alberta government is committed to the closure of its coal generating stations which contribute a significant portion of its carbon emissions. Alberta boasts the first wind farm to be developed in Canada and the dry, sunny southeast portion of the province offers one of the best locations for solar farms in Canada. But Alberta is in the fossil fuel energy business and right now, oil and natural gas are the energies in demand. Obviously, this will change over time and, in preparation, the province is at the forefront of a number of developments. Beyond wind and solar farms, below are a few of the wide range of initiatives, large and small, currently being undertaken (Fig. 5.10).

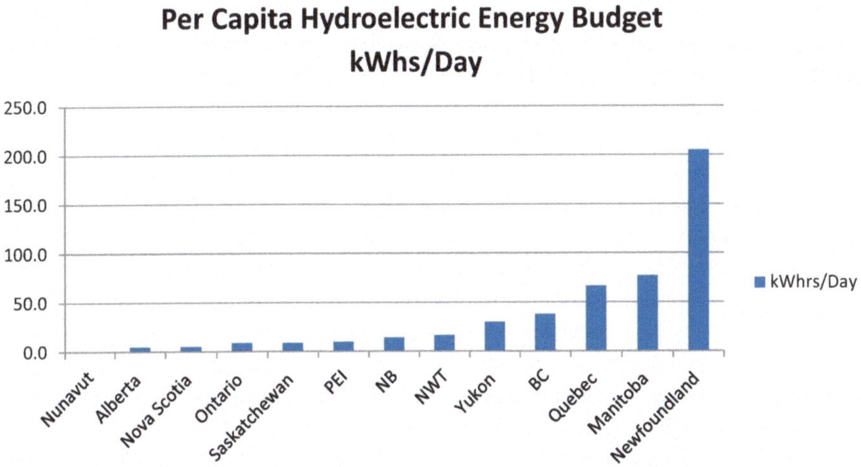

Fig. 5.10 Hydroelectric production per day per capita by Canadian province (Statistics Canada 2017)

Carbon Sequestration via Next-Generation Bioreactor Technology—This process maximizes the quantity of fuels produced per ton of wood waste in an integrated biorefinery.

- **The Shell Alberta Bioenergy Project** diverts organic wastes, avoids methane emissions, and transports them to a primary anaerobic digester facility to produce biogas and digest.
- **Saddlebrook Solar and Storage** has created a utility-scale, solar plus storage electricity which takes advantage of Alberta's climactic conditions.
- **Battery-Based Bus Charging Routes is developing** solutions to electrify Alberta's transit with a DC microgrid and on-route charging stations.
- **ENMAX E-Fleet** electric vehicle technology applied to two medium-duty transports to test the technology in all weather conditions, including harsh winters.
- **These are promising ideas but their energy costs and EROI have not been calculated or at least made public.**

Hydro Have and Have Not Provinces

There are essentially four "have" provinces in Canada with respect to hydroelectric resources. The economic "have" provinces of Ontario, due to its manufacturing and finance, and Alberta due to fossil fuels are both deficient in hydro capacity. Alberta has very little and Ontario outgrew the resources it has. Alberta has natural gas to fall back on as a path to much lower greenhouse gas emission future. Also, it has substantial wind and solar potential.

Ontario, however, by virtual of its population size, almost twice that of the next most populous province, Quebec, has electricity demands which exceed what it could be reasonably expected to generate from solar, wind, and water sources. Nuclear power currently contributes 58% of its electricity generation and the last nuclear plant is scheduled to close in 2037. This means the province has to more than double either its generation from other sources or its importation of electricity over the next 17 years.

Viewed on paper, on its own, Ontario would appear to be an energy powerhouse, but when its numbers are viewed on a per citizen basis, Ontario appears much more vulnerable in a world transitioning to renewables. If it continues to follow a "damn the torpedoes" obsession with simple growth, it may follow the path that California is charting of rapid population growth and failing social structures and rising greenhouse gas emissions despite progressive carbon reduction programs, only without having the tremendous solar potential that state possesses.

Quebec is a net exporter of hydroelectricity but it can't fill that gap. To the south, Ontario is staring at the US states which are net electricity importers themselves, and have little in the way of solar or wind potential that, if developed, would allow them to become renewable energy exporters.

Quebec

Quebec is a large province with very large hydroelectric resources and progressive climate policies. Given the huge hydroelectric capacity bestowed on it by the province's massive river systems, Quebec has been a leader in Canada in clean energy, low greenhouse gas emissions and EV penetration. The large hydroelectric projects harnessing the power of the province's many rivers have allowed its electricity rates to remain the lowest in North America while the greenhouse gas emissions associated with that power are also extremely low. Quebec has enough clean and cheap power to drive its de-carbonization faster and further than any other province in Canada with the least amount of social and economic disruption.

Clearly this also applies to Newfoundland and Manitoba which both produce more hydroelectric power per person than does Quebec.

California

Over the past 30 years, California has proven to be a global leader with its very progressive and often ground-breaking environmental and climatic standards regardless of which of its two parties was in power. California standards require that 33% of the electricity being put into the grid be from either solar or wind sources. In addition, California has very strong incentives for the adoption of electric cars and reduction of gasoline consumption.

Yet, despite all of its strong steps, California's greenhouse gas emissions went up slightly from 1990 to 2017. What happened? Over the same period, California's population increased from 30 to 39 million. This increase in the number of consumers overwhelmed the decrease in its per capita emissions.

Despite surging social problems, a fleeing middle class, housing inflation, burgeoning cheap labor rates, and crumbling infrastructure, California is still staying the course on its renewable energy targets. Whether, it is residential solar systems, vehicle emissions standards, or incentives for wind power, this state is at the forefront of efforts to transition to clean energy. It has significant wind potential and very high solar PV potential given its southern location and many arid regions.

European Union

For at least the last century, Europe has been a net importer of energy and raw materials. Most of Europe can be considered to have fairly high energy demand to meet the basic needs of its citizens. Some southerly regions like Spain and areas bordering the Mediterranean have moderate energy demand while regions to the north and east have substantially more. The demand for this energy is satisfied by imports from the Middle East and Russia. In particular, Russia supplies a great deal of oil and natural gas to Europe.

The European Union has been strongly in favor of addressing the issue of global warming. Crop failures, forest fires, and freak weather events are becoming more frequent on the continent and Europeans are solidly behind greenhouse gas emission reduction programs. Despite this, the commitments and actions of the EU members have fallen short of what is needed to meet the targets identified with 1.5° and 2° warming limits.

The estimates from CAT indicate the level of currently implemented climate action is not yet compatible with the Paris Agreement's 1.5 °C limit. The European Parliament has called for increasing the EU's NDC emissions 2030 reduction goal to 55% below 1990 levels which would come close to being "2 °C compatible" but far off what is needed for 1.5 compliance.

In November 2018, the European Commission presented its proposed 2050 climate strategy which had as its goal emissions neutrality by 2050. Despite the awareness and strong efficiency and emissions reduction goals of the EU, their emissions went up in 2017 driven by the use of low grade coal (lignite) and natural gas for electricity generation. This step back underlines the need to accelerate the coal phase-out and development of renewables.

The EU Emissions Trading System is pushing coal into the realm of uncompetitiveness by adding the carbon cost of 28 Euros per ton which, when combined with stiffer air pollution standards, will actively support the push to renewables or a changeover to natural gas.

In terms of the displacement of fossil fuels the EU trails behind Norway, China, and California in support for EVs, the number one means by which to reduce oil consumption. Further, progress in improving the efficiency of buildings is taking place at the very low rate of 1–2% annually.

Clearly, the EU recognizes the importance of addressing climate change and is also sensitive to its position of being dependent on foreign sources of fuel but the complexities of implementing progressive policies are resulting in halting progress.

The share of renewables in gross final consumption of energy is one of the headline indicators of the Europe 2020 strategy. The target to be reached by 2020 for the EU is a share of 20% energy from renewable sources in gross final consumption of energy. However, renewables will continue to play a key role in helping the EU meet its energy needs beyond 2020. For this reason, member states have already agreed on a new EU renewable energy target of at least 27% by 2030.

Solid Planning, Clear Achievements

There may be many examples of inaction and green energy failures but progress on renewable energy infrastructure and the reduction of fossil fuel consumption initiatives is occurring in virtually every part of every country. Renewable technology is progressing rapidly, and, more importantly, societies are engaged in the learning process of how to adapt technology and their own demands to support sustainable living. The recognition that humans must address the issues of climate change and resource depletion is now entrenched in the world's consciousness. Trying and failing is not failure, ceasing to try is. As we fail and try again, our knowledge toolkit is expanding and is enabling us to deliver an increasing trickle of renewable energy and sustainable society successes (Fig. 5.11).

Global Good News

Battery electric vehicles are the low hanging fruit in energy conservation and the transition to renewables. If Elon Musk woke up one morning and found a battery-powered magic wand on his bedside table, it is safe to assume that the rest of humanity would wake up the following morning to find their gasoline-powered vehicles replaced by EVs.

Replacement of the fossil fuel light transport fleet of cars and light trucks, with vehicles using electricity as their motive power, is the fastest and cheapest way to reduce carbon emissions and fossil fuel use for any country. It is also the most effective step that can be taken with the least amount of disruption. Once people discovered these electric cars in their garages, there would be some unease at the prospect of getting used to something new. There would also be indignation on the part of a few at the thought of this change being foisted on them.

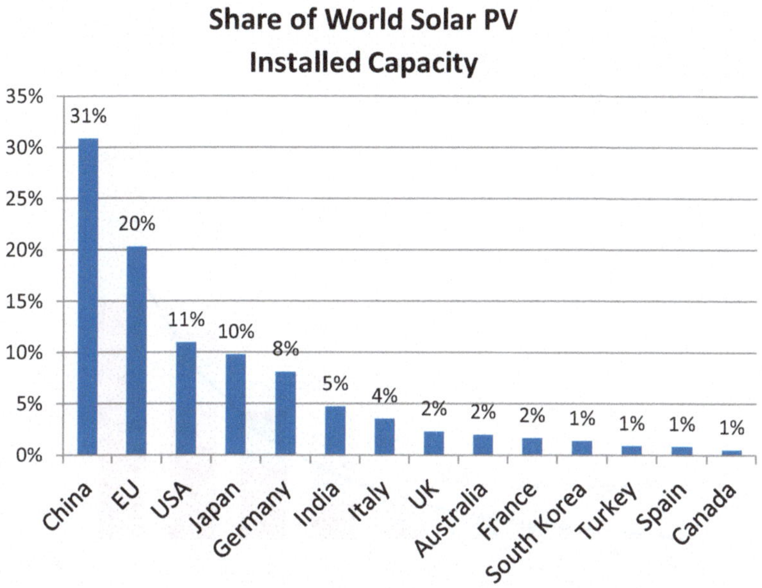

Fig. 5.11 Country share of world installed solar pv capacity (IRENA 2018; BP 2018)

But all this would become history once people experienced the EV driving characteristics of abundant torque, instantaneous throttle response and the ultra-luxury car levels of smoothness, quietness and refinement. The appreciation of the lower running costs of these vastly simpler vehicles and the convenience of home charging would come later. In several weeks, the thought of going back to internal combustion engine (ICE) cars would seem laughable to most.

Not everyone would be happy of course. The smell of gasoline and the bracing roar of an ICE car under full throttle are a substantial part of the visceral appeal of the driving experience for many. Electric car races have all of the drama of fossil fueled events but lack the physical intensity of roaring engines and belching exhaust that are part and parcel of piston vehicles being pushed to their limits.

And then there is the range and recharge time issue. Most people will never experience a limitation with vehicles now offering 400 km ranges and 30 min time to 80% charge, but a cross country jaunt will probably take about 15% longer than it would with the 500 km range and 5 min 100% "recharge" of ICE vehicles. In ICE cars, human bladder capacity is more of a limiting factor than range.

EVs solve a number of problems. They are vastly more efficient than ICE cars which send two-thirds of their energy up in heat. EVs can use nuclear, solar, wind, and hydroelectricity making their carbon footprint extremely small. Even natural gas plant-fed EVs have small wellhead-to-wheels carbon emission levels and, in the worst-case scenario of coal-fired plants, an advantage is still there, but only quite a narrow one. The life expectancy of EVs should be longer than their ICE forbearers.

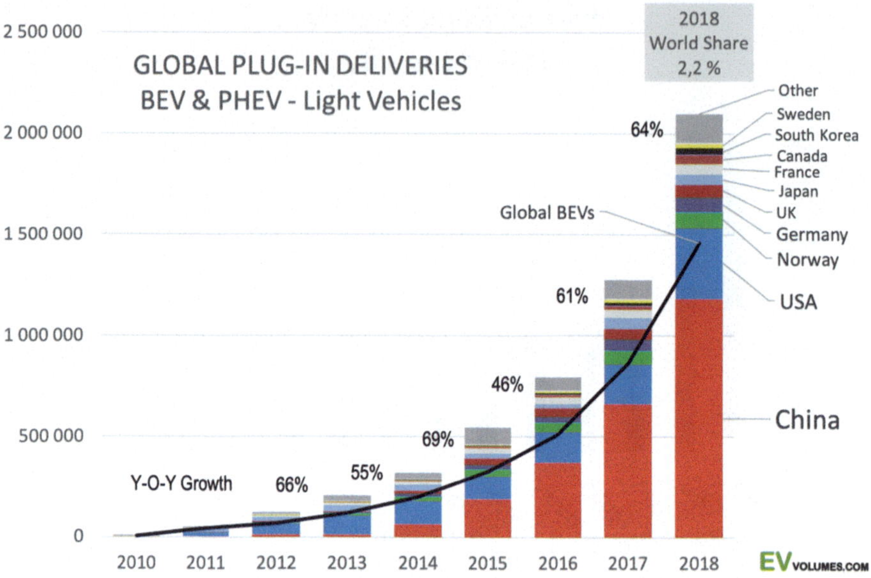

Fig. 5.12 Annual production of fully electric and hybrid electric vehicles (Irle 2019)

The first car Ferdinand Porsche designed in 1898 was an electric vehicle using lead acid batteries with a fairly impressive range of 70–80 km (Pruit 2018). This vehicle was parked in a Vienna warehouse in 1902 and sat for 108 years until purchased by the Porsche company for their museum. During the restoration, the batteries were, of course, dead but when power was applied to the electric motor, it turned over. This would not happen with an internal combustion engine. Their complex valve trains, water cooling systems, exhaust and oil circulation systems would have to be very delicately brought back to life or replaced entirely. Then they would have to be maintained several times a year including fluid changes.

These systems do not exist in an electric motor whose only points of friction are two bearings. EV owners will simply make far fewer trips to the garage (Fig. 5.12).

The graph above from EVvolumes.com shows the rapid growth of sales of battery electric vehicles (BEV), terms EVs in the rest of this book and plug-in hybrid vehicles (PHEV) which have a battery large enough to travel 20–80 km but would use their internal combustion engines for longer trips. But by comparison, there are still more than one billion internal combustion engine passenger vehicles on the road in 2019.

Consumer demand is rapidly increasing for EVs which, as we go to press, has motivated Toyota to announce it is moving up the target dates for its electrification plan. The company's previous target for producing 5.5 million EVs per year was 2030, but that's now been moved up by 5 years to 2025 (Auto123 2019).

Conservation

In Ontario, it is estimated that the cost of adding electrical capacity is 1.7 cents per kWh whereas the cost of adding additional nuclear capacity is 16.5 cents per kWh. Unsurprisingly, once past the heady days of the fossil fuel explosion of cheap energy, saving energy is cheaper than finding and making more. This math works for virtually every country and hence, simple conservation measures like time of use billing, smart meters, LED lights, improved insulation, and upgraded windows are being encouraged through a variety of incentives, worldwide.

Once past EVs and conservation, the effort required to reduce energy consumption increases, and that will be discussed in a later chapter.

Extreme Heat vs Extreme Cold: Different Countries, Different Energy Demands

The challenges presented by between extreme cold and extreme heat are quite different. As the climate warms, many hot regions of the world are becoming even hotter and the heat may be accompanied by an increase in humidity. This combination can produce conditions in which the ambient temperature and level of humidity make it impossible for the human body to shed heat through sweat.

A temperature of 34.4 °C with 80% humidity yields a temperature that feels like 53.9 °C which is considered extremely dangerous without some way to cool down. Unlike the progressive stress of increased cold, heat stress has a knife-like edge, beyond which danger to life increases rapidly. In a temperature of 35 °C and 85% humidity, the human body cannot cool itself enough to survive more than a few hours. One study in Bangladesh showed that extreme heat is more likely to drive migration than flooding is, so strong is its threat to life. The "wet bulb" heat discussion (the ability of the human body to cool itself) will only expand in the coming years (Wehner 2017).

In extreme cold climates, locking oneself outside without the proper clothing in −30 temperatures with a strong wind can prove fatal in under an hour. In the case of livable temperatures in the living space, the temperature must be increased by at least 45 °C to afford any measure of safety. In cases of extreme heat, it may only be necessary to reduce temperatures by 4 °C or 5 °C to assure long-term survival. But artificial cooling can be achieved only through the use of electricity, whereas heat can be provided by fossil fuels and biomass as well.

Currently, extreme heat can persist for periods of weeks in some regions. These periods will undoubtedly become longer and more frequent. However, extreme cold can persist for months in the far north and therein lies the difference in the amounts of energy required for survival in the far north vs the hottest regions on earth. It can't be a surprise that vastly more people live in regions of extreme heat than live in regions of extreme cold. However, over time, as global warming progresses, the heat extremes will likely worsen while the cold extremes will moderate.

Structural Social Problems

The record of almost all nations in reducing greenhouse gas emissions and tackling climate change is chequered at best. Clearly, if we were going to succeed or fail based on our efforts to date, we would fail very badly indeed. There are many reasons for this but they can be categorized as follows:

- Failure to recognize the difficulty and complexity of the necessary changes and thus setting optimistic goals implemented by simplistic policies.
- Resistance of powerful groups to changes which will reduce their wealth, power, and influence.
- Resistance of the poor and middle class on whom the majority of the burden will fall and who attach some of the blame for a large part of the societal failures they see occurring to the impacts of renewable energy policies.
- Failure of the governments to set long-term goals and implement permanent policies that become part of the culture rather than just short-term special initiatives.

Detailed Look at Complex Issues

The work by David MacKay, a preeminent British physicist, in detailing the use of energy and its sources in the UK is unmatched in its scope and depth. From the energy it takes to ride a bike to heat your home, Prof. MacKay has it quantified in his book "Sustainable Energy: without the hot air" which can be purchased or downloaded for free in any of 12 languages. His YouTube video on the realities of renewable energy breaks the energy issue down into some resolutely clear and manageable math.

Prof. MacKay looks at the UK's energy situation, many would use the term "predicament," in terms of types of energy uses, types of sources and offers the actions which could be taken and lead to the best outcome. Watching this video is highly recommended as Prof. MacKay breaks down the energy question into harvestable energy by land area, but anyone doing so should keep in mind the following points. MacKay's analysis is based on gross energy, not net energy. In other words, he does not allow for the energy input it takes to produce the energy output. This would eliminate biofuels at the latitude of the British Isles and would substantially reduce the available energy for end uses from the sources of solar panels and wind farms.

He mentions, but does not quantify, or include the embedded energy in UK imports which increase the energy consumption side of the budget by up to 35% according to some estimates. Population size got the same treatment. Although implied by the area shown on the video, he also fails to explicitly state that biofuels would take more than all of the crop area of the UK. In conversation with the author, Prof MacKay was keenly aware of these issues but didn't include them in his TED Talk in the interest of maintaining both an accessible level of complexity and an air of optimism.

David MacKay's work on renewable energy in the UK has yet to be matched by any individual in any other country in terms of its scope, detail, and frankness. Here is a brief summary of the issues in dealing with the transition to renewables as he presented them in his book which can be downloaded at no cost.

David MacKay "Sustainable Energy: Without the Hot Air"

> The red stack (demand) in figure 18.1 adds up to 195 kWh per day per person. The green stack (production) adds up to about 180 kWh/d/p. A close race! But please remember: in calculating our production stack we threw all economic, social, and environmental constraints to the wind. Also, some of our green contributors are probably incompatible with each other: our photovoltaic panels and hot-water panels would clash with each other on roofs; and our solar photovoltaic farms using 5% of the country might compete with the energy crops with which we covered 75% of the country. If we were to lose just one of our bigger green contributors—for example, if we decided that deep offshore wind is not an option, or that panelling 5% of the country with photovoltaics at a cost of £200,000 per person is not on—then the production stack would no longer match the consumption stack.
>
> Living on other countries' renewables?
>
> "Whether the Mediterranean becomes an area of cooperation or confrontation in the 21st century will be of strategic importance to our common security."
>
> *Joschka Fischer, German Foreign Minister, February 2004*
>
> We've found that it's hard to get off fossil fuels by living on our own renewables. Nuclear has its problems too. So what else can we do? Well, how about living on someone else's renewables? (Not that we have any entitlement to someone else's renewables, of course, but perhaps they might be interested in selling them to us.)
>
> Most of the resources for living sustainably are related to land area: if you want to use solar panels, you need land to put them on; if you want to grow crops, you need land again. Jared Diamond, in his book Collapse, observes that, while many factors contribute to the collapse of civilizations, a common feature of all collapses is that the human population density became too great.
>
> Places like Britain and Europe are in a pickle because they have large population densities, and all the available renewables are diffuse—they have small power density (table 25.1). When looking for help, we should look to countries that have three things: a) low population density; b) large area; and c) a renewable power supply with high power density.

A net renewable energy country will have to be one on which either a great deal of rain falls or a great deal of sun shines. And it will have to have its energy generation substantially exceed its domestic demand. This points to the likelihood of green energy export nations being much closer to the equator than to the north or south poles.

Potential "Green Donor" Countries vs Recipient Countries

MacKay's discussions call out a wealth of questions on our options but the need to transition to renewables was always made crystal clear in his publications and presentations. It should be noted the list of countries which are now in the position of living on 100% renewable energy with the ability to export their surplus of renewable energy numbers precisely zero. That list cannot be expected to grow beyond a

Table 5.13 Land area per capita

Region	Population	Area (km^2)	Density (persons per km^2)	Area per person (m^2)
Libya	5,760,000	1,750,000	3	305,000
Kazakhstan	15,100,000	2,710,000	6	178,000
Saudi Arabia	26,400,000	1,960,000	13	74,200
Algeria	32,500,000	2,380,000	14	73,200
Sudan	40,100,000	2,500,000	16	62,300
World	8,000,000,000	148,000,000	54	18,500
Scotland	5,050,000	78,700	64	15,500
European Union	496,000,000	4,330,000	115	8720
Wales	2,910,000	20,700	140	7110
United Kingdom	59,500,000	244,000	243	4110
England	49,600,000	130,000	380	2630

handful of countries in the future, hence the need for self-sufficiency and the crunch on small, densely populated regions. And it must be reiterated, on particularly, small, densely populated countries in northern regions (Table 5.13).

Ingredients of Successful Policy

Nothing is more important to the implementation of a successful transition to renewable energy than starting with a full set of books, which is to say, a very thoroughly described energy budget. Knowledge of what is necessary and what can be delivered is critical to a smooth shift away from fossil fuels and along the path toward a state of sustainable energy and resource balance.

In the end, to replace fossil fuels in the UK, MacKay outlines the need to implement heavy conservation measures, an increase in wind farms and nuclear plants as well as the importation of about 25% of energy requirements from solar farms "in someone else's desert."

Once all of the implications of the recommendations made above were put on the plate of any government, it might appear as a very unpalatable dish. However, any policy initiative is much more likely to succeed if it is spawned from a clear picture rather than from murky suppositions of what the forces at work might be.

National Incentives

Only a handful of nations are energy self-sufficient. The majority import a goodly percentage of their energy with most of that being some form of fossil fuel. For the nations which do depend on imported fuel, there is a clear incentive to transition to home-grown renewables. Increasing their renewables production has the benefits of reducing trade imbalances, increasing domestic employment, creating a healthier environment and reducing dependence on foreign countries whose political objectives might not be wholly compatible with their own.

A broader based economy and a higher percentage of energy produced at home will give any nation a higher degree of resilience and make them more immune to the shocks inherent in complex trading networks and changing weather systems.

In the process of transition to renewable energy, the technical issues and the social issues have to be seen in their own right. Nations may not be at the stage of being able to confidently implement solutions, but many have gotten to the point where they understand what the problems might be and solutions can only be achieved after the problems are clearly identified. And during the implementation of a transition to new energy systems, lots of problems will present themselves to be identified.

References

Agora (2018). https://www.agora-energiewende.de/en/service/use-of-agora-materials/

Auto123 (2019). https://www.auto123.com/en/news/toyota-panasonic-electrification-solid-state-batteries-china/65978/#vpaQWLLTPjUKh9oO.99

BP (2018) Statistical Review of World Energy. https://www.bp.com/en/global/corporate/energyeconomics/statistical-review-of-world-energy.html

CAT (2018). https://climateactiontracker.org/

CIA (2018) World information handbook

Glen B (2018) Orphan wells: Alberta's $47 billion problem, https://www.producer.com/2018/03/orphan-wells-albertas-47-billion-problem/

Global Carbon Project (2018). https://www.globalcarbonproject.org/

Han A (2014). http://sites.lafayette.edu/egrs352-sp14-pv/technology/history-of-pv-technology/

IEA (2018). https://www.iea.org/newsroom/news/2018/october/finland-shows-how-bioenergy-and-nuclear-can-drive-the-energy-transition.html

International Energy Agency (2018) Fuel economy in major car markets

IRENA (2018). irena.org

Irle R (2019) Global EV sales for 2018—final results

Luft S (2016). http://coldair.luftonline.net/2016/03/ontarios-curtailed-wind-energy.html

Ministry of Environment, New Zealand (2019) New Zealand's Greenhouse Gas Inventory. http://www.mfe.govt.nz/climate-change/climate-change-and-government/emissions-reduction-targets/about-our-emissions

Natural Resources Canada (2018). https://www.nrcan.gc.ca/forests/topics/state-canadas-forests-report/how-does-disturbance-shape-canadas-forests/indicator-carbon-emissions-and-removals/16552

Prieto P, Hall CAS (2012) Spain's photovoltaic revolution: the energy return on investment

Pruit S (2018). https://www.history.com/news/ferdinand-porsches-first-car-was-electric

Statistics Canada (2017) Statistics Canada Archives Series Q19-25

Wehner M (2017) The impact of moisture and temperature on human health in heat waves. https://oxfordre.com/naturalhazardscience/view/10.1093/acrefore/9780199389407.001.0001/acrefore-9780199389407-e-58

WRI (2017) CAIT World Resources Institute

Chapter 6
Renewable Energy in a Spectrum of Countries

Abstract Most countries have recognized the need to increase their production of renewable energy. The acknowledgement that renewables will one day replace fossil fuels may not have been made across the board but broad initiatives are underway to change consumption habits and reduce reliance on stored energy. Now that energy is becoming less easily available, it is critical to measure how much energy we use to produce the energy we consume. The level of energy a nation requires depends on its geographical and climate realities, not just on its level of technological achievement or wealth. How easily a nation is able to produce energy and how much energy is required to maintain itself are two key metrics of prosperity and social stability going forward. Every nation faces unique challenges and unique opportunities with social structures which range from very open to highly resistant to the transition to renewable energy.

The transition to renewable energy is the process by which we phase out our dependence on fossil fuels until all of our energy requirements are ultimately met by renewable sources. This complex process involves modifications to the way we live and do business and perhaps even where we live. Humanity just climbed on the fossil fuel energy horse little more than a century ago and we now find ourselves having to change horses again in midstream. This will require many steps and involve different energy mixes at all stages of the transition. Although biofuels are not a net source of energy in most countries, perhaps they will have to be pressed into service to keep a critical part of our infrastructure functional for years or even decades, despite a very high energetic cost.

The timing, infrastructure investments, skills development, regional circumstances, and adaptation of demand are complexities best illuminated by mathematical models which can lay out the range of options before policy makers (Day and Hall 2007). These models take years and decades to develop. WhatIf Technologies, an Ottawa-based group, has produced perhaps the most advanced model for this specific kind of analysis, certainly in Canada. Solid models are built on the level of the understanding of the processes involved and the quality and uniformity of the data available. Timeline data takes years to develop. Just as importantly, public

J. E. Meyer, *The Renewable Energy Transition*, Lecture Notes in Energy 71,
https://doi.org/10.1007/978-3-030-29115-0_6

access will have to be built into these models to prevent them from becoming just another glass tower blaring its superiority and demanding compliance.

No matter what time frame one might see as the one which ends in fossil fuels being reduced to small single digit percentage components of our energy diet, the endgame, which is clearly on the horizon, is for a de facto conversion of our carbon based societies to de-carbonized societies. The expected timeframe may be the extremely optimistic 10 years of the American Green New Deal, which calls for a complete decarbonizing of the economy by 2030. The motivation for this extremely ambitious plan is the threat of catastrophic climate change, while the means to implement it in a decade are less clear.

Both climate optimists and deniers are aware that fossil fuels are finite and an extreme view here might be that the severe scarcity of oil and natural gas will not occur for another 80–120 years. Others, such as Mohr et al. (2015), project a severe decline in the availability of each fossil fuel within one or two generations led by the collapse of Chinese coal production. But after that point, the forced increase in renewables in the energy mix is inevitable. So the choice to decarbonize is not a matter of "if," but of "when." Once this has been realized, "how" and "how best" to accomplish the transition over what time period are the nuts-and-bolts questions decision-makers will have to grapple with.

Certainly there will be no cookie-cutter solutions which can be applied across swaths of the world's surface. It is far too varied for that. Many smaller regions in many different countries may share similar energy demands and generation profiles but few national governments will be able to simply adopt another nation's strategies with a reasonable expectation it will work well for them. National governments will need to incorporate a number of regional strategies into one coherent national strategy in order to make reliable and sufficient energy available to all of its citizens.

This chapter examines the possible paths and options for various countries on their journey toward 100% renewable energy. Models which can reliably chart the course of the transition or predict the outcomes for any country are in short supply at this time. But they are desperately needed if the conversation on global warming and resource scarcity is to transform into positive action.

This chapter looks at some of the relative strengths and weaknesses of a cross section of countries with dramatically different geographic, climatic, and energetic realities. Gross national statistics may be useful for some elements of society but a nation stands or falls based on the well-being of its individual citizens. Having a large amount of energy available may be one thing but the critical consideration is over how many consumers this energy has to be stretched. Therefore, this comparison will be done in units of resources on a per capita basis.

The process of transition will be one of moving from mining extremely concentrated and dense materials containing high EROI energy in the form of coal, oil, and natural gas which are available all of the time, to harvesting relatively low density energy flows in the form of sunlight, wind, and water, which are available only some of the time.

The amount of energy consumption, and when it occurs, ultimately has to comply with what the energy supply system can deliver. This puts the issues of storage, time of use, and the total amount of available energy into perspective. In looking at the circumstances of modern nations, it might be helpful to briefly examine the environment in which previous great civilizations were spawned when fossil fuels laid untouched in the ground. The energy flows they made use of, and the energy demands they had to fill, will once again become prominent considerations as modern societies step back into the world of renewable energy stocks and flows.

Pre-fossil Fuel Empires

Rome

Romans are famous for the roads they built. The road network of the Roman Empire comprised almost 400,000 km, 80,000 km of which were carefully engineered, excavated, and paved roadways to accommodate the movement of raw material, finished goods, food, and the rapid deployment of troops. This network stretched from modern day Scotland to Iraq and south to Ethiopia. The Romans knew the value of transport and they were well placed to exploit the natural super-highway system at their doorstep—the Mediterranean Sea. This 4000 km long body of water was completely surrounded by rich farmlands, dense forests, and productive peoples, which were easily within a few days to 2 weeks sailing of Rome.

Rome could exploit or defend any point of its great empire very efficiently. But, as Thomas Homer-Dixon points out in energy analysis of Rome, "The Upside of Down," once its boundaries became too far flung, the energy required to maintain it exceeded the energy that could be sustainably extracted from it. Rome might be militarily able to take ground but after that, the ground it took had to be able to pay the cost of its own maintenance and defense. If no tribute flowed to Rome, what use was a territory? Roman expansion reached its limits as lands became more sparse and less rich.

Hadrian's Wall in Britain is an excellent example of this. Beyond the wall to the north lay what is now Scotland, a country that was thinly populated and with no significant natural wealth, either mineral or agricultural. The effort to establish Roman control over this region would have required more energy and material expenditure than the Empire could ever expect to recoup.

In its day, all roads led to Rome and so did all energy and material flows. Energy flowed into Rome and military power, the civil order of strong public administration and culture flowed out of it. Rome was in a temperate region with excellent farmland, rich fisheries, extensive forests and, above all, moderate energy demand. Togas weren't a fashion statement; they were an indicator of the hospitableness of the Roman climate and of gracious living.

From this favorable environmental base, Rome could stretch out and extend its power into well-endowed neighboring regions and thus expand its empire and build its size and power. The wealth that poured into Rome subsidized arts and learning in a wide variety of fields from medicine to engineering. Unfortunately, the Romans became highly dependent on cheap labor in the form of slaves and they therefore failed to apply their advanced learning potential to making either their agriculture or manufacturing processes more efficient.

As the productivity of the over-worked land declined, the ability to both maintain a large slave population, and defend a very extended border while producing a surplus to maintain the one million Romans declined as well, thus gradually destabilizing the labor-intensive and top heavy socioeconomic structure. The sprawling Roman Empire effectively expired in 476 AD, 500 years after it began. The more compact Eastern Empire collapsed 1000 years later.

The Roman Empire exploited natural and human created wealth in a wide geographic area made much more compact by the ease of transport. Coal was used extensively for heat in the northern reaches of the Empire and bronze and iron tools and weapons were well developed. But fundamentally, the Roman Empire operated on current solar input.

At its peak, the Roman Empire extended over 5 million square kilometers and had a population of 60 million people for an average density of 10 people per square kilometer. At its height, around 210 AD, the population of Rome itself reached 1 million people. Several hundred years later, that had fallen to just 30 thousand and it was a full 1700 years, in the 1930s, before Rome once again boasted 1 million inhabitants. The expanse of the Empire may have been gone but the dense energy from the newly available fossil fuels more than made up for it with the energy needed to once again sustain such a large, dense population.

Persia

The First Persian Empire from 550 to 330 BC ranged from Eastern Europe proper in the west to the Indus Valley, in present day Pakistan in the east, a span of some 4500 km encompassing almost 6 million square kilometers at its peak. It developed a central government, a single official language, roads, and a postal system. Populated by some 49 million people representing as much as 45% of the world's total 110 million figure and the population density averaged 9 people per square kilometer.

The progressive Persian Empire was the glittering prize Alexander the Great seized after his victory in the 331 BC Battle of Gaugamela in which his outnumbered Greeks defeated a vastly superior force under Darius the Second on the strength of brilliant leadership.

Mayan Civilization

No bronze, no iron, no wheels, no great sailing ships. Yet despite failing to develop the technologies which had proven key to other great civilizations, the Mayans developed sophisticated mathematics and astronomy. Their postal system serviced a network of some 40 cities with populations between 5000 and 50,000 spread over a compact portion of Central America with very fertile land. The Mayan civilization lasted for some 3000 years until a series of droughts over several hundreds of years eroded its social and economic structures. With an overall population of 12 million, the density of the major areas would have been perhaps 20+ people/km^2.

Egypt

Egyptian civilization rode the immense power of the self-renewing agricultural engine of the Nile River for thousands of years. Much of the 500,000 km^2 region Egypt controlled was desert and the two million Egyptians averaged out to a density of four inhabitants per square kilometer. But the great majority lived within a 30 min chariot ride of the Nile. The extremely fertile soils of the 1400-km-long Nile Valley and delta were flooded and renewed annually and in its 25,000 km^2 area, the population density might well have exceeded 100 per square kilometer.

China

Many dynasties rose and fell during the course of Chinese history, and there were many famines, wars and plagues, but China's population hovered around 50 million for thousands of years. These dense populations, concentrated in its regions of rich soils and powerful rivers, produced many technological breakthroughs from paper to gunpowder to mechanical clocks. Presaging many of the modern world's banking collapses, when the first printing press was invented, one of its first applications was to print paper money, replacing the physical wealth holders of cowrie shells and iron coinage in use at the time. When money first met the printing press, Chinese government officials felt the same urges as our modern bankers and triggered the world's first case of hyper-inflation.

Common Circumstances

Sophisticated cultures sprouted and decayed throughout the regions of the world blessed by mild climate, fertile soil, and adequate rainfall from Japan to Indochina to the Yucatan.

These empires and populous regions shared a number of characteristics. They were generally centered in earth's most favorable climatic zone between the equator and 20° north in the Americas and between 20° and 40° north in Europe and Asia. They all had extensive agricultural lands and the highly developed irrigation systems to feed them. Many empires and dynasties rose in these regions, pushed their

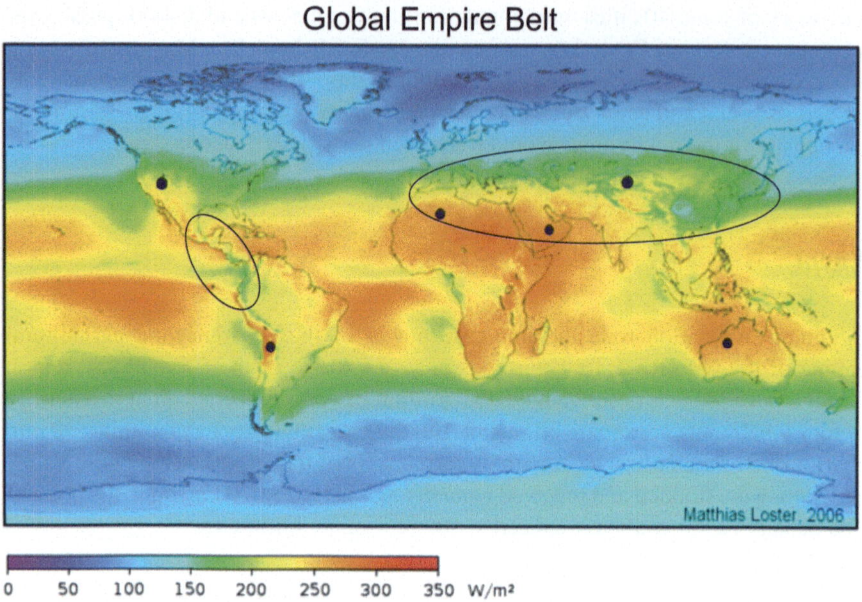

Fig. 6.1 Pre-industrial age regions spawning empires (Loster 2006). Pre-fossil fuels, the Great Empires of the world flourished in the most agriculturally richest regions, typically the belt of mid-latitudes north of the equator. Peoples living outside of these accommodating regions, effectively the Empire Belt, were not able to build up the population density, specialty of trades or free time necessary to develop advanced technologies and social structures

influence outwards, and then fell with some leaving a legacy of improved technology, laws, or social advances. Whatever their legacies, they all grew out of the same favorable conditions. The key requirements were low energy demand, rich soils, moderate and consistent rainfall, and access to fisheries and forest resources (Fig. 6.1).

Why Not Southern California or Australia?

Many areas of the globe might seem well suited to support a complex society, but lacked one or more critical components. California would seem to have everything, but its regions with its best soils receive only 10″ of rain annually, and most crop lands require 20″ to be viable. California is now one of North America's largest food producers, but only because an extensive system of dams and irrigation channels store and distribute the high seasonal flows of its northern rivers for use at the right time, in the right regions. Agriculture in Australia also faces water constraints, as does much of Africa.

Thomas Homer-Dixon made the point very well that Empires are really energy systems in his "The Upside of Down." Numerous other authors have illustrated that same point about most living things. Whether an empire, a nation, a fish, or amoeba, every dynamic organism requires energy to function. How the level of its energy intake compares to the amount it consumes determines the following:

- An even balance between energy-in and energy-out means survival.
- A surplus of energy intake over energy required means enhanced development and the possibility of expansion of empire.
- A deficit of energy intake over the amount needed to survive means decline and the eventual collapse of vital systems.

Empires grew in regions and eras that provided the energy and resources necessary to expand, and they declined when the marginal regions turned into energy sinks. This could be the result of an actual erosion of the regions' productive capacity, or of increasing demands for more "tribute." If this process of decline was not stabilized, the drain on the Empire's core would eventually end in disarray and collapse.

Modern Empires

The more modern Empires which spread out of Europe were fueled, to a large extent, by the resource bases of the nations they conquered, rather than from their own expansive and deep wealth of natural assets. The Europeans' advantage lay in their technology, which grew out of the coal-fired industrial revolution. This technology, expressed in arms, logistics, and mobility, was vastly more advanced than that of the indigenous armies they faced. Ancient Empires usually faced adversaries with roughly equal weapons and numbers. But in the modern era, advanced weapons took the place of large numbers and allowed smaller, more mobile, highly coordinated forces to defeat much more numerous and highly motivated opponents.

Although the force of arms is still used to project power, most empires now use technology, finance, and information to exert control over resources. Energy networks transmit power to every point on the globe, underwriting the fullest exploitation of resources which were simply inaccessible to societies before fossil fuels made their entrance.

Before fossil fuels, the world ran on a renewable energy budget. Today, that energy budget has been multiplied to the point where 80% of the energy we harvest and apply comes from stored carbon. Virtually every modern nation has been built on the assumption that energy supplies will continue to be abundant. These modern nations stretch well outside of the rich "Empire Belt" and occupy far less hospitable regions of the planet with far higher population densities than were possible before fossil fuel. How will these sophisticated societies adapt to the withdrawal of the very energy supplies which made their existence possible?

Renewable energy is the future. How kind will that future be to countries born and bred of mined energy? The options are very different for every nation with the options for different regions within many countries being almost as divergent.

A Nation's Necessities

In the ancient world, societies had to be able to provide adequate food for their population or social chaos would ensue. Little has changed as the book "Failing States, Collapsing Systems; Biophysical Triggers of Political Violence" clearly details (Ahmed 2017). In the modern world, a nation must be able to provide its citizens with both affordable food and affordable energy to remain stable. Doing so, while replacing 80% of its energy supply in the fossil fuel-renewable switch, will make for, as the ancient Chinese used to say, "interesting times" across the globe. Success will depend on how well nations understand their current energy budget realities, and how well they are able to implement the changes necessary to adapt to a huge shift in the very underpinning of their societies.

What energy consumption is crucial and what can be termed discretionary? How much savings can conservation programs be counted on to deliver? What increases in the supply and storage of energy will flow from technological improvements? Which energy harvesting infrastructures will be workable in which regions? How thoroughly are the citizens invested in the recognition of the problems and in the success of the transition?

An election promise of American President Herbert Hoover's campaign was that he would provide "a chicken in every pot and a car in every garage." for every American. A year later, in 1929, the collapse of the inflated financial markets made fulfillment of that promise impossible. However, after three decades of what would have to be considered "the oil boom," that promise was largely made real by 1960. And subsequent decades have seen ever higher levels of consumption to the point where most developed societies, including Canada, are now over-fed and over-mechanized, resulting in growing obesity and health problems.

Renewable Energy Realities: Canada

Society needs to deliver energy when it is needed, where it is needed. Fossil fuels excel in those areas, while wind, sun, and water power do not. In order to make renewables function in a complex society, their diffuse energy has to be gathered, stored, and then released as required. Sun and wind are low grade, less concentrated, energy sources compared to coal, oil, and natural gas. Large areas must be committed to use as solar and wind farms. Large river systems must be dammed and controlled for hydroelectricity to be made practical. One nuclear plant with four Candu reactors generates more power from a 1 km^2 footprint than is available from water of the 700,000 km^2 catchment area of the Niagara River as it thunders into the Niagara Gorge.

In Canada's case, huge areas for farms of both solar panels and wind turbines will be required to replace the fossil fuels in our energy diet. Shifting from dense to diffuse energy is a matter of area. How large an area will these farms need to cover?

In a 2016 study by WhatIf? Technologies, titled "Current and Future Canadian Energy Demand Expressed in Hectares of Renewables," the area require to replace fossil fuels in the Canadian economy with both photovoltaic and wind turbine farms was calculated. The CanESS model used data from Statistics Canada, the Office of Energy Efficiency, Natural Resources Canada, Environment Canada and other sources and was calibrated in 1 year increments from 1978 to 2013.

This small study, funded by Canadians for a Sustainable Society, was the first of its kind in Canada and indicates the scale of infrastructure needed to replace fossil fuels over a period of decades. It found the following for energy demand based on the current level of population:

- Solar farms would need to cover an area equivalent to a square block of land 142 km on each side or 20,000 km^2.
- This represents an area eight times the size of the cities of Calgary, Edmonton, Montreal, and Toronto combined.
- Wind farms would cover an area equivalent to a square block of land 438 km on each side or 192,000 km^2.
- This is almost five times the area of Canada's Class 1 farmland.

What is the magnitude of wind turbine infrastructure necessary to populate 192,000 km^2 of wind farm? It would involve the erection of well over 100,000 2MW wind turbines and the expansion of the electricity grid to connect them.

The above figures assumed that 22% of the nation's energy demand would be filled by hydroelectricity and wood scrap from logging operations and pulp and paper mills. Wind and solar would make up the other 78%. Further, the study also assumed the energy required to power the nation's passenger fleet was cut by 50% to allow for the much higher level of efficiency of electric vehicles compared to their internal combustion engine predecessors.

Biofuels: Green on the Outside but Black in the Inside

Liquid biofuels were addressed, but given their very low EROIs, very nearly 1:1 at Canadian latitudes, they simply don't represent a viable option. For instance, the area required to generate Canada's energy requirement previously provided by fossil fuels is on the order of 860,000 km^2, which is greater than the farm area in the country. Not only would biofuels displace all crops grown in Canada, but also they would not provide the energy needed by end users, since the figures used in these calculations are gross energy output, rather than net energy. A net energy calculation would deduct all of the energy inputs required to grow and harvest the crops from the gross energy and produce a much smaller figure for energy output. Murphy (2011) and others found that generating ethanol for fuel in the USA from maize required about as much fossil energy input as was produced (Murphy 2011).

In the Second World War, Germany developed several breakthrough technology weapons. The most spectacular of these were the V1 and V2 rockets. The V1 was the forerunner of today's cruise missiles and the V2 was the first missile to reach space and acted as the template for future intercontinental and space missile development. The bewildered Allied intelligence agencies knew of their existence, but had no idea of what these weapons were, even though they had received detailed reports on them from various underground organizations (Keegan 2004). It wasn't until the V1s started falling on London that their purpose and capability was finally understood.

Although a huge technological achievement, these weapons were incredibly inaccurate and could do little better than hit a target 10–15 km in diameter from 200 km away. Hence, they were of little use to the German war effort and served only as terror weapons aimed at very large cities and were far more likely to kill innocent civilians than do any actual damage to combatants or their infrastructure of airfields and weapons manufacturing.

These missiles ran on a fuel mixture including ethanol distilled from potatoes. Being chronically short of oil throughout the war, the Germans used crops to fuel their wonder weapons. It took the annual output of about 6 ha or 15 acres of potatoes to power the launch of a single V2 rocket. Germany launched about 15,000 V1 and V2 rockets and their demand for fuel seriously crimped domestic food supplies late in the war.

If we attempted to power the modern space program with biofuels rather than fossil fuels, the challenge would be daunting. As an extreme example, the heaviest lift rocket in the American space program was the Saturn V. If ethanol from potatoes was used to power it, the annual output of 8 km^2 of cropland would be required to lift it and its payload into space orbit. The low biofuel productivity of northern climates takes it out of the running as a mainstream energy source. But for specific uses and certainly in some equatorial regions, Brazil for instance, liquid biofuels from crops can be a viable mainstream part of a renewable energy strategy.

Renewable Intermittency

Power from sun, wind, and water occurs only when the sun is shining, the wind is blowing, or the water is flowing. In the case of the sun, except above the Arctic Circle for a few weeks of the year in the summer, the sun does not shine 24 h a day. In that same region, effectively, it does not shine for months at a time in the winter. Over the rest of the globe, power from the sun goes off and on through the 24 h day just like clockwork. The sun-up period of the day can also be subject to high humidity, clouds and during winter, snow can cover solar panels for days and weeks at a time.

The wind can blow or not blow with sufficient strength to turn the blades of a wind turbine at any time of the day or year. Water flows from rivers can fluctuate greatly seasonally through the year and from year to year. Damming rivers to create storage can provide much more consistent flows.

There are three means by which the inherent intermittency of renewable electricity sources can be addressed. They are:

- Storage by electrical battery, hydrogen battery, or pumping water into reservoirs.
- Fast response base generation like natural gas or hydroelectric plants.
- "One huge copper plate": a term to describe a very large electrical grid into which every energy source and every energy consumer is plugged. This completely integrated grid would extend over thousands of kilometers in a country or region and be built on the assumption that the greater the number of sources, the greater the consistency of supply. If the wind is not blowing in one region, it certainly is blowing in another if the area covered is large enough. One side of a continent might be getting sun while the other side remains in darkness.

Renewable energy is intermittent and consumer demand is variable. Below is a chart showing how the output from all sources is layered to conform to the end user demand through an average day with significant variability of sun and wind sources (Fig. 6.2).

Storage

With fossil fuels, producing them and storing them are almost one and the same because coal, natural gas, and oil merely require containment. They are self-storing. Coal can sit in a pile on the ground and oil requires only the most crude of containers. The USA uses huge caverns as storage for their strategic oil reserves. Natural

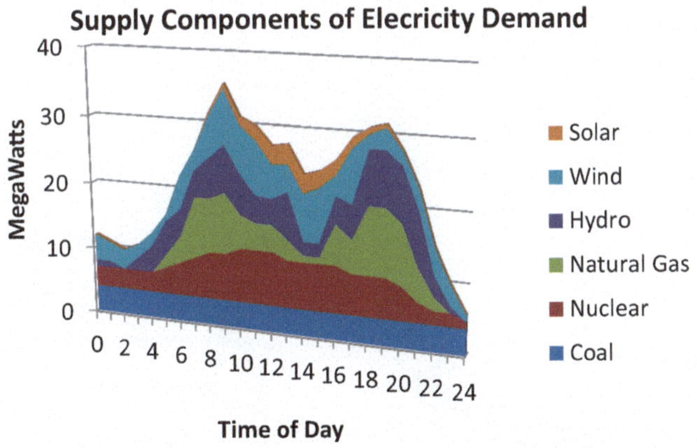

Fig. 6.2 Hypothetical makeup of energy supply during a 24-h period (author). Consumer demand peaks in the morning and in the evening and solar power peaks in the middle of the day. The wind is variable throughout the 24 h day. The coal plant output is an unchanging base load while the nuclear output changes slowly. Output from both natural gas and hydro plants reacts quickly to the consumer demand and renewable fluctuations.

gas does require basic tanks and compression, but sun and wind generated energy comes only in the form of electricity, and its storage can be considerably more complex and expensive.

The saying of "Lightning in a Jar" illustrates the difficulty of electricity storage as opposed to holding a lump of coal in one's hand or 20 L of gasoline (200 kWh of energy) in a jerrican. It can be achieved through electrical batteries and pumping water into large reservoirs but batteries are expensive and there are few locations in the world suitable for large-scale reservoir or "pumped" storage. Since the time and output level of renewable energy is rarely going to match the time and level of demand, storage is a necessity in an all-renewable energy systems.

If storage is a necessity, then the size of the solar and wind generating capacity has to be increased because it will have to have the ability to both produce the amount of energy to match the current demand and simultaneously produce even more energy to be stored when its output falls below the amount of energy required. This is called "overbuild," the factor indicating how much extra capacity has to be built to "both fill the tank and run the engine" at the same time. Overbuild is another factor which will vary enormously from region to region based on the level of demand and the vagaries of local renewable production.

Storage is a vital consideration in the inclusion of increasing amounts, or degree of "'penetration,' of renewable energy into the world's energy networks. The need for it can vary from a few hours" worth of low-level demand in a warm country with high levels of sun or wind throughout the year to cold, northern locations where energy demand is very high and the wind it intermittent and the sun doesn't shine strongly for months on end.

Thermal storage is another type of energy storage but it doesn't always involve heat and it isn't a new concept. "Cold batteries" in the form of ice blocks made the 1000 km journey from Norway to London in the 1850s (Fig. 6.3).

Fig. 6.3 Storing ice blocks cut on Lake Simcoe, Central Ontario (Barrie Archives 2019)

"In 1880, it was reported that the Northern Railway had a contract to deliver 300 tons of Kempenfelt Bay (central Ontario) ice to the city of Cincinnati, Ohio." Ice was shipped even further. "Carlo Gatti brought his first consignment of ice from Norway to London in 1857, of 400 tons, and one of the two ice wells at 12–13 New Wharf Road was almost certainly dug to receive it and store it until it was wanted by customers. Customers wanted ice for food preservation, for making ice cream, and for medical use. In the last 40 years of the century Norwegian ice dominated the market in London" (London Canal Museum 2019).

Mohammed and the Mountain

In the renewable energy world, human demand is Mohammed and it must acquiesce to the intractability of "the Mountain," which is the variability of energy production from solar and wind sources. Time of use must be made flexible if frequent, large-scale, interruptions of planned activities are to be avoided. There must be lifestyle changes and possibly some sacrifices on the part of the general population to harmonize their demands with the realities of energy supplies, which have their own timetable.

The Look of Renewable Energy Systems

But the conversion to 100% renewable energy is a long way off and it will have to be achieved in a process of many small steps over time. Perhaps it is useful to look at a median point where the economy is largely weaned off its oil dependence and carbon emissions are greatly reduced. Such a point would occur when the economy is largely electrified. Natural gas generators would ramp up electrical production while adding vital quick reaction time to the grid to smooth out the inevitable peaks and valleys of sun and wind. With that additional capability, the electrical system would then be capable of powering a 100% electrified passenger and freight ground transport system.

Assuming some sectors such as agriculture, mining, military, some emergency services, and aviation would be downsized but still remain largely fossil fuel powered. Oil would still be produced, but in far smaller quantities. Heavy conservation programs would have had a dramatic effect on demand levels as would significant lifestyle modifications. There would still be a long way to go to achieve 100% decarbonization, but the potential for energy supply shocks would have been substantially mitigated, and the driving force behind climate change would have been greatly reduced.

In the discussion below, the circumstances of different nations with respect to the energy transition will be examined at two points. The first will be where they sit circa 2019. The second will be the point at which they reduce their carbon emissions by upwards of 80%, in compliance with the Paris Targets. A final stage, achievement of nearly complete decarbonisation, is left for future generations to ponder. It can be safely assumed that technology will improve and that we will become more efficient at both using energy and in harvesting it. But this discussion assumes there will be no breakthroughs in the next 100 years which would re-establish an era of safe, affordable energy-abundance-for-all comparable to the oil energy boom of the middle of the last century. Humanity has already had its free lunch and we can't expect another in the near future.

Comparison of Current Circumstances

Following is a comparison of the energy circumstances of a number of disparate nations using data from the Energy Information Agency for the year 2015. These range from large-scale energy producers in northern regions to net energy importers in warm climates. There is a wide range of standards of living and necessary energy demand levels. On the supply side, the potential for sun, wind, and hydroelectric power is a different mix for each. They all face the same problems, but the degree of the problem is different for each as are the opportunities afforded by their renewable potentials.

Their current energy budgets are represented by these graphs and each country will be discussed just enough to outline the uniqueness of its situation. Several countries will be examined in more depth but the same questions apply to all. The metric of energy per capita per day in kilowatt-hours is used because any national energy strategy has to be built from the ground up, and per capita measurements illuminate real consumption patterns and policy options that aggregated national statistics simply cannot. The size of the national energy budget does not matter, but rather it is the energy available to each citizen that is the critical.

Modern economics sees only aggregated dollar flows for national commercial markets. In that light, all nations are comparable on the same basis. In the physical world of energy, nations face radically different demand and supply realities and their citizens will experience radically different outcomes (Figs. 6.4 and 6.5).

To contrast the extremes, compare the high consumption, northern nation of Canada to the low consumption country of India, positioned just above the equator. Canadians expend three times more energy heating their homes than do Indians for all uses combined.

South vs North: It's More Than Just the Heat

The further north one goes, the fewer local resources are available all year round and therefore supply is more sporadic and more transport is required to retrieve any particular good. The more sparse the population means the economies of scale and

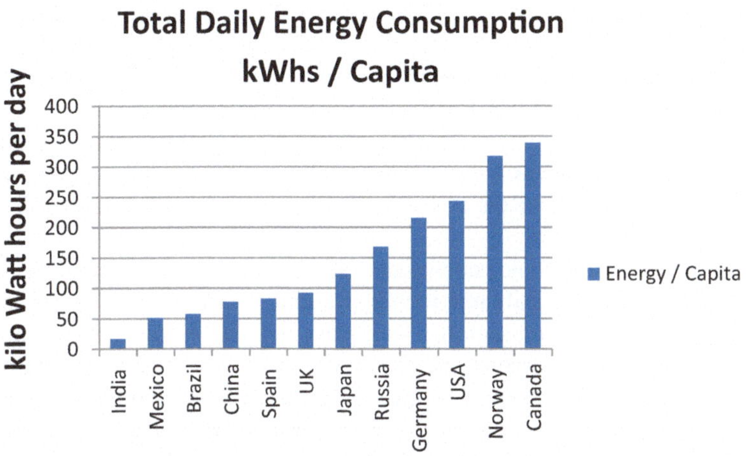

Fig. 6.4 Daily energy budget per capita by country (BP 2018). The southern nations consume far less energy per capita than do the northern nations, with the northern resource extraction based economies of Canada and Norway having the highest energy demand

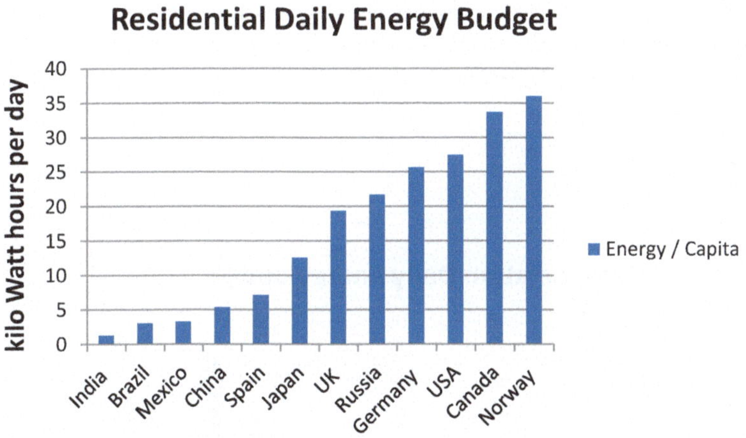

Fig. 6.5 Daily residential energy budget per capita by country (EIA 2015). Northern nations use many times the amount of energy to heat their homes than do residents of more moderate regions. In fact, many countries close to the equator have little need of energy for heat or even heating systems themselves

density at play in the southern regions is eroded and energy consumption rates climb for the progressively smaller and more northern centers.

In Guadalajara, Mexico, it is possible that daily needs, including food, services, and work, may be within walking distance. Given the mild climate, the practical walking and biking radius might be quite large. In contrast, Sudbury, Ontario must have all of its food transported over considerable distances to distribution centers most of which have to be accessed by automobile. Walking and biking long distances is far less practical or even next to impossible for 5 months of the year due to snow ice and cold.

Greater distances and more energy intensive activities can explain why a 24 pack of 500 mL bottled water can sell for $2.80 in Toronto but demands $29.95 in Igloolik, a small community north of the Arctic Circle. The clothing "toolkit" is larger in northern climate as is the variety of implements necessary to perform routine tasks in and outside of the house for four completely different seasons every year.

In geothermal rich Iceland, sidewalks in some towns are heated to eliminate the need for snow shoveling and large indoor swimming pools are maintained in communities to help people through the long, dark months of winter. In Croatia, the Adriatic Sea swimming season runs from April to October. Conditions are different for Icelanders where Reykjavik provides access to ocean swimming for those willing to take the plunge. On an inlet on the south side of town, Nautholsvik Geothermal Beach provides a heated pool open to the North Atlantic which allows a swimmer to gradually ease into the ocean. The "heated" pool is a bracing 7 °C and the ocean is a double-cappuccino replacing 2 °C. It takes a great deal of energy to offer this pool heating but, thanks to abundant geothermal energy, the Icelanders are uniquely positioned to do it.

In northern communities, energy has been expended to not only allow basic survival but to provide many of the activities and conveniences that are taken for granted and made naturally available at no energy or resource cost in southern climates (Fig. 6.6).

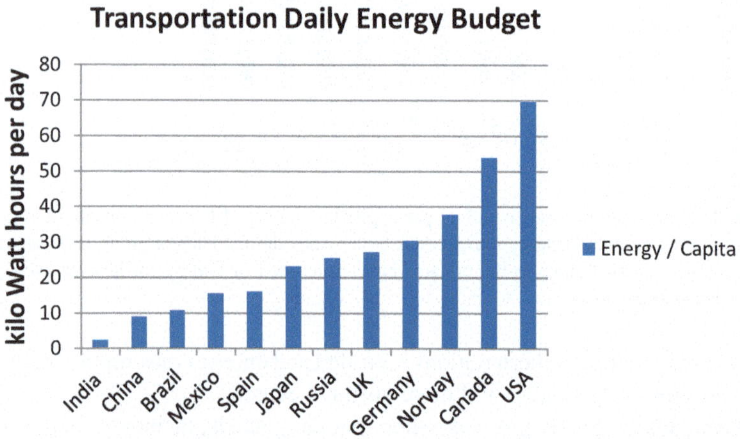

Fig. 6.6 Daily transportation energy budget per capita by country (EIA 2015). More northern countries expend more energy on transportation both out of necessity and for leisure pursuits. "Car culture" is more dominant in the USA than in any other country. Also, the US train transport network is less comprehensive than in most countries and rail is four times more efficient than truck.

Country Energy Profiles

Brazil

Brazil straddles the equator and a very small percentage of its citizens require heating systems to be built into their homes, places of work, or means of transportation. A wide variety of food is available from local sources and the Brazilian wardrobe and domestic toolkit has to accommodate one season a year. Sugarcane can be planted during any month of the year and is used in the country's extensive production of ethanol as a transport fuel (Fig. 6.7).

Canada

Canada is the second largest country in the world geographically, but its population centers do not extend very far north of the limits of its agricultural zone. Consequently, most of its large expanse has only a scattering of small communities. All areas of Canada require significant energy consumption to heat buildings and provide all-weather transport. It has a very large resource extraction sector, including some of the world's largest energy producing fields. It takes a great deal of energy to exploit and transport oil and gas from the generally low-grade plays (Fig. 6.8).

Canada's energy consumption almost doubled from 1990 to 2017 due to the exploitation of oil and gas in the western provinces, mainly Alberta, and due to a high rate of population growth. Being an energy powerhouse, Canada has not focused on renewable energy, which only generated 3% of its consumption as of 2017. From the earliest days of colonization, to the current sophisticated society,

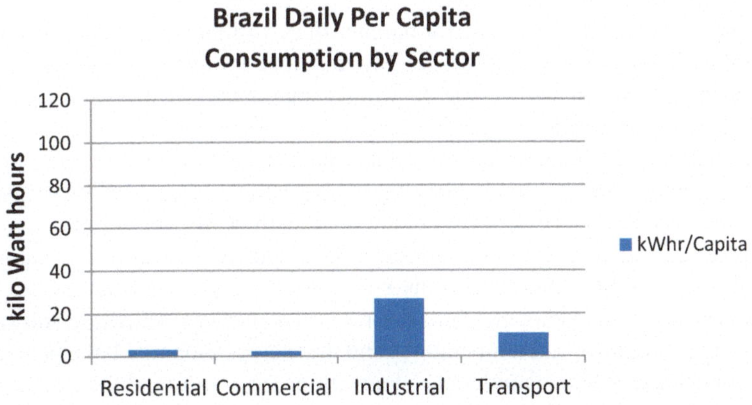

Fig. 6.7 Daily energy per capita budget by sector Brazil (EIA 2015)

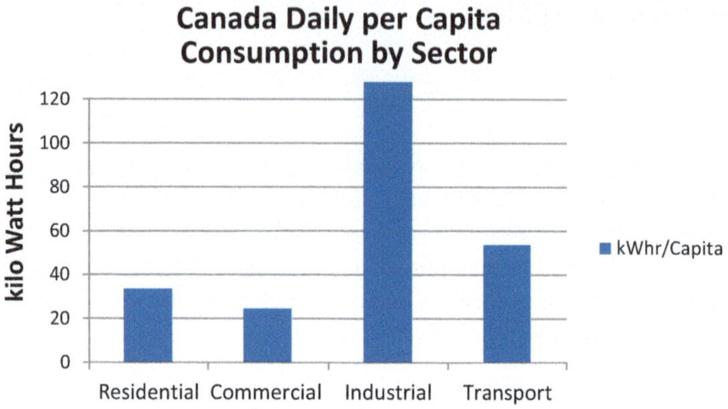

Fig. 6.8 Daily energy per capita budget by sector Canada (EIA 2015)

Canadians have been among the highest energy consumers, some of which is due to choice, but a good deal of which is due to necessity.

China

After the Second World War China was faced with the task of transforming a nation, now numbering 1.4 billion souls in 2019, from a fragmented agrarian economy into a cohesive modern society. Along the way, the government had to find jobs for ½ billion farm workers displaced by more efficient farming methods. The means by which these problems were solved was by turning China into the factory of the world. This created hundreds of millions of jobs and gave China the funds to acquire much needed foreign resources. This incredible metamorphosis was made possible by the exploitation of China's large coal reserves. Between 1990 and 2017, China's energy production tripled (Fig. 6.9).

China's coal reserves will last for another three to four decades and it imports large amounts from other countries, notably Australia. Its very small oil reserves are depleting rapidly, and it has become the world leader in the size of its solar and wind power installation growth.

The chart above illustrates how dominant industry is in China's energy profile, using over eight times as much energy in its industrial sector as in its residential sector, and over 20 times that of the commercial sector. Only South Korea remotely approaches this singular focus on industrial development. For all of its coal plants and industrial might, Chinese citizens still consume less than one-fourth of the energy Canadians do. Although China has regions with severe climate, a substantial portion enjoys semitropical temperatures and these areas support a large percentage of the population. China's unique development process will be discussed in more detail in following chapters.

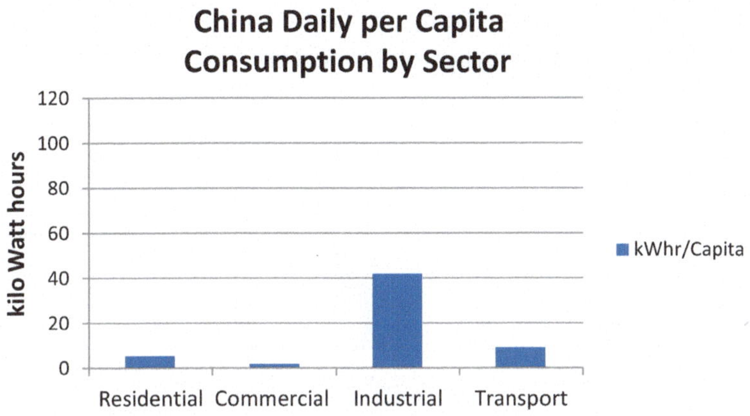

Fig. 6.9 Daily energy per capita budget by sector China (EIA 2015)

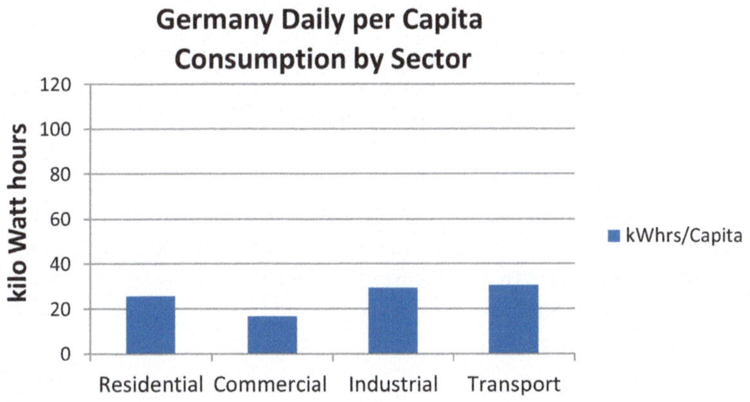

Fig. 6.10 Daily energy per capita budget by sector Germany (EIA 2015)

Germany

China may be the workshop of the world but Germany is Europe's manufacturing powerhouse. It still has large coal reserves (although mostly lignite) and is rapidly developing solar and wind resources. It must import all of its oil and most of its natural gas. Energy self-sufficiency is not in its immediate future despite being a leading nation in solar and wind development. Over 13% of Germany's power comes from solar and wind but its hydroelectric resources are relatively small (Fig. 6.10).

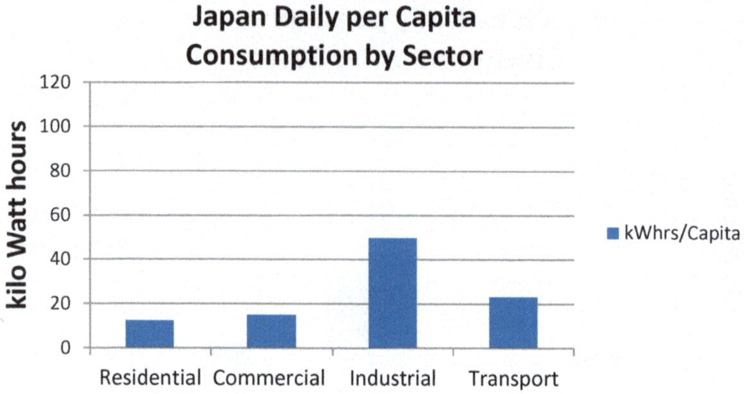

Fig. 6.11 Daily energy per capita budget by sector Japan (EIA 2015)

Japan

Despite its status as an energy have-not country, Japan has developed a very large and sophisticated industrial sector. It imports large amounts of coal and also has a network of 54 nuclear reactors, of which only 6 are back in full production after the 2011 Fukushima disaster. The goal is to have 30 in operation by 2030. Japan has climate zones extending from semitropical to Nordic and it has a very highly advanced industrial infrastructure yet manages to generate only 5% of its budget from renewables while importing almost twice as much energy as Canada exports (Fig. 6.11).

https://oilprice.com/Alternative-Energy/Nuclear-Power/Nuclear-Energy-Stages-A-Comeback-In-Japan.html

Mexico

The gas and oil reserves of Mexico are depleting rapidly and it has once again become a net energy importer. Much of Mexico enjoys very moderate temperatures as their very low residential energy consumption indicates. As in Brazil and Spain, living well in this climate does not require a great deal of energy. Despite having many high and arid regions, nearly ideal for solar installations, Mexico's renewables contribute only 2% of its overall energy mix (Fig. 6.12).

Norway

Norway is another northern energy powerhouse. The scale on every other chart in this section has been maintained at 120 kWh for ease of comparison. However, Norway has such a comparatively large and energy intensive resource sector, mainly

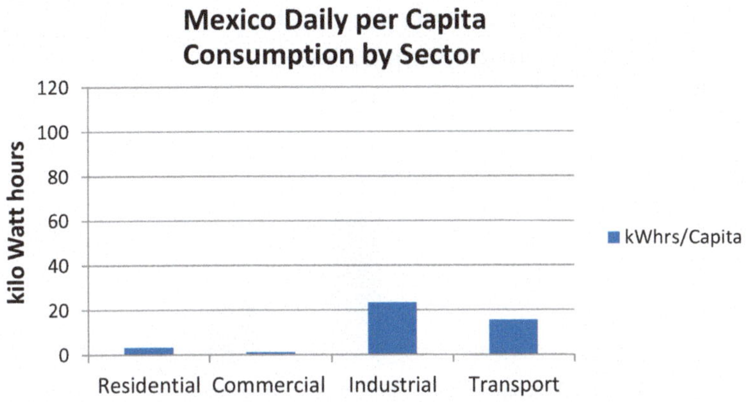

Fig. 6.12 Daily energy per capita budget by sector Mexico (EIA 2015)

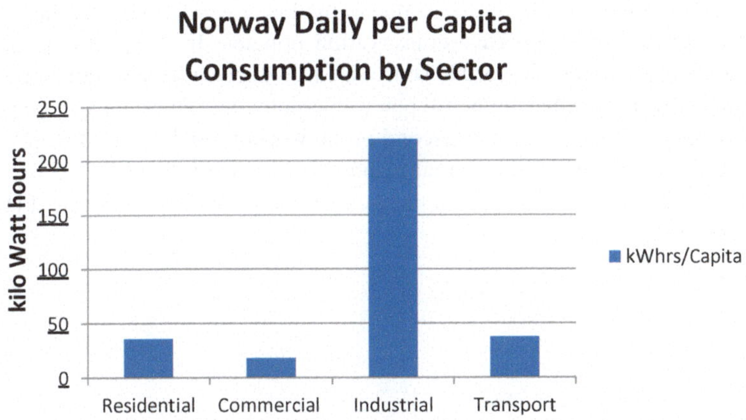

Fig. 6.13 Daily energy per capita budget by sector Norway (EIA 2015)

in the form of oil and gas extraction, that its industrial sector consumption required a doubling of the scale. Transport and residential consumption remain high in this progressive country and in the future, its huge hydroelectric capacity, at 67% of its energy budget, will be even more important as it transitions to a no-carbon future (Fig. 6.13).

Russia

Russia, the third great northern energy powerhouse remains one of the world's largest energy exporters and the country upon whose natural gas Europe in particular depends. Citizens of the world's largest country use slightly less energy than their

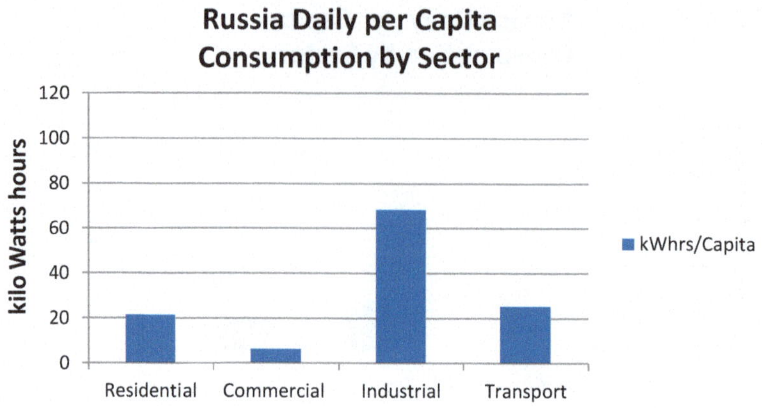

Fig. 6.14 Daily energy per capita budget by sector Russia (EIA 2015)

counterparts in other northern nations due to the lower level of development of the economy rather than progressive conservation policies. In 2017, Russia derived 0.04% of its energy from renewable sources. Clearly, given its size and geographic and climate diversity, it has immense renewable potential. The same can be said of its hydro capacity, where it ranks second in the world in undeveloped hydropower resources. However, these substantial resources are mostly located in Siberia, far from the large population centers in the western end of its 9000 km extent (Hydro Power 2019) (Fig. 6.14).

Spain

Spain features a sophisticated society with a high standard of living and a very small energy budget for a western nation. Credit the Mediterranean climate in which it finds itself. Spaniards use less than 25% of the energy Canadians do in their homes. Over 11% of their energy comes from solar and wind, the dry portions of the country providing excellent solar sites which have double the output of the ones in more northern and more humid Germany (Fig. 6.15).

The United Kingdom

The original coal fueled energy/industrial super-power, the UK has had to rely on imported energy for most of the last century. It has few good hydroelectric sites and solar, in this humid, northern country, holds less promise than it would in more southern, arid regions. However, wind potential, particularly off-shore, is fairly high. The UK is now densely populated and is a net importer in most categories

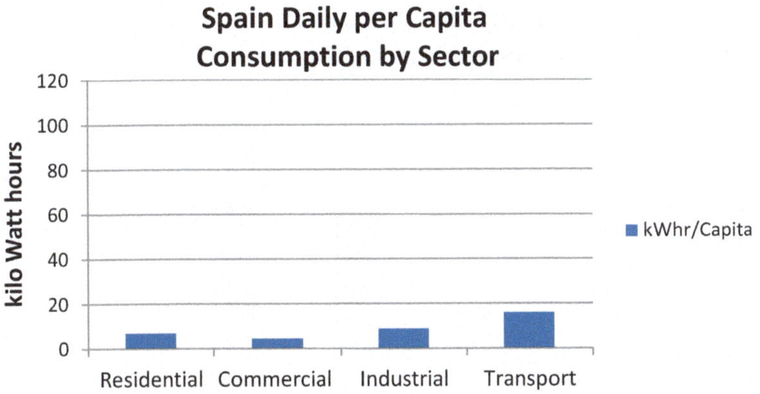

Fig. 6.15 Daily energy per capita budget by sector Spain (EIA 2015)

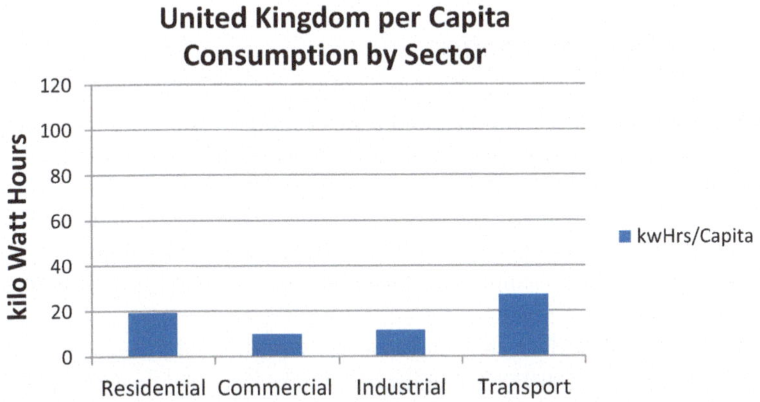

Fig. 6.16 Daily energy per capita budget by sector UK (EIA 2015)

from energy to raw materials to finished goods and food. Its industrial sector has greatly diminished in importance from 150 years ago, when it was the world's industrial powerhouse and the center of the world's largest empire (Fig. 6.16).

Where once it poured high technology steel implements and finished goods into the colonies and other nations around the world, now the UK is a net importer of embedded energy representing roughly 20% of its directly consumed energy budget. For reference, Switzerland's imported embedded energy is actually greater than the amount it directly consumes. Predictably, both China and Russia are both huge embedded energy exporters. (Kejun 2008) The UK gets almost 11% of its energy from renewables, but less than 1% from hydro.

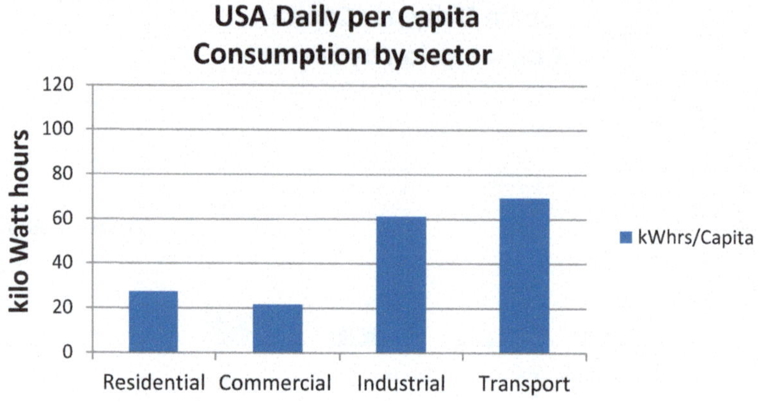

Fig. 6.17 Daily energy per capita budget by sector USA (EIA 2015)

The United States of America

The USA is both a super-consumer and a super-producer of energy. Fracking has produced a gusher of new energy by adding almost 6 million barrels per day since 2010 in oil alone, and has raised the prospect of possible oil independence. However, even if that hope were realized, the very high depletion rates of tight oil wells would mean the celebration will likely be short lived. Seven thousand additional new wells are required every year to simply offset the rapid decline in existing tight oil output (Fig. 6.17).

However, if the Americans reduced their world leading oil transportation budget through downsizing, electrification of the automotive fleet, and increased public and rail transportation, dependence on foreign energy could be eliminated for decades. Even climate and distance-challenged Canada has lower transportation expenditures than the USA. America has excellent sites for solar in the arid Southwest and wind installations in many coast, plain, and mountainous regions. It currently gets 4% of its energy from non-hydro renewables. Hydro capacity from its long and powerful, yet mostly slow moving rivers is, at 3% of total energy, fairly limited given the size of the population.

Ideal Circumstances

A nation with an ideal balance of resources and demand would find itself in a southern climate with very large hydroelectric potential, a very modest yet concentrated population, fertile soils, a sophisticated rail network, geothermal springs in abundance, and high and dry mountains on which to place large solar arrays and wind turbines. If Iceland or New Zealand were blessed with the Great Lakes and Niagara Falls, and moved to a location slightly south of Bermuda, they might fill the bill as the first nations with broad based, energy self-sufficient economies, substantial manufacturing capability, a strong social safety net, and very high educational attainment to progress to 100% renewable energy societies.

On the other side of the coin is a nation facing insurmountable hurdles in achieving decarbonisation. Imagine a country with very little hydroelectric capacity, no natural gas, very high population density, and rocky topography. If England were stripped of its rich farmland and towed slightly further north, it would have no chance whatsoever of being able to provide enough energy for any sort of sophisticated life for its population. Very possibly, even bare survival might be out of the question for a large majority.

As it is, every nation will have to work hard to transition to renewable energy. How hard depends on their individual mix of circumstances. One asset every nation should covet is that of hydroelectricity. Large-scale hydro plants have long term, very high energy returns for the energy that was invested to create them while using an extremely small amount of other resources for maintenance. Assuming the hydro resources go hand-in-hand with large-scale reservoirs, pumped storage enables the highest rate of solar and wind penetration in the system, and at the lowest cost (Fig. 6.18).

Along the path toward 100% renewable energy, natural gas can drive electric generating plants with far lower emissions than any other fossil fuel and provide quick response time for integration into a renewable energy grid. The most efficient societies will use electricity for all of their end use consumption. The technologies which enable both rapid adoption of renewables and rapid displacement of oil and coal consumption should be valued highly. Hydro may be the ideal, but natural gas fits the bill as a solid interim step.

Nuclear power can provide a solid base for a fossil fuel-free energy system but it, like fossil fuels, is fired by mined minerals which will ultimately deplete to the

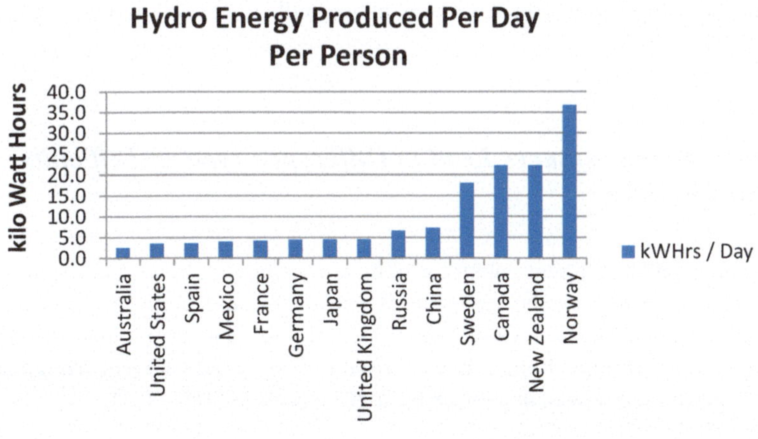

Fig. 6.18 Hydroelectric daily per capita budget by country (BP 2018). Hydro capacity is a key resource in the renewable transition

point of inaccessibility. Its future is clouded at best, but, if small, safe, quickly deployable nuclear reactors became practical, they could quickly replace natural gas as the transition stepping stone of choice. Nevertheless, both China and India are actively pursuing nuclear plants as a critical part of their energy infrastructures.

The Green Energy Transition Cards Nations Have Been Dealt

The table below puts many of the factors involved in the transition to renewables in one place. It also leaves out a great many. Agricultural potential, fisheries, recycling capability, and many others would all have a significant impact on the process and end result. Furthermore, although not random, the values in the table are arbitrary and many of the categories are simply value judgments. Some though, are based on hard numbers. Energy demand, population size, planned growth, and hydro capacity are all based on real data. Hydro capacity is largely fixed as most of the best sites in the world have been dammed.

Energy demand due to northerly location and hydro capacity, both have a huge influence on both the need for energy storage and the ability to store energy. The capacity to produce all of one's own energy is also quite easy to quantify, but some would hold the view that in a stable, globalized world, it is unimportant. Consumer aspirations and technological capability are purely value judgments and, in fact, population growth policies can be reversed very quickly. Public support for the greening of the energy system is not represented in the table, but is critical. Fill in your own values! More details on the variables can be found below the table.

Finally, even countries with the highest ranking in this comparative exercise may, in the real world, have a very difficult time succeeding and possibly, even the lowest ranked might succeed and find the whole process rather unremarkable. Whatever the positive and negative surprises in the process, early in the next century certainly, and very possibly within 40 years, every nation will count itself as a largely renewable energy society.

Different Circumstances Lead to Different Levels of Difficulty in Transitioning

In the above table, different countries, with very different circumstances, are compared to give an idea of the ease with which they could ultimately transition to renewable energy. The scales used are crude estimates and are not meant to represent linear relationships. Rather they are used to simply indicate whether a country is better positioned now than another for a given criteria (Table 6.1).

Table 6.1 Ease of transition

	Energy Demand	Pop growth Policy	Consumer Aspirations	Tech Base	Hydro Potential	Energy net Imp/Exp	Solar/wind Potential	Storage Required	Ease of Transition
Brazil	−2	−2	−3	5	3	8	4	−1	12
Spain	−3	2	−1	5	2	3	6	−2	12
New Zealand	−4	−6	−3	6	8	3	5	−3	6
Australia	−4	−8	−8	7	2	8	8	−1	4
Japan	−5	7	−2	8	2	−3	1	−4	4
Norway	−8	−5	−4	7	10	8	3	−8	3
China	−6	5	−1	9	2	−5	2	−4	2
Mexico	−2	−4	0	3	2	0	4	−1	2
France	−5	0	−2	8	2	−2	3	−4	0
Russia	−10	2	−3	6	5	6	4	−10	0
Germany	−6	0	−3	9	2	−3	3	−4	−2
India	−2	−4	−1	4	1	−1	1	−1	−3
Finland	−9	0	−2	8	3	−2	2	−6	−6
Sweden	−8	−6	−3	7	6	2	3	−8	−7
The USA	−5	−4	−8	8	3	−2	2	−5	−11
Canada	−10	−8	−8	7	7	8	2	−10	−12
The UK	−6	−4	−4	6	2	−8	1	−6	−19

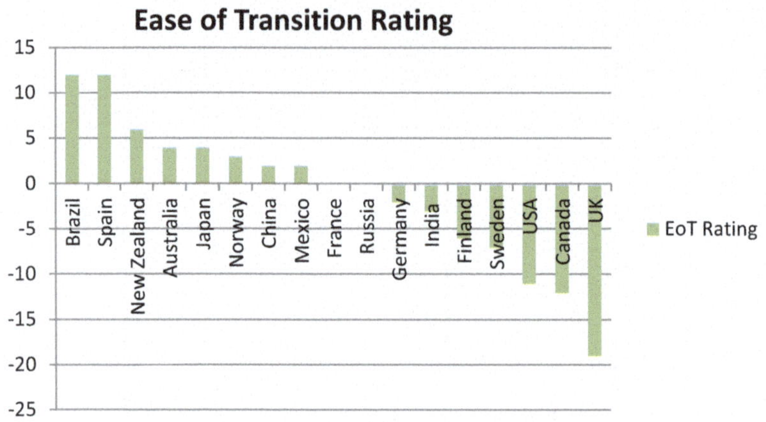

Fig. 6.19 Ease of transition to renewable energy by country (author)

All energy demand will be seen as a negative, but the demand (that is, the requirement needed to sustain life) in southern countries is much lower than the essential energy demand in Canada or Russia.

If a nation is fixed on high consumption of material goods, then this makes great demands on the renewable energy available.

Ease of Transition

If a nation has a strong base of advanced industrial capacity and a population with abundant scientific and technical education, rather than depending on foreign sources for technology and components and possibly even installation and maintenance, the ability to transition to renewables will be made a great deal easier (Fig. 6.19).

Hydroelectric power is a huge asset in making the transition to renewables of solar and wind practical. The higher the hydroelectric power capacity per person, the more seamless, with fewer sacrifices the transition will be.

If a nation is dependent on foreign sources of energy, it will be increasingly limited in what it can do and will be vulnerable to disruptions in energy supply. The potential of a nation's landscape to produce sufficiently high level of renewable energy per capita will have an immense impact on its ability to prosper as the opportunities to import green energy will likely be extremely limited.

The storage requirement is really a function of seasonal demand and the seasonal fluctuation in energy production. These are compounding factors. The higher the fluctuation in demand, it is likely the higher the fluctuation will be in energy production. Locations in the far north see these two factors multiplying the amount of storage required.

The Ease of Transition index suggests that the UK will have the most difficult time and Brazil the easiest. The UK is grossly overpopulated given its resources and particularly its renewable energy potential. It has very low hydro capacity and it is a net importer of all basic resources as well as finished goods.

Brazil might have the easiest time transitioning as it has very low energy demand necessary for survival, significant hydro capacity and strong solar potential along with very substantial ethanol production from sugarcane.

The transition to renewables will involve many steps involving all areas of the economy and society as a whole. Below is an examination of some of the steps which several countries might take on the way to a very low or no-carbon economy. In the charts below, all countries do the following:

- Replace oil and coal with natural gas
- Reduce consumption of finished goods
- Implement heavy conservation measures
- Modify lifestyles
- Continuously develop more advanced and efficient technology

Seven Large Caveats

There are several large unknowns which must be mentioned at this point. They have the potential to impact many countries significantly and in different ways. These caveats are not included in the graphs and tables above but may be mentioned in the discussion on some countries below.

- China and its manufacturing output.
  - A great deal of the material wealth enjoyed by people around the world originates in China and was made with coal-based energy. Chinese coal reserves currently have an estimated lifespan of 38 years. Once China chooses not to be the factory to the world, production of finished goods will, at least partially, flow back to the countries where they are consumed. This process will have a significant impact on both the energy budgets of the Chinese people and those of its former customers.

  China, manufacturing and coal are discussed in more detail in Chap. 8.
- Embedded Energy
  - When one good travels across a border, the energy consumed in its creation flows with it. This is called embedded energy and the degree to which nations are net importers or exporters can only be determined when not only the flows of raw energy are counted but when the flows of embedded energy are also included. China's exports of embedded energy are huge while both the UK and Switzerland for example, import a large quantity of finished goods and

raw energy. When this energy is compared to the energy content they export, the size of their "energy trade deficit" becomes apparent.

- Electricity Bonus.
  - Electrical processes are always more efficient than fossil fuel-fired processes. The most common example is internal combustion engine (ICE) vs the electric motor as found in automobiles. The typical ICE vehicle converts 35% of the energy in the fuel in its tank to motive power at the wheels. Electric motors in EV—electric cars—are closer to 90%. When the supply chain is included, the EV advantage is 3 or 4:1. (more on this in Chap. 7).
  - If a nation which used 30% of its energy, in fossil fuel form, to power its automobile and truck fleet, converted to EV mobility, its energy use might drop by over 20%. This drop in the energy budget would not represent a loss of utility for end users, as they would still have the same mobility. That is, the same amount of power would be available at the wheels.
  - From the US government website:

    "EVs have several advantages over conventional vehicles: Energy efficient. EVs convert about 59%–62% of the electrical energy from the grid to power at the wheels. Conventional gasoline vehicles only convert about 17%–21% of the energy stored in gasoline to power at the wheels." https://www.fueleconomy.gov/feg/evtech.shtml

- Population
  - The bottom line for economic well-being is the amount of disposable energy available to each individual. That determines the work an individual can do whether it be expressed in terms of travel or production of goods or food.
  - Obviously fossil fuels are finite but so too are rich sites for the production of renewable energy. EROI and scarcity will be discussed in depth in the next chapter but the greater the demand the more we are forced to exploit progressively poorer sites. A small nation with large hydroelectricity capacity is far more likely to produce positive outcomes for its citizens than one which has a large and even growing population depending on relatively limited hydro power.

- Natural Gas Bonus
  - By replacing coal and oil with natural gas, carbon dioxide emissions decrease dramatically for the same energy produced. Compared to natural gas, oil (gasoline) has 35% higher emissions and coal 80% (EIA 2015).
  - In the figures in this chapter, the goal is to outline some of the mechanics involved in transitioning from a society which gets 85% of its energy from fossil fuels to one that depends on fossil fuels for 15% of its energy. While the numbers accurately represent the fossil fuel energy reductions, the reduction in greenhouse gases may be even greater.

- Renewable Energy Infrastructure Overbuild

- The solar and wind generating energy structure has to be sized to not only deliver the power that is needed at the time but to be able to generate excess power for storage and use later. The degree of overbuild is related to the amount of storage available, the reaction times of the base loads, (hydro, gas, nuclear, and coal) and to the extensiveness of the grid. Somewhere the sun is shining and the wind is blowing but is that energy potential connected to a grid you can access?
- On the consumption side of the issue, how quickly can consumption be cut without disruption? At night, when demand might consist of charging EVs and possibly storing heat in thermal batteries, maybe there is no need to make up for a decline in output for a number of hours.

  Smart meters, appliances, and charging strategies can be made very flexible in their demand for energy, thereby lessening the need for base loads to ramp up to fill gaps.

- Energy Storage
  - Energy storage requires a great deal of infrastructure investment and it reduces EROIs dramatically. Northern countries have a much larger need for storage than do southern nations. Storage considerations are not included in this simple comparison.

In developed societies, food on the table, heat in the house, juice in the car, and technology and luxury items aplenty are now fused into our way of life, both in fact and in expectation. How the lifestyles many covet can be preserved while undertaking the largest transition in human history is discussed in the section below.

A Plan for Energy Conversion: The 85% Transition

In 2017, 85% of the energy for our group of countries came from fossil fuels, which equals the world average. At the point where we are able to use hydro and natural gas as fast reaction base loads without any assistance from nuclear power, it will be possible to completely invert the energy mix of the fossil fuel age and rely on renewable energy, including hydroelectric, for 85% of our energy supply. The 90+% greenhouse gas reduction which would accompany this shift will be sufficient to meet the Paris 1.5 °C target.

A few questions come up. How long will it take to implement these changes? A question for further down the road will be, how long can natural gas supplies last because it will be replacing both oil and coal. Despite the efficiency improvements reaped by electrification and conservation measures, demand for natural gas will increase by 45%.

Steps in the process toward 85% renewables.

- First series of steps.

- Oil and coal eliminated and replaced with natural gas which requires 53% less energy overall.
- Renewable output assumed to increase 400% over current levels.

  This creates a large bias in favor of the countries which have gotten off to a strong early start like the UK and Germany.

  It puts at a seeming disadvantage countries currently with very low renewables but very high potential like Mexico and India.

- Conservation assumed to cut overall end user energy demand by a further 30%.
- Greenhouse gas emissions drop by 72%.
- Natural gas consumption increases 45%.

• Final series (no nuclear).

  - 80% reduction in natural gas
  - further 50% increase from renewables,
  - 20% further increase in conservation.

Lifespans of Fossil Fuels

The lifespan of reserves of fossil fuels is represented by dividing the known reserves by the current production levels. The estimates of the lifespans of coal and conventional oil have been fairly consistent. Those for tight (or fracked) oil and tight natural gas have not. Despite high production in the short period of time we have exploited them, their lifespans do not seem to decline according to the BP Energy Outlook. Given the very high depletion rates for these fuels, the concern is that once absolute reserves do start to go down, they may do so very rapidly. Fracked resources are found in small amounts and deplete at 70% annually. Conventional, large reservoir reserves, once they reach their peak, can deplete as slowly as 3% annually. Others can deplete at an annual rate of 7%.

How long will it be before we develop a fuel to replace natural gas? 10 years? 50 years?? Can such a thing as "quick and clean coal" be perfected? The vast coal remaining reserves could hold out for 160 years at the current level of consumption and over 200 years if used at the same level as natural gas in the 15% baseload application. Its greenhouse gas emissions would be over twice as high and a means would have to be found to improve its reaction time. Possibly a new boiler process or a secondary storage medium like hydrogen could be developed. Hydrogen is a very lossy process but, like coal, if it allows the overall system to function, perhaps it could find a place (BP 2018).

There are a great number of unknowns in this process but unknown now does not mean unknownable in the future, once humanity decides it needs to develop the technical and social skills to successfully manage large-scale energy change.

The end point below does not represent a 100% renewable energy state but rather one in which greenhouse gas emissions are over 90% lower than in 2017 and over

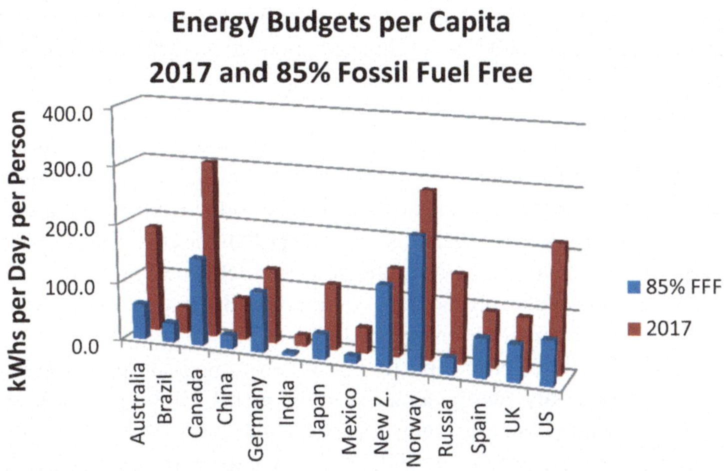

Fig. 6.20 Current per capita energy budgets by country vs 85% reduction in fossil fuel use (BP 2018). 85% Fossil Fuel Free involves a large reduction in energy production but not necessarily a large reduction in the final output of the economy or of the quality of life. Depending on the country!

85% of energy comes from renewable sources. Nuclear energy has been replaced by renewables at this stage.

What Does This Mean for Each Country?

In the chart above, perhaps Americans experience the biggest declines in their energy budgets followed by Canadians, Russians, and Australians. But this does not mean certain disaster or deprivation for any of their populations. The USA has extremely inefficient transportation and will realize huge cuts in energy consumption by electrifying and modernizing its passenger and freight infrastructure. In Canada's case, replacing almost all oil with natural gas will eliminate the huge amounts of energy currently consumed by the oil sector. This is also true for both Russia and Australia whose largest energy consuming sectors are fossil fuel extraction. As well, both have very large and untapped potential for renewable energy going forward (Fig. 6.20).

Canada in a Bit More Depth

In the analysis above, Canada starts in 2017 with a daily per capita budget of 301 kWh derived from 41% oil and 15% coal and arrives in the post oil, coal, and nuclear era with a budget of almost exactly half that at 150 kWh per day. Since Canada will have stopped large-scale exports of fossil fuels and the energy it took to extract and refine them, they will be in a net embedded energy deficit position if they do not increase the production of finished goods to replace those made in other countries.

In fact, this is a situation that faces all developed countries that have lost manufacturing to China and other cheap energy, cheap labor countries. If China is either unable or unwilling to continue to supply other countries with huge amounts of manufactured goods, those countries will then rebuild their moribund manufacturing sectors. Energetically, this is the most efficient path if the cost of energy increases and it begins to be seen as more of a strategic good that is best applied to domestic uses rather than sold off cheaply as a raw commodity. But manufacturing can easily eat up 15% or 20% of a nation's energy budget.

A high level of hydroelectric capacity is a very strong asset in offsetting one of the highest basic necessity energy demands in the world. High ratios of hydroelectric capacity, especially when accompanied by natural gas plants will reduce the need for storage and capacity overbuild. High capacity with quick reaction times can also reduce the end user compromises that are inherent in a system dependent, to a large extent, on renewables.

To illustrate the magnitude of the infrastructure needed to fulfill the renewable component of this budget, here is some math illustrating the number of wind turbines required to make up the difference or shortfall between the output of natural gas plants and hydroelectric facilities and the daily end user budget of 150 kWh in the era of an 85% reduction in the use of fossil fuels. Hydro output doesn't change and in the 2017 budget it constituted 22% or 66 kWh of the energy supply. If the 85% fossil fuel free budget is 150 kWh, the natural gas contribution is 15% or 22 kWh (Table 6.2).

So simple math provides the scale of the renewable energy infrastructure required which is 137,000 2 MW turbines with no overbuild and no storage losses which would push the number of wind turbines needed up significantly. Solar PV is not included in the above exercise.

Solar PV installations can work in Canada but their capacity factors (effectively the amount of energy produced annually) is between half and one-fifth of that of

Table 6.2 Estimate of number of wind turbines required in 85% FFF era

Daily per capita Budget			150 kWh
Less natural gas base capacity	Minus 22 kWh	=	128 kWh
Less hydro base capacity	Minus 66 kWh	=	62 kWh
Daily budget shortfall		=	62 kWh
Canada's total annual energy budget shortfall is daily budget shortfall × 365 days times 37 million Canadians	62 kWh × 365 × 37,000,000	=	8.4 trillion kWh
A 2 MW turbine has the potential to produce 6.1 million kWh[a] annually in Canada	8.4 trillion kWh/6.1 million kWh	=	137,000 wind turbines

[a]A 2 MW wind turbine has a capacity of 17,500 MW annually. In Canada, with a capacity factor of 35%, it would be expected to produce an average of 6.1 million kWh annually

installations in the US Southwest or many parts of Australia. And the timing of production and of demand are reversed for Canada compared to Australia. In Australia, peak demand comes during heat waves, when the sun is strongest, reducing their need for energy storage. In Canada, peak solar generating periods occur during the summer, and peak heating demand occurs during the winter, a time shift of several months, requiring a huge amount of storage.

Germany's Steps to Clean Energy

Germany has a large manufacturing sector but this was primarily enabled by the energy from its extensive coal reserves. A Germany running on renewables sees its energy budget fall from 127 kWh per day per person in 2017 to 104 kWh in the final stage of the transition. Germany has made a very strong start on renewables but the difficulty it faces is the lack of high levels of hydro and also the common problem of storage.

Like many other nations, Germany has looked to other nations to solve its resource supply problems in the past. For future energy storage, many European nations look to Norway's hydro and dam capacity. But even with its prodigious (on a per capita basis) resources, these can only address a small part of Europe's issues and, in fact, during the droughts of the spring and summer of 2018, electrical energy rates increased in Norway to dampen demand to match declining electricity production.

China's Steps to Clean Energy

Population growth dramatically decreases the ability of a nation to make a successful transition, whereas a declining population improves their ability via higher ratios of largely fixed renewable resources per person. Canada is attempting to triple its population over the next 100 years, which will quickly negate the advantage of its large hydroelectric resources. On the other hand, China, as well as Japan, can expect to see reductions in their populations of close to a third by 2100, giving them an improving relationship with their renewable base. China had an energy budget of 71 kWh per capita in 2017 and this would to 24 kWh in the 85% formula. However, the large hydro resources, falling population and its ability to equip itself with all of the renewable energy generation infrastructure it deems necessary, would indicate it will be able to provide sufficient energy for the government's goal of a semi-sophisticated lifestyle for its people.

China is electrifying rapidly and well into building a network of high-speed trains and public transport. It is the world leader in both production and installation of solar panels, and has the largest fleet of EVs. It should be noted that Norway currently has the highest number of EVs per capita.

The degree to which China will divest itself of its energy intensive manufacturing export sector is an open question but it is a decision that will be made by the Chinese government in what they see as the national interest, and not simply something that occurs due to external market forces. Other countries must recognize that much of their renewable energy infrastructure is currently being made using cheap Chinese coal, just as most of their existing social infrastructure was made by cheap fossil fuels. They must also recognize that they may be left to build their green infrastructure with their own energy supplies, and paying full domestic labor rates accompanied by higher environmental standards rather than those existing in the third world. Dollar costs will vary accordingly.

America's Steps to Clean Energy

Despite the political ups and downs in American green energy programs, the country is moving forward in many areas. One notable slippage though has come in the form of the energy consumed per passenger vehicle mile. Already the world's highest, over the past several years, consumption has actually gone up while it has declined in virtually every other country. Americans currently have an energy budget of 216 kWh per day which falls to 76 kWh post carbon, post nuclear.

American hydro assets are not impressive, but the areas of southern California and the Southwest offer excellent solar potential, with almost double the capacity factor of installations in higher humidity areas further north and along the east coast. There are large temperate areas in the USA which make little in the way of energy demands, yet are capable of providing substantial amounts of year-round renewable energy.

Conservation and support for renewables is inconsistent but California for one, is one of the leading jurisdictions in the world.

Britain's Steps to Clean Energy

Without an Empire and compliant states to feed it resources and energy, the UK faces very difficult choices. From the reference 2017, its 71 kWh per person slides slightly to 63 but, from this, it likely would have to now supply a great deal more of its own food and manufactured goods, greatly reducing the level of "disposable energy." The UK has a very large population on a very small resource base and it will have to be very careful and determined in the choices it makes. The moderate northern climate does make significant energy demands but is not as harsh as that in other more northern nations.

Britain has implemented strong energy conservation measures in the past decade and its consumption has gone down from 8% since 2005 despite a population increase of 6% over the same period. The country also has gotten off to a strong start on renewables, which currently supply 11% of its total consumption.

David MacKay, in his YouTube TED Talk (MacKay 2012), implied that the UK needed to make arrangements to be able to use someone else's energy resources, perhaps those of a dry and sun-drenched African nation willing to supply upwards of 25% of Britain's energy requirements. It would seem that, despite large investments in conservation and wind power, it might be prudent for a substantial percentage of the country's population to move to more temperate regions.

In the post-fossil fuel era, this is a consideration many northern nations with modest natural energy supplies might be forced to entertain.

Norway Steps to Clean Energy

To counter the very large energy demand of its northern location, Norwegians have more hydroelectric power per person than anyone in the world, and stand to transition to clean energy more easily than any northern country. Their biggest energy sink is that of oil exploration and development, which disappears in the near elimination of oil consumption. Currently Norwegians are using 284 kWh a day but, post oil, they still have a world beating 223 kWh to deal with the heavy requirements of their very northern and transport intensive country.

Too far north for solar panels to be broadly useful, Norway can get by on its hydro, and when they are ready to install it, wind power. As with every nation, consumers will have to adapt to the variability of renewable supplies, with the very high level of hydro penetration, variability might actually be very small. Without the export of large amounts of oil, to subsidize the importation of high levels of consumer products, Norway will almost certainly have to increase its domestic manufacturing which will eat into its disposable energy budget but broaden its economic base, and increase their level of resiliency.

Spain's Steps to Clean Energy

Spain's daily per capita energy budget drops from 94 to 68 kWh but, given their moderate climate and potential for very large solar energy production, Spanish society has considerable flexibility in determining what level of consumption they feel is necessary and desirable.

Brazil's Step to Clean Energy

Energetically, Brazil seems to have it all, with significant hydro capacity, very high solar potential, in addition to one of the few large scale viable sources of biofuel, which comes in the form of sugarcane. Its current budget of 45 kWh declines by formulas used here to 33 but Brazil has the capacity to exceed that level with sufficient investment in solar infrastructure.

Mexico's Steps to Clean Energy

Mexico has relatively low-energy demand given its moderate climate and very strong renewables potential, but despite a moderating rate of growth, its population is still increasing by 1.3% annually, an even faster rate than that of Canada. Oil production has declined 50% since 2005 and natural gas imports are now larger than its own slowly declining production. Mexico spent its oil bonanza quickly and, at the same time, failed to invest in renewables despite its immense potential. It has many arid and desert regions and is approximately 20° further south than Spain, a proven solar powerhouse.

The average Mexican energy budget is currently 43 kWh per day, and by the formula used here will fall to 12 kWh, which is a huge and socially destructive drop. However, given its high solar potential, only national will stands in the way of increasing this substantially.

Real World Shock Experience

Changing from one primary fuel to another and modifying the way life is lived and work is done in a country is a very complex process. Having experience to draw upon is very useful in gaining perspective and there have been two very different "energy shock" events on a national level which might provide insight.

The first is the crumbling of the Soviet Union, an event that caused Cuba's oil subsidy to come to an abrupt end. Cuban agriculture had to convert from mechanized and chemically fertilized methods to organic almost overnight. The population was forced to walk and bike more, and there was a cut in food calories from 3000 to 2000 per day accompanied by a 40% drop in protein. Cubans lost an average of 12 lb and it took a full 5 years to bring agricultural production back fully with organic methods. But deaths from cardiovascular disease fell by a third and adult-onset type 2 diabetes by a half, respectively while overall mortality rates went down.

Cubans ate more fruit and vegetables as these are what they could grow themselves with. An energy shock such as the one Cuba experienced was manageable and even provided health benefits. But Cuba is a semitropical country. Would a sudden loss of oil be quite so manageable in a northern country, where crops cannot be grown all year long and walking and bike riding long distances is problematic for at least 4 months a year (Schiffman 2013)?

Japan imports 90% of its energy requirements and in 2011, nuclear energy supplied 30% of their electric capacity. After the Fukushima disaster, a complete shutdown sent nuclear output to zero and it recovered to only 4% by 2017. By 2019, only 9 of the 37 reactors had been reactivated. The Japanese electrical system is self-contained and is not connected to any other grid. How did Japan cope with such a large portion of its energy being removed?

Coal plant output was increased by 100 TW and the renewables output increased by 45 TW but overall electricity output still dropped by 15%. This shortfall was dealt with in several ways. Demand was reduced through efficiency improvements, lower service levels, (lower voltage in the system is maintained). Also, end users accepted higher indoor temperatures during the air conditioning season and lower indoor temperatures during the heating season. In the end, increased fossil-fuelled generation made up more than one-third of the nuclear power lost while renewables and reduced demand compensated for the rest (Zissler 2017).

Cuba and Japan experienced sudden structural changes to their power systems. Although the effect on Cubans was undeniably larger, both they and the Japanese had to put strong compensating policies into effect very quickly. The public in both countries understood the nature of the crises and cooperated with the new measures the governments implemented.

From Analysis to Planning

Clear visualization of the need to change is central to any successful implementation of an energy transition. Also, fairness in spreading the load equally, is key to the public acceptance of any perceived inconvenience or hardships involved in the process.

Whatever emergency measures planning Cuba and Japan had in place before their respective crises, they were probably not adequate to deal with events as they unfolded. Given the nature of the crises, citizens were willing to accept hastily put in place measures. But over the long haul of the transition from fossil fuels to renewable energy, planning has to be complete enough to avoid resentment and fatigue stemming from poorly thought out, half measure changes that require constant readjustments.

National energy accounts must be made complete enough to provide an accurate vision of the structure of each nation's relationship with energy. There has to be international standardization of energy metrics to allow accurate comparisons among nations and the varying results of their different programs. This needs a full spectrum of comparable data which does not exist at this point. However complete and internationally consistent the information is, each nation will have to decide on its best course based on their own interpretation of their own data and their own priorities.

An outline of basic issues may help in developing perspective and laying a foundation for discussion but for actual operational planning, nothing replaces a working model that actually represents the way current systems work now. In the simple analysis above, we concluded that Canada would have to install 137,000 wind turbines to provide all of the energy hydro and natural gas did not. No other form of energy was included; neither solar farms nor tidal nor nuclear.

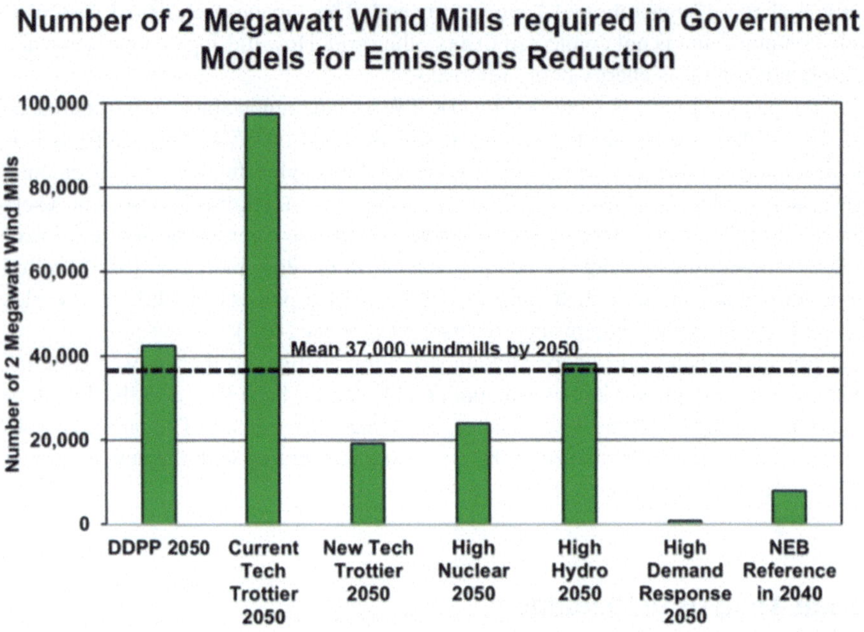

Fig. 6.21 Range of wind infrastructure forecast by different models (Hughes 2016) from David Hughes presentation "Canada's Energy Future: The Path to Transition"

The models cited below are vastly more sophisticated and may have included overbuild, storage and many other sources of energy but there are substantial differences in their conclusions. How would policy makers deal with variances of that magnitude? Each country needs to have all of its planners reading off the same sheet and at least having their assumptions laid out for all to examine. Each country needs a fully fleshed out energy model with access open to all experts and interested parties (Fig. 6.21).

From Planning to Implementing

It's reassuring to see figures on paper indicating possible paths to the successful transition to green energy and a sustainable and climatically stable world. However, once pen leaves paper and we start to cut steel, we need to understand exactly what the scale of the undertaking is. It took over 150 years to create the fossil fuel energy networks which propelled us to the level of consumption we now enjoy.

There are thousands of coal and gas generating plants, nuclear stations, and refineries spread over every country. There are hundreds of thousands of kilometers of various pipelines and perhaps millions of oil derricks dotting many landscapes. Almost all of the heating systems in our homes, factories, and offices use fossil fuel as do the automotive and freight transport fleets. Presently, these all work. Further, almost all of this infrastructure was built using fossil fuels.

We need to replace a sizable part of this network with new infrastructure designed to generate, store, and utilize electrical energy. We need to do this before the fossil fuel tank runs dry or even before the engine starts sputtering and we need to do it before the damage we've done to the planet reaches a level that proves to be fatal for high level human societies. How much infrastructure we need can only be determined once we understand the basic principles of energy measurement and how we humans, and our energy budget, interact with the biophysical world around us. Asking our questions using new terms like net energy, EROI and understanding the benefits and costs of converting most of our energy systems to electric will allow us to ask the right questions.

References

Ahmed N (2017) Failing states, collapsing systems; biophysical triggers of political violence. Springer, Cham

Barrie Archives (2019). https://www.barriearchive.ca/piece/ice-harvesting-on-kempenfelt-bay/

BP (2018) Statistical Review of World Energy. https://www.bp.com/en/global/corporate/energy-economics/statistical-review-of-world-energy.html

Day J, Hall CAS (2007) Revisiting the limits to growth

EIA (2015) Energy consumption by sector by country. https://www.eia.gov/tools/faqs/faq.php?id=73&t=11

Hughes D (2016) Power Point Presentation data from Canada's mid-century long term low greenhouse gas development strategy

Hydro Power (2019). https://www.hydropower.org/country-profiles/russia

Keegan J (2004) Intelligence in war

Kejun C (2008) Embodied carbon in traded goods. https://www.iisd.org/pdf/2008/cph_trade_climate_carbon.pdf

London Canal Museum (2019). https://www.canalmuseum.org.uk/ice/iceimport.htm

Loster M (2006). http://www.ez2c.de/ml/solar_land_area/

MacKay D (2012). https://www.ted.com/talks/david_mackay_a_reality_check_on_renewables

Mohr SH et al. (2015) Projection of world fossil fuels by country. https://www.researchgate.net/publication/267870440_Projection_of_world_fossil_fuels_by_country

Murphy D (2011) New perspectives on the energy return on (energy) investment (EROI) of corn ethanol

Schiffman R (2013). https://www.theatlantic.com/health/archive/2013/04/how-cubans-health-improved-when-their-economy-collapsed/275080/

Zissler R (2017) How Japan has managed to substitute nuclear power and the trade balance consequences. https://www.renewable-ei.org/en/column/column_20170308.php

Chapter 7
Choosing the Right Metric for the Job

Abstract Failure to be able to measure progress toward a clear goal is a sure sign of inevitable failure. When goals are so ill-defined that a crude ancillary measure, over time, becomes the goal, then progressive management becomes impossible. For the most part, through history, elites rose to power on the waves of economic dominance of particular sectors. When those sectors began to decline and be supplanted by other sectors, the elites tied to those sectors typically held on to power as long as possible, employing all manner of devices to retain their grip on privilege. The measures that best represented their interests were the ones they championed. But as economies and nations transform, metrics which once represented national benefit or at least coincided with the rise of national welfare may become obsolete or even misleading, hence the need to be clear on the objectives, rather than be tied to one simple metric to the exclusion of all critical thought. In the past, nations have only measured gross energy or the total energy produced. This was not a problem in an era of energy abundance, but policy based on gross energy, rather than net energy and EROI, is like judging the health of a company on the basis of gross sales rather than net profit.

In Communist Poland, furniture factories were judged by the number of kilograms of furniture they produced each month. Since physical units rather than money were used by central planners, and since measurement in tables, chairs, and sofas would simply not fit on any ledger, weight was chosen as the yardstick of output. Consequently, furniture production targets were set in tons. The result: Communist era Poles had the heaviest furniture on the planet.

Over time, metrics often become the goal and that is why Poland came to produce furniture designed by weight. Factory managers made their targets and earned their privileges not by turning out high volumes of pieces of furniture or improving quality but by dreaming up ways to make their products heavier. The original intent undoubtedly started off as an attempt by central planning bureaucrats to quantify output and simplify accounting, but it came to dominate operational strategies and

J. E. Meyer, *The Renewable Energy Transition*, Lecture Notes in Energy 71,
https://doi.org/10.1007/978-3-030-29115-0_7

ended up distorting the very fabrication process and function of the products they were designed, just simply, to measure.

Just as furniture should be measured in its own specific units most useful to the plant producing it, energy and resources are physical entities and have to be measured in their own physical units rather than be "captured" or represented by a proxy valuation in a system of measurement designed to suit another purpose.

Very often, monetary units (dollars or fiat currency) are used to represent not only a commercial economy but more broadly the health of the entire society and all of the processes within it. Why? Because measuring in fiat currency is easy, and it allows everything to fit into the same set of books or spreadsheet. Not only is measurement in money an accounting convenience, it also best represents the interests of specific groups who benefit from ever greater levels of money turnover rather than the creation of real product with real value.

Currencies facilitate efficient transactions which see the transfer of goods of all manner and size all over the world from one party to another. Money does not make the world go round for most people, but it certainly does grease the cogs of the productive economic engine and allows it to spin as efficiently as possible to the benefit of all. Without money, we would be left trying to barter arrow heads for beaver pelts or cell phones for cabbages. Money provides an extremely efficient interface between all parties in all commercial transactions.

Facilitating transfer is what currency is designed to do. It is not designed to represent anything else. Money is a claim on real goods or resources but is not real wealth itself. Gold has come to be a global fiat currency over thousands of years. It represents real wealth to probably every person on the planet, yet it has little intrinsic value. It is attractive and feels good, and it certainly is useful in some electronic applications, but, in the end, gold is only important so long as people with real goods are willing to trade those real goods for the gold. Once it loses the aura of being a store of buying power, it becomes almost useless.

Energy is of critical use to all people at all times, but it is not so easily measured, handled, and distributed as gold or printed currencies. Otherwise, it might be the final currency in the chain of real value proxies through human history which spanned seas shells, beads, arrow heads, knife blades, and bushels of wheat or rice, all of which had intrinsic value. Those awkward currencies were finally supplanted by more convenient and scalable proxies with some intrinsic value like gold and silver coins which were followed by items with no intrinsic value but high enough confidence levels to maintain their value over time, namely copper and steel coins and printed paper and finally numbers in a computer. It should also be pointed out that the adoption of fiat currency (artificial currency, usually printed by governments, which have no commodity or real value backing) also gave certain groups the ability to manipulate valuations and currency flows and thereby make money with money. The result is more claims on wealth than the amount of wealth created leading to inflation, in effect an invisible tax by money manipulators on the real wealth producers of a nation. In the end, confidence in a currency does not necessarily mean that currency is an accurate representation of real cost and real value.

Gross Domestic Product: The Health Indicator of the Commercial Economy

GPD is a cash flow metric that is designed to capture all of the paid activities in the commercial economy. This system of accounts is dollar based and originally designed to provide insight into the workings of the increasingly complex economies of western countries, so that a balanced tax system could be put in place. As it was developed through the late 1800s and into the mid-1900s, gross domestic product was never envisioned as a social indicator or as a tool of national policy development. But it came to assume that role as can be witnessed hourly in news coverage calling for increases to the GDP and endless growth to solve all manner of social problems. GDP, the simple measure of paid activity, has become the shining light by which public policy is examined. Growth of the GDP, not the health of the economy, is held as an unassailable national goal with small variances being anxiously reported by intense media commentators.

But in that role, one for which it was never designed, it produces many serious distortions of perception and therefore many outright damaging decisions. By using GDP as our measure, GDP growth has become the fundamental goal, and the singularly goal-focused national policy system has delivered it by the ton. But just what are we measuring?

- GDP is not a financial statement. It is not netted out for changes in man-made or environmental assets, or net external debt. GDP is not a measure of real product. It is not a costing system describing mechanics of the wealth-creation process.
- It has no ability to represent the value of the environment for the destruction of environmental assets or even man-made assets.
- Another misleading aspect of GDP is that it counts only *paid* activities. A great deal of GDP growth has come from including previously unpaid productive activities into the paid category. An example is women leaving the home to work for pay cheques. The domestic work women did in the "old days" account for about 40% of all productive efforts. As women transferred into the paid work force, the paid activities of day care, restaurant meals, cleaning services, extended care homes, and a wide variety of social, medical, and educational programs moved to fill the void. GDP growth rates soared, but real output did not.
- Again, when natural disaster strikes, repair activity boosts GDP despite the destruction of assets. Neither the economy nor individuals are better off, since the work merely replaces lost personal goods and infrastructure.

Clearly, we need to move beyond a prime indicator that represents activities contributing to the destruction of the environment as productive activity and its consequences as economic stimuli. Many people have pointed out the folly of using a commercial market indicator as the prime metric to represent individual well-being and socio-economic progress or to use as a working tool to determine future policies.

Calls for a Better Way

Despite the constant media barrage of the need for growth and a larger economy and GDP, the shortcomings of GDP as an indicator of social welfare or environmental health have been well understood since its inception. Below are observations from a broad range of people who had reasons to question the prominent place GDP holds in our vision of society.

> The gross national product does not allow for the health of our children, the quality of their education or the joy of their play. It does not include the beauty of our poetry or the strength of our marriages, the intelligence of our public debate or the integrity of our public officials. It measures neither our wit nor our courage, neither our wisdom nor our learning, neither our compassion nor our devotion to our country, it measures everything in short, except that which makes life worthwhile.
>
> Bobby Kennedy on GDP: Kennedy Library

> At present we are stealing the future, selling it in the present, and calling it gross domestic product.
>
> *Businessman, Environmentalist, author* Paul Hawken

> We are consuming the past, present, and future of this biosphere, our only home, in an unthinking rush for profits and GDP that we call 'progress'.
>
> Pavan Sukhdev is an Indian environmental economist

> For 60 years Gross Domestic Product, or GDP for short, has been the yardstick by which the world has measured and understood economic and social progress. However, it has failed to capture some of the factors that make a difference in people's lives and contribute to their happiness, such as security, leisure, income distribution and a clean environment–including the kinds of factors which growth itself needs to be sustainable.
>
> Joseph Stiglitz, Economist in *Beyond GDP: Measuring the impact of Operational Programmes in relation to economic growth and sustainable development at a local/regional level*

> We deserve better indicators of societal well-being that extend beyond mere economic growth. Many economists and social scientists are proposing such indicators. … Whatever we come up with, it has to be better than GDP with its absurd emphasis on endless growth on a finite planet.
>
> DavidSuzukihttps://www.straight.com/news/593761/david-suzuki-measuring-progress-gdp-gross-mistake

Finally, the individuals who created the system of national accounts in the USA and Britain were clear on their objectives and the capabilities of the systems they were developing. They were creating a system to develop a comprehensive and fair tax system, not a national planning metric. They foresaw that their measures might be misused as some sort of policy guidance tool and were emphatic and explicit that they were unsuitable for any such purpose. Simon Kuznets was the chief architect of the American system and summed up his thoughts on the subject in the following quote:

> The welfare of a nation can scarcely be inferred from a measurement of national income as defined by the GDP.
> Simon Kuznets

No matter how complete the GDP accounts become, they will still suffer the limitations of their printed currency base and therefore will never be able to represent a wide spectrum of physical processes. If the mechanism or even part of the mechanism of any process lies outside the commercial market, in which it would receive a dollar valuation of some sort, GDP is unable to represent it. Also, assets only begin to have "value" to this system when they are processed and not when they are lying dormant or doing their job without paid human interaction.

GDP continues to be the excellent tax and commercial market size tool it was designed to be. But the physical realities upon which commercial economic activity is based are best examined by counting those physical assets and their flows in their own physical units (Fig. 7.1).

One Metric to Rule Them All

The human toolkit includes measurements for many different things, from the healthy functioning of the body to the price of goods to energy flows, but there is no uber-metric that represents everything. Unfortunately dollars, euros, yen, and gold, so vital in the realm of the commercial economy, have been adopted as prime metrics far beyond their useful ranges and have come to dominate the field of public policy.

Just as businesses would face insurmountable problems trying to function using the Bhutanese Index of Happiness as their prime metric instead of dollars, the health of a society, and the well-being of its citizens and the environment they live in, cannot be represented by cash flow. Understanding the measurement of energy and physical and social processes for most people must involve the un-learning of some concepts baked into our national consciousness. One of the principle areas of these

Fig. 7.1 Nal cartoon, survival is bad for business (unknown) "What you environmentalists have to understand is the destruction of the planet may be the price we have to pay for a healthy economy"—Quote from NAL

is the notion that money can reflect true cost or illuminate dynamics of natural systems and predict either the priorities or the outcomes of the land or the people.

Money-Based Decisions Under an Energy Microscope

Once past the illusion of money metrics, it is possible to grasp the physical distortions produced by using a monetary process to make energetic, environmental, and social decisions.

Wrong Metrics, Wrong Decisions

- Ethanol from corn has been added to gasoline in an attempt to reduce carbon emissions.
- Legislation for mandatory minimum of 5% ethanol in gasoline was passed, and various government subsidies and incentives were built into the production chain from corn to its processing into liquid biofuel. The program works monetarily due to financial subsidies but not energetically because ethanol, at northern latitudes, uses almost as much energy as it provides, with an EROI of between 0.7:1 and 1.4:1.
- This means it is not green but rather just as black as the oil and gas which are its feedstocks.
- In 2016, Canada produced 1.8 billion L and imported another 1.1 billion L from the US. (Haney 2017)
- Fracking or tight oil and gas have launched American energy production back into a growth mode and have contributed to an increase in world production of almost six million barrels of oil per day as the output from other conventional producers such as Saudi Arabia remained stable.
  - Early studies indicated that the fracking process with its intensive well drilling and water and chemical inputs had a very low EROI of 5:1, in the range of the Alberta oil sands. However, some recent studies have put the figure above 10:1.
  - Whatever the energy economics, the monetary economics have been distorted by essentially free investment money. Hundreds of billions of dollars "quantitative easing" money has been invested and not recouped. The surge in oil production has depressed oil prices below production costs for many producers. Low gas prices have encouraged the purchase of very large passenger vehicles leading to soaring and unsustainable energy use.
  - The end effect is to supress oil prices with surging production but with little or no return to the hundreds of billion dollars (Fig. 7.2).
- Solar systems in the Canadian Arctic work with large government incentives but with capacity factors well under 10%, and no output for months at a time, they cannot be said to be energy effective. Systems near the Arctic Circle may produce as little as one fifth the energy as similar sized installations in the sunny,

Fig. 7.2 Terraforming earth, fracking wells in Texas (EcoFlight 2012)

arid portions of the American Southwest. Or possibly only one third of those in southern Alberta. Subsidies rather than energy rationality drove northern solar PV system installations.

- Chinese researcher Yan Jun (Yan et al. 2019) compared decisions based on currency with those based of energy for a large corporation and found they produced different results.

  - Her paper on a Russian natural gas company that found that, over several years, decisions based on energy proved to be the better ones.
  - Also, after a period of 10 years of upgrading the dollar decisions to be more energy rational, the dollar decisions yielded the same results as the energy decisions.

  Energy analysis has been added to Russian gas companies' annual reporting system as per government requirement. Energy analysis is intended for external users, and therefore, indicators must be standardized.

  Data gathering was broken down into specific areas including preparation and pumping of oil, production service, maintaining reservoir pressure, and oil and gas production.

  The breakdown by type of task allowed the overall system efficiency to increase by 16% over 10 years which resulted in a higher level of profitability by the company.

- Energy intensity is a ratio that compares the amount of energy used in an economy to the size of the commercial economy as represented by GDP. Energy per dollar ratios over the past 40 years indicate most economy are becoming more efficient since cash flow has increased more quickly than energy consumption. There may be some use for this metric if it could be effectively decoupled from currency inflation and from all of the problems inherent with the GDP metric. Asset inflation and commercialization of previously unpaid work in addition to the growth of the largely unproductive, or transfer based, FIRE sector (finance, insurance, and real estate) would indicate efficiency improvements where none exists in reality.

False Metrics: Usurious Promotion

Politics and money can serve up a confusing mish-mash of conflicting information which leads to poor decisions being made by well-intentioned people. And then there are the outright efforts to mislead. The American Motors 1975 model Pacer was an attempt by the company to offer the benefits of a small car in the form of a large car. In the car culture of North America in the late 1900s, people tended to measure gas mileage by how far they could go on a tank and how often they had to refill. There were no "Miles per Gallon" or "Liters per 100 km" readouts on the dashboard in those days.

The AMC Pacer was sold into the market as a vehicle that could go further than their competition on a tank of gas. It was also sold as being larger than the competitive models from Ford and GM, the Pinto and the Vega, respectively. The Pacer was a lot larger, being, in fact, a mid-sized car, whereas the Pinto and Vega were genuine compacts and had the fuel efficiency of compacts.

The Pacer had the fuel efficiency of a mid-sized car because that is what it was. The Pinto weighed 2350 lb and the Vega 2200 lb, while the Pacer tipped the scales at a full 3000 lb and was almost as wide as a Cadillac. But yes, the Pacer would get more to the tank because its tank was 83 L, at least 1/3 times larger than the 61 L of the Vega and 49 L of the Pinto. AMC was not lying, but the deception of confusing miles per gallon with miles per tank was deliberate.

In terms of critical analysis and perspective, GDP mistakes size for importance. Farm income in Canada only accounts for about 2% of GDP, an insignificant amount. Clearly, the economy would be fine without it, as it amounts to maybe one-tenth the economic contribution on paper that the finance industry makes. However, Canadians could live without a finance industry but could not live without food. Similarly, energy generation constitutes a relatively small part of the Canadian economy, but without it, all economic activity would cease.

Policy makers have to recognize that national decisions made on the basis of GDP, rather than real metrics, will not only be misguided but, in the long run, disastrous. Unless decisions are based on the right metrics using good data, the wrong decision can be arrived at despite the best of intentions. Confusion and poor data and lack of clear goals enable self-serving decisions promoted through manipulation by parties with an agenda other than that of the greater good.

What Is Needed

Coherent national policies, policies which are tuned to work as a whole, rather than as disconnected and independent parts, require a full set of energy-based books. This means energy accounting, with enough detail to illuminate stocks and flows. It is also necessary to maintain communication between energy metrics and the commercial economy, and this can be done by identifying the "energy intensity" of various sectors at various points of the economy to clarify the energy/economic relationship.

If a society's costs increase due to the higher costs of energy, a strictly financially based metric management system will most likely raise interest rates to combat the resultant inflation. This is exactly the wrong measure as it will hamper investment in renewables and conservation.

- Monetary tools should be applied to monetary problems and physical tools should be applied to physical problems.
- Without an energy rational approach, based on physical reality, today's monetary policy-focused governance will eschew specific structural change actions, and in the face of real energy cost increases, simply dampen broad demand rather than channeling investment to where it is most needed, in the energy production and conservation sectors.

GDP, EROI, and Growth

In the industrial boom of the mid-twentieth century, it was said that "what is good for General Motors is good for the country." This was probably true for the majority of people. Manufacturing produced real goods and real salaries, and at the time, these goods were well distributed among the citizenry resulting in an improved lifestyle for all.

In the late twentieth century, manufacturing was transferred out of North America, and the economic void this created was filled by services, banking, and investment. At that point, the welfare of the leading economic sectors diverged from that of the average person. What is good for the developers, cheap labor employers, traders, and bankers could no longer be said to be good for the country. The size of the commercial economy expanded but the real output of goods did not and the benefits which accrued with this kind of growth did not end up in the pockets of the vast majority of the population, but rather in those of a very narrow segment of the society. Even further, the means by which this group enriched itself is directly responsible for the growing debt, job losses, and declining equality levels which have reduced the quality of life for the broad majority of citizens.

But GDP growth is well served by asset inflation, ever higher consumption, and growing debt, and these do indeed constitute the lifeblood of a sector of the economy whose only source of income is ever increasing cash flow. This was noted by

Stephen Chu, former US Secretary of Energy, 1997 Nobel Prize winner in physics, in his 2019 speech at the Energy Policy Institute of Chicago as reported by Forbes Magazine. "Depending on a pyramid scheme or a Ponzi scheme, there's no such thing as sustainability,"

Money may be the currency of human wealth transfers but energy is the currency of human wealth creation.

The Prime Commodity

Despite the vital nature of energy, it is generally placed into the same basket as other commodities by policy makers and commercial economists who deal with money flows rather than units of physical input. But energy is, in fact, the prime commodity by which all other commodities are produced. Understanding how energy is produced and how it is used and ultimately its value to society is critical as we face a changeover of energy sources.

Technically, many scientists know very well how to measure energy in raw units and mostly understand how it flows through the economy and society both in its raw form and in the forms in which it has become embedded. However, the national conversation has yet to acknowledge its ubiquitous presence and importance in our society.

Since energy is a critical foundation of our society, we need to have a clear idea of the effort required to obtain it. In other words, how much energy is required to acquire more energy. The metric for this is the previously mentioned Energy Return on Energy Invested (EROI). This metric or ratio, which essentially represents the ease, or lack thereof, with which we acquire more energy, was devised by a systems ecologist and energy analyst, named Charles Hall while studying the migration of fish (Hall 1972).

Prof. Hall, like many other biologists, and especially his PhD advisor Howard Odum came to view the lives of fish and the animals and organisms they studied as a series of flows of energy which represented how the fish was able to survive. It had to generate net positive energy to become healthy enough to invest in activities such as migration that were not directly energy-acquiring activities. He transferred the energy investment concept to other animals and finally humans and their social and economic systems. It became apparent to him that not only was energy vital to every living entity, but the pursuit and extraction of energy, being a universal process, needed to be represented by its own metrics. He finally made the leap to be able to represent this process by simply comparing energy returned to energy expended for both fish and societies (Cleveland 1984; Hall et al. 1986).

When Prof. Hall developed the Energy Return on Energy Invested yardstick, he dispensed with the specificity of how energy and its users were related in innumerable individual processes. This new ratio freed up the study of energy to enable it to be addressed as a completely independent field, yet common to all fields. He used that fundamental principle to synthesize a much better understanding of all life processes as well as the vital stocks and flows of energy which are literally the life-blood of every life-form, organization, and physical process. This new EROI metric enabled the relationship between energy and the economy to be brought into much sharper relief. A synthesis of his life's work on this concept in the biological and economic world is given in Hall's recent book: Energy Return on Investment: A Unifying Principle for Biology, Economics and Sustainability (Hall 2016).

What Is in a Number?

Monetary measures cannot capture the physical realities on which our lives and societies depend. But once the focus is put on measures like EROI and scarcity indicators enumerated as "yield in ounces per ton," etc., what do these numbers actually mean? An EROI for oil of 40:1 means that for every barrel of oil used in an oil extraction project, the yield is 40 barrels with a net energy yield of 39 barrels. This is a very healthy return. But it is a single number for one project for one point in time. When all energy projects are averaged together, it is possible to come up with an average for that form of energy, in a region or country or the world. That begins to show what humanity's relationship is with that energy source. Long time series analyses have been done, and at least one study indicates that EROIs peaked in the 1930s for oil and gas (Fizaine and Court 2017). In many ways, EROI is a societal and economic bottom line metric which does reflect real outcomes and potential. It has also spawned numerous other means of measuring the relationship between energy, society, the economy, and the environment (King 2015).

With very high EROIs, the difference between 100:1 and 150:1 would not be felt by the society, but only the firms involved in the energy extraction sector. Even at a 50:1 level or a 30:1 level, the society still has quite easy access to abundant energy. Once down below the 20:1 level, real prices begin to go up, and there may be some constraints in the supply of some goods for some income groups. But at that level, all elements of a sophisticated society could be maintained in place, from broad-based manufacturing to science to education, health, and social support (Fig. 7.3) (Table 7.1).

A 20:1 ratio means that society is only spending about 5% of its entire economic activity acquiring the energy it needs to function. Most would view a net energy gain of 95% as entirely manageable. But the EROI metric can be used for more than just an occasional static snapshot showing the real cost of energy. The field of energy production is an extremely dynamic one, and EROI can be used to illustrate a trend.

There may not be much in a one-time number, but there is a wealth of information in a series of numbers showing a clear trend over time. EROIs for world oil production have fallen from perhaps 70:1 in the 1960s into the same region as that of the combined Canadian oil and natural gas ratio of 17:1. Note in the graph below that output from the oil sands drags the ratio down dramatically. Given that the oil sands come in at around 4.5:1 for mining operations and 3.5:1 for in situ, and their

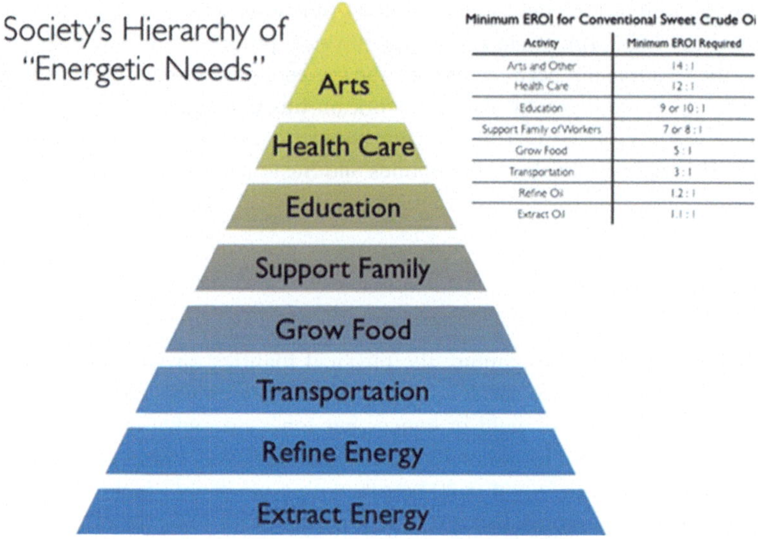

Fig. 7.3 EROI pyramid of potential social sophistication (Lambert 2014) as a function of EROI of society's main fuels. The lower tier estimates are fairly solid (Hall 2009) and more elevated values increasingly speculative and dependent on externalities such as government and culture

Table 7.1 EROI for social development

Minimum EROI for conventional crude oil	
Activity	Minimum EROI required
Arts and other	14:1
Health care	12:1
Education	9 or 10:1
Support family of workers	7 or 8:1
Grow food	5:1
Transportation	3:1
Refine oil	1.2:1
Extract oil	1.1:1

share of energy production keeps on rising, a steep decline in Canada's overall EROI can be no surprise.

At an overall EROI of 17:1 or 15:1, we can see perhaps a little tightening in the affordability of energy for many citizens. Hydro rates go up a bit, gasoline prices are dynamic but generally twitching upward. Taken as a snapshot, EROI ratios give a good indication of where we sit but taken as a trend over decades, it provides a clear indication of where we are headed. Clearly, energy costs are not going down as they were in the 1950s when EROIs were much higher. Once EROIs sink below a certain level, and energy either is distinctly more expensive or even begins to be less available, several processes begin to engage. Either activities are cut back or activities are made more efficient. For instance, if Canada was forced to turn to the oil sands,

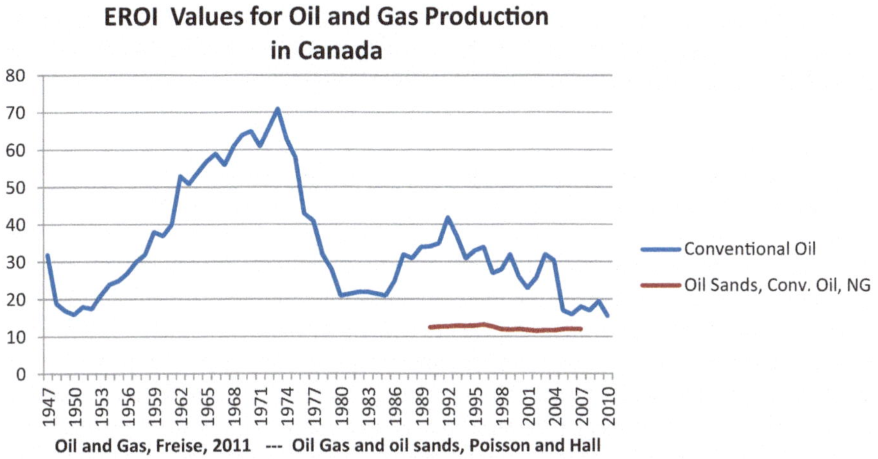

Fig. 7.4 Declining EROI of conventional oil and gas and with oil sands (Poisson 2013)

with its EROI of around 4:1, then economic output would be much lower as perhaps one quarter of all economic activity would be needed to get to run the other three quarters of the society.

When EROI sinks below 10:1, efficiency becomes a paramount social priority and activity cuts become inevitable. Just as Canadian society and industry became more efficient over 300 years of its early history, its capability increased immensely yet the energy consumed on a per capita basis stayed roughly even from the mid-1600s to 1950 (Fig. 7.4).

The necessary energy demand of higher latitude and harsher environments inevitably results in the effective EROI being reduced. An EROI of 5:1 for the Inuit above the Arctic Circle means bare survival, whereas at a seacoast in Mexico, it could mean the development of a sophisticated society. Perhaps with solar and wind energy with an average EROI of 10:1, in a Mediterranean climate, a very high standard of living might well be possible, whereas in most of Canada and northern Europe, it might mean bare subsistence. Efficiency and necessity demand modulate the EROI so that specific EROI ratios have different meaning for different societies in different regions of the globe.

But the trend is clear. The end of cheap and abundant energy is within sight, and we have to manage our economy and infrastructure to prepare for a large change in our relationship with energy. We will have to change from consuming energy stocks to using energy flows.

Any conflict between commercial market metrics and physical metrics should be put into the perspective that recognizes they are counting two entirely different things. Strong arguments may come from the interest groups who benefit from an ever larger commercial market place. Once the focus of policy attention is shifted away from GDP, the beneficent metric that implicitly promotes growth and therefore narrows interests, a challenge to the existing order will have occurred. Like any good costing or financial system, our next generation of indicators must illustrate real productive activity and steer us toward sustainable balance while flagging problems at the same time. Information must trump special interests.

Lifecycle Examples and EROI

Energy moves from source to end use where it dissipates back into the environment. The more complex the society, the more complex the "lifecycle" of energy and the more difficult it is to determine the true EROI. In an Inuit community, the hunter invests energy to harvest a seal which he then brings back to share with the rest of the community. Failure to share with the community would mean the hunter would have an abundance of energy for a short period of time but would soon be unable to continue due to the loss of the critical support structure which provides him with clothing, tools, cooked food, shelter, and ultimately children to carry on. So the lifecycle of energy would have to include the energy required to support the hunter's critical infrastructure, the rest of the band (Fig. 7.5).

The average food energy consumption for an Inuit band member might be 3.4 kWh per day or 100 kWh per month. An average 130 kg harp seal might yield 116 kWh or 42 days of food meaning one person would require nine seals per year. But the hunter must be able to feed both himself and four band members, so that the output of his efforts has to be five times what he himself consumes or invests in the hunt. Therefore the minimum EROI required for survival of the Inuit in the Arctic could be considered to be 5:1.

EROI is a ratio which indicates how much available net energy a society has to use for various activities. The very simplified lifecycles described below give an indication of how energy efficient those activities and their energy delivery systems are (Fig. 7.6).

The process of producing gasoline for the automotive or truck fleet consists of extracting natural gas to use in the mining and processing of oil sands bitumen into oil which is then refined into gasoline. Finally, the gas is burned up by the internal combustion engine in a process which delivers mechanical energy to the wheels of the vehicle.

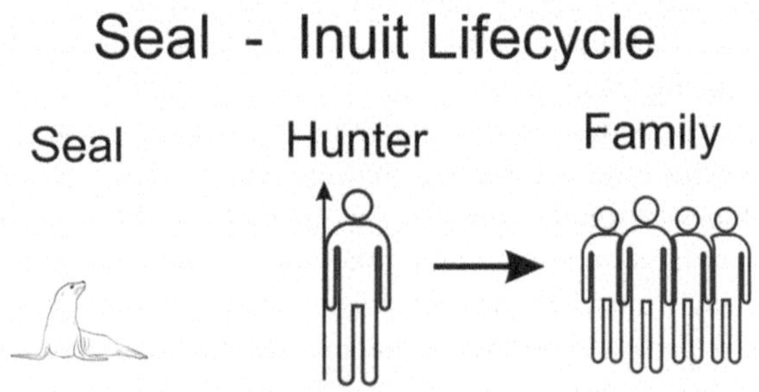

Fig. 7.5 Inuit EROI for community survival (author)

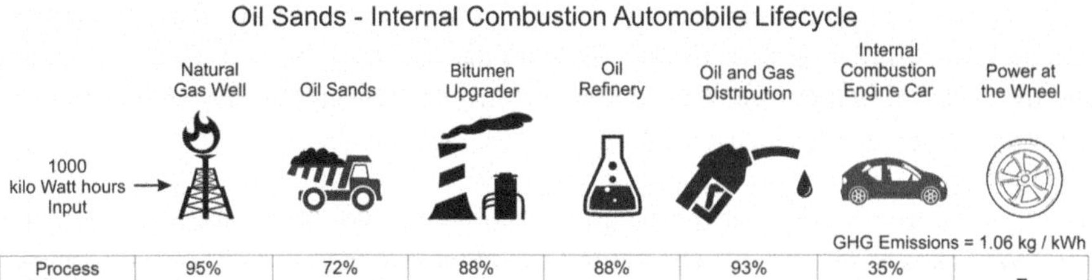

Process Efficiency	95%	72%	88%	88%	93%	35%	-
Energy Remaining kWh	950	3325	2927	2575	2394	-	838

Fig. 7.6 Oil from oil sands, energy, and emission lifecycle (author)

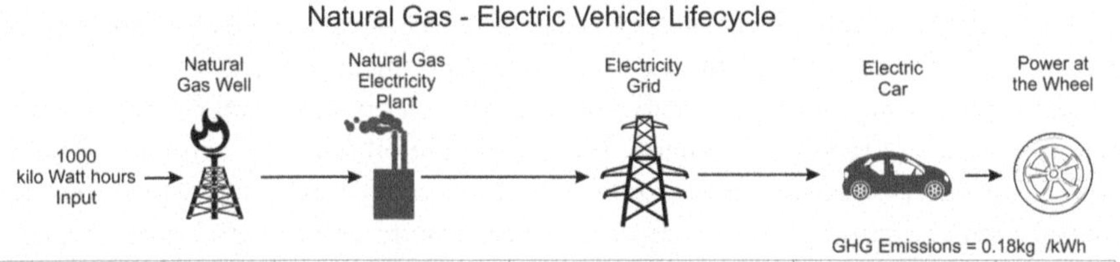

Process Efficiency	95%	70%	94%	95%	-
Energy Remaining kWh	950	665	625	-	594

Fig. 7.7 Natural gas-driven electric automobile energy and emissions lifecycle (author)

The starting energy investment is 1000 kWh equivalent of natural gas in the ground. Natural gas extraction has an EROI of around 20:1 in Canada which means it is 95% efficient and is able to use only 50 kWh while producing 950 kWh for other uses such as oil sands extraction. Oil sands in situ extraction has an EROI of only 3.5:1 which means it consumes 28% of the energy fed into it. Using all of the remaining natural gas to extract oil produces 3.5 times the 950 kWh input. The result is bitumen with an energy content of 3325 kWh after the extraction.

An upgrader turning the bitumen into oil is 88% efficient as is the refining process which turns oil into gasoline leaving 2575 kWh worth of gas ready for transport. Transport and distribution require 7% of the energy leaving 2394 kWh worth of gas to put in the tank of the internal combustion car.

Internal combustion engines are very inefficient and convert only 35% of the energy in fuel into mechanical energy at the wheels. This energy comes at a total resource cost of 4275 kWh, 950 kWh worth of natural gas and 3325 kWh worth of bitumen. The greenhouse gas emitted from the burning of this amount of fossil fuels is 1.06 kg of CO_2 per kWh of propulsion at the wheel. Since the average car might require 0.2 kWh to travel a kilometer, the energy process that powers internal combustion engine automobiles down the road results in the emission of about 0.2 or 1/5th of a kilogram of CO_2 for every kilometer traveled.

Overall, an oil sands powered internal combustion vehicle has a lifecycle efficiency in the neighborhood of 20% (838/4325), a level 1/3 that of a natural gas/electric system (Fig. 7.7).

The process of producing electricity from natural gas for the electric vehicle fleet is somewhat simpler than producing oil from the oil sands. Natural gas is extracted and then burned in a gas plant producing electricity. The electricity charges the EV batteries which power electric motor driving the wheels.

The starting energy investment of 1000 kWh is put into natural gas extraction which uses only 50 kWh of the potential 1000 kWh worth of natural gas leaving 950 kWh ready to be burned. The natural gas plant is able to turn 70% of the energy in the remaining natural gas into electric power and sends it down the line to the waiting EV which converts 95% of the power it receives into motion.

Electric motor engines are very efficient and convert 95% of the electrical power into mechanical energy at the wheels. The greenhouse gas emission from burning the natural gas to power this process is 0.18 kg of CO_2 per kWh of propulsion at the wheel. At a consumption level of 0.2 kWh to travel a kilometer, the emissions would be 0.03 or 1/30th of a kilogram of CO_2 for every kilometer traveled.

The electric route does not involve the oil sands or conventional oil deposits and therefore does not receive the energy boost from the oil which is extracted. The oil process resulted in an increase of power at the wheel of 40% (838/594), but this energy increase came at the expense of a much more complicated process and a quintupling of greenhouse gas emissions to do the same work. It should also be noted that natural gas will be the last resort supplier of electricity with hydro and renewables being the main sources. Since hydro and renewables involve vastly lower greenhouse gas emissions than natural gas, the real-world greenhouse gas emissions involved in electrifying the ground transportation fleet would be much smaller than the 0.03 kg/km given above.

Overall, a natural gas to electric car system has a lifecycle efficiency in the neighborhood of 60% (594/1000), a level triple that of a typical gasoline powered car. If the electricity was generated by solar, wind, or hydro sources, the efficiency level would approach 90% (Fig. 7.8).

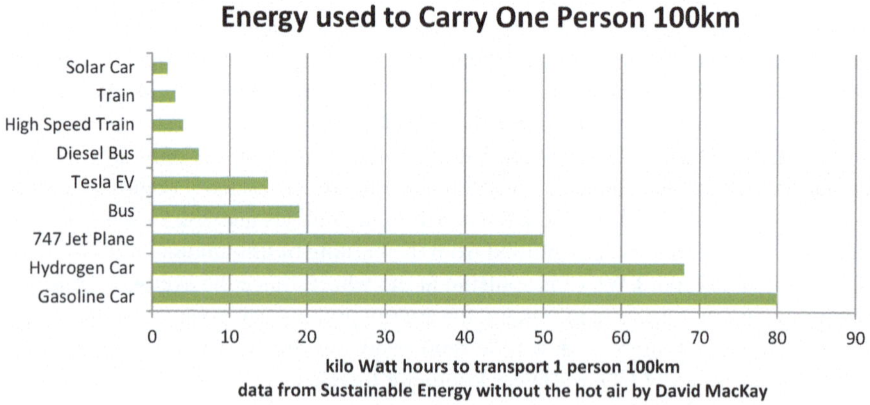

Fig. 7.8 Energy intensity of various methods of transport (MacKay 2009)

Lifecycle of Hydrogen Automobiles

Hydrogen is not a source of energy. but it can be used as a means to store energy. Hydrogen (H) is a very light gas which, when exposed to oxygen (O), produces the by-products of electrical current and water (H_2O). Hydrogen does not occur naturally in nature since it rapidly bonds with other elements to form other compounds, but it can be created by applying electrical current to water, thus separating the oxygen and hydrogen atoms. The resulting hydrogen gas can then be compressed into manageable volumes allowing it to be stored for future use. It can be transferred rapidly between holding tanks, and this rapidity of transfer, about equal to gasoline, and emission-free "burning" are the characteristics which make it attractive for large-scale use as solar PV and wind curtailment storage and as EV batteries.

Additional features of hydrogen storage are the potential for large-scale tanks, the lack of requirement for rare earth minerals and the very low "long-term degradation" of its charge. Loss of electric battery charge over time is a well-known issue as is the evaporation of water from large reservoirs.

Hydrogen would seem to be the answer to several large problems facing renewable energy networks. But predictably, hydrogen systems have their own significant problems which are laid out in the lifecycle pictogram below (Fig. 7.9).

The process of using electricity to convert water into hydrogen and then converting hydrogen back into water involves many energy losses. Heat is lost in the initial process of electrolysis and also during compression of the very slippery hydrogen gas in a process that is 58% efficient (Bauer and Simons 2015). When hydrogen is "burned" to produce electricity and water, only 40% of its energy can be captured resulting in a system lifecycle efficiency of 23%. This level of efficiency is less than half that of a natural gas plant-powered EV system and close to

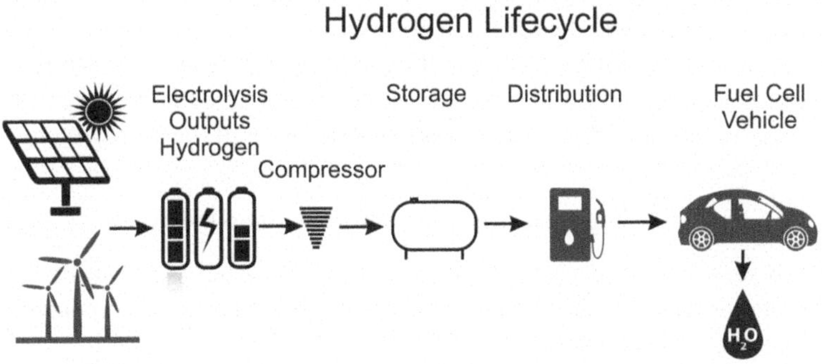

Electricity Generation	Electrolysis & Compression	Storage	Fill-up Station	Coversion to Electricity & H2O	Overall System
Efficiency	58%	-	-	40%	23%

Fig. 7.9 Solar- and wind-fed hydrogen automobile lifecycle (author)

an internal combustion engine. One advantage however is the very low greenhouse gas emissions produced by this all electric system.

To be practical for a transportation application, hydrogen must be compressed to between 350 and 700 atmospheres or 5,000 and 10,000 psi. The small hydrogen molecule is hard to contain and can leak easily. Very high-quality standards must be maintained throughout the handling system, and it also dictates the use of heavy, well-made tanks for containment. For grid level storage, underground salt caverns have been used, obviously at much lower pressures but the "round-trip" efficiency is low—approximately 40% compared to pumped hydro storage of 75–80% (Hyweb 2019).

If hydrogen systems were to use nothing but curtailment energy perhaps they could play a positive role in the complex energy systems which will evolve from the transition away from fossil fuels. However, for mainstream transportation systems or core storage applications, efficiency will have to increase for hydrogen to become practical.

Solar PV with Storage: EROI Impacts

Sunshine and wind are intermittent. Solar may be slightly more predictable than wind given its daily and seasonal cycles, but nevertheless, when sun has the potential to shine strongly, it often does not due to clouds, humidity, and snow cover. Winds may blow or not at any time of the day or season and wind turbines, like their solar panel cousins, produce energy when conditions are favorable, not when people would prefer to it. Rather than waste the energy produced when it is not needed (curtailment) or go without when energy is required, but not being produced, storage capacity can be added to energy systems to allow smoothing out differences and time delays between output and demand (Fig. 7.10).

Given the level of inefficiency of storage, it has a very large impact on EROI of any system into which it is built. But other realities come into play even in a world which, after the elimination of fossil fuels, has to replace 80% of the energy it once used.

- Pumped storage requires huge reservoirs into which to pump the water.
  - Possibly a sufficiently large holding capacity for freshwater is not available, meaning the process is not possible or salt water must be used, introducing environmental consequences.
- Electric batteries are expensive and require a great deal of raw materials to be extracted from the ground and processed and finally to be fabricated into their final form.
  - Ultimately electric batteries must be extremely recycleable, or else they simply assume the scarcity/depletion problems inherent with mining.

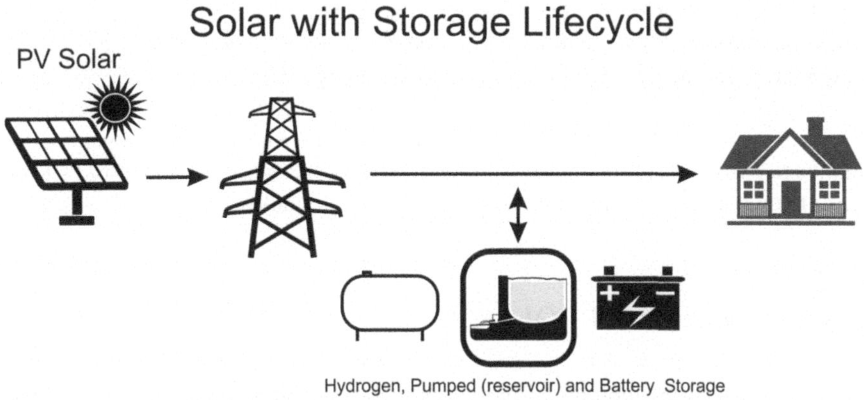

Transmission Efficiency	Storage Efficiency	System Efficiency
94%	none	94%
	Chemical Battery 95%	89%
	Pumped 80%	75%
	Hydrogen 40%	38%

Fig. 7.10 Solar lifecycle efficiency with various storage methods (author)

- Hydrogen is abundant, and the steel from which hydrogen storage tanks and piping are made is highly recycleable.
  - Will these strengths one day make hydrogen a core means of storage?
  - Graham Palmer demonstrates that storage has a large impact on the EROI of energy systems when the penetration (i.e., use) of renewables is "high" (above 20%) in his study, "A Framework for Incorporating EROI into Electrical Storage" (Palmer 2017).
  - Storage can decrease the EROI of renewable energy systems from 25:1 to 7:1.
  - The need for storage increases in more northerly locations due to higher demand and lower solar PV production for longer periods of time.

Hydro EROI

Sun-powered water cycles are able to lift water from the oceans and elevate it high into the atmosphere before dropping it onto prominent geological formations where the water begins to flow back to the ocean. If the terrain is high and rugged enough, significant drops in altitude can create a potential source of energy if the water can be harnessed.

Large hydroelectric plants have very high EROIs because their lifespans extend potentially into hundreds of years, and their output is generally continuous with very high-capacity factors. It is possible to speculate that if there were 10 or 20 times the number of excellent hydroelectric sites on the planet, coal, oil, and gas would have remained very minor players in the energy field.

Long life coupled with high, 24/7 energy output makes EROI values for hydro very high, and at the same time, remediation or downstream impacts are very small relative to emissions producing fossil fuels, the remediation costs of their old site or the unknown de-commissioning and removal costs of nuclear plants.

Key Components of EROI

In simple terms, the amount of energy that comes out of a process divided by the amount of energy that went in determines the EROI value. But many decisions are required on what should be included and on how long the life of the installation might be. What about maintenance costs, storm damage, and theft? It is a lot easier to secure a 1 km^2 nuclear site than it is to secure its photovoltaic equivalent of scores of site, large and small, in fields and on rooftops. The numbers generated by theoretical or laboratory-based researchers are often considerably different from those generated by researchers who study actual field operation. The empirical estimates of EROI (Prieto and Hall 2013; Ferroni and Hopkirk 2016) are often much lower than what is measured in the "nameplate" (rated output specification) based research (Breyer 2016).

Should all roads used be included in the energy cost of a project or just those built specifically for the project? What about the food to feed the workers and their families or the heat for their homes or for the schools in which they learned their trades? The broader the scope of inputs included in an EROI calculation, the lower the EROI is going to be.

Virtually all EROI calculations will include the energetic costs of all machinery and energy that went directly into the project but how broadly the net is spread to include social capital, and other inputs can have a very large impact on the results. Maintenance will be included in all field-based estimates but perhaps not in all theoretical ones. Remediation costs are rarely included given the time frames of decades for many projects. But they are real and are the ones the public purse must increasingly deal with. For instance the cost of "fixing" the old and abandoned oil wells in Alberta is estimated to be \$40–\$70 billion. The energy cost is unknown. Currently no estimates are available for the remediation of the thousands of square kilometers disrupted by the mining and upgrading operations of the Alberta tar sands.

In accounting terms, EROI is like a profit and loss statement and the data modeling that forms the basis of its math which directly compares to a costing system used to describe processes in a business model.

- EROI declines with a broader base of processes included beyond just direct "wellhead" costs:
  - If social infrastructure is included, then EROI is reduced.

  - If living costs of workers and their families and their support systems are included, then EROI becomes smaller.
  - Long transmission lines to smaller energy sources.
  - "Depletion" of good hydroelectric sites due to "shadowing" as too many wind turbines are installed, the output per turbine drops in a given area.

- EROI improves with:
  - Increased recycleability
  - Higher efficiency
  - Extended lifespans
  - Optimal placement

Despite these and other problems with measuring EROI, the numbers generated by quite different methodologies are broadly the same for the same situation (e.g., Fizaine and Court 2017).

Conflict Between Cash Flow Growth and Sustainable Progress

- There is a fundamental conflict between sustainable balance and GDP growth, and it manifests itself in many forms.
  - Lifespan of durable goods vs. planned obsolescence
  - Conservation vs. immediate exploitation
  - Recycling vs. minimizing short-term costs
  - Valuation of natural assets and human priorities
  - Immediate benefit now vs. larger benefit later
  - Personal profit vs. welfare of the community (Garrett Harden, Tragedy of the Commons)

Ultimately the metric of GDP promotes ever higher consumption without regard for material or social efficiency while the EROI metric promotes efficiency and asset management.

Critical National Metrics

Tax departments and corporations are guided necessarily by monetary measurements which indicate the size of the flows of revenue they can expect in their business model. Nations have to be guided by other more basic physical metrics which can guide policy to minimize problems. This suite of real unit metrics would telegraph the real state of society's underpinning before the symptoms showed up in monetary or social yardsticks or in the streets.

- Energy used per calorie of food produced

- Energy used per person for heating
- Energy used per person for cooling
- Energy used per unit of mined mineral
- Energy used per person kilometer transport
- Energy used to produce more energy
- Overall ecological footprint (Rees and Wackernagel 1995)
- A Human Effort Index? EROI × resource richness × resource requirement for survival (or latitude)

The work done by researchers Prof. Uwe Schneider, University of Hamburg, and Pete Smith, University of Aberdeen, on the energy needed to produce our food supply is critical to anticipating social shocks. Up to 2008, western countries had fair-level energy inputs per calorie of output, but in less developed countries, energy input was increasing as they adopted more mechanized production, higher levels of fertilizer application and irrigation. As climate change accelerates, are these figures changing? If we require more energy per unit of food output, then, in a world of energy transition, surely we need to track changes in this critical area.

EROI is being widely studied, but the subsets of this key energy metric should also receive close attention as they directly impact the necessities of life and social stability (Lambert and Hall 2014).

Resilience

No system is going to prove 100% reliable or stable. It will encounter problems of some sort over time, and the degree to which these problems can be tolerated by the user base will determine how resilient the system is. Resilience is the metric which defies quantification. How much abstinence do we endure in order to assure stability? How risk-averse are we? Immense stores of energy and stable climate are the best guarantors of a high level of resilience. In other words, if we had just maintained the level of consumption we had in the 1950s, and started the transition to renewables at that point, we would not have had to deal with large fossil fuel cost increases or rapid climate change. Human progress could have continued without the impacts or disruptions of climate change and energy cost shocks.

Countries living a hand-to-mouth existence are extremely sensitive to large supply shocks, whereas those who have a surplus of supply from different sources are subject to fewer crises. Surplus is required for stability, whereas pushing the limits of resources at hand invites a cascading and unpredictable series of shocks. These shocks play out as social turmoil as evidenced all over the world on an unfortunately frequent basis (Ahmed 2017). Countries with larger surpluses and capacity overbuild have fewer crises than those with smaller or no surpluses.

Accurate Terminology

The EROI ratio has gained prominence in energy and national welfare conversations because it most accurately describes or illuminates the physical workings of different energy systems and of nations as a whole. Accurate, comprehensive terms are needed if we are to develop a thorough understanding of the issues before us. We need to use common terms which are widely understood.

Communication counts and using terms correctly is vital for effective communication. A number of terms have been used to describe economic and social conditions but have been used to mislead or obfuscate. The terms of "efficiency" and "competitive" are often used as part of the discussion on the benefits of globalism and free trade. They are repeatedly misused in that context.

Cheaper does not mean more efficient or productive or competitive, it means lower monetary cost. Competition implies a level playing field.

– Pre-free trade and globalism, Canadian autoworkers, who were paid approximately $35 an hour plus benefits, like most of their counterparts in the developed world, were more productive and produced a higher quality product than did those in China.
– But as China began to land foreign manufacturing trade deals, their wage rate, in the controlled exchange rate in the late 80s and early 90s was under $1 a day.
– It has improved greatly to $4.10/h in early 2019.
– In Mexico, the minimum wage is $2 a day, and the average manufacturing wage is $2.30 an hour.

Energy is another important input, and China had a great deal of cheap coal and very little in the way of environmental regulations to interfere with its wholesale use. Given cheap labor, cheap energy, minimal environmental or health and safety standards, and a rigged exchange rate, China vacuumed up a good portion of the worlds' manufacturing capacity by producing, initially at least, very low quality goods incredibly cheaply.

"Cheap" applies but the terms "competitive" and "efficient" are simply misplaced and misleading, very often deliberately so, in a discussion on trade. China had the energy supplies and 400 million people displaced from agriculture who had to be employed along with a burning desire to upgrade its technology base. The west got cheap goods, the loss of millions of highly paid, very productive jobs, tax base erosion and a slide into higher inequality, and debt at all levels. The world got much higher greenhouse gas levels than if the same goods had been manufactured more efficiently in countries with much cleaner energy sources. More on China's rise is in the next chapter.

Misleading Terms

Measuring critical systems with the right metric is critical but so is using descriptive terms accurately. In public policy there are a number of terms which are used in a misleading fashion which serves to drive constructive conversation off course. Often these terms are used interchangeably but may mean exactly the opposite.

- Growth—progress
- Bigger—better
- Efficiency—cheaper
- Competitive—cheaper
- Well-being—higher income, higher consumption
- Clean growth—an oxymoron at our current level of technology
- Sustainable growth—growth in the use of earths resources simply is not sustainable (Fig. 7.11)

At the beginning of its modernization, Chinese industry was inefficient and energy intensive. As companies acquired more advanced technology, energy intensity declined, although the very energy (and carbon emissions) intensive coal-generating plants remained.

- In a renewable energy economy, efficiency is critical, and it must be very well understood.
- In a national policy context, efficiency does not mean cheap in dollar terms, it means making the most of the available resources. This can be better determined using physical units than monetary ones.

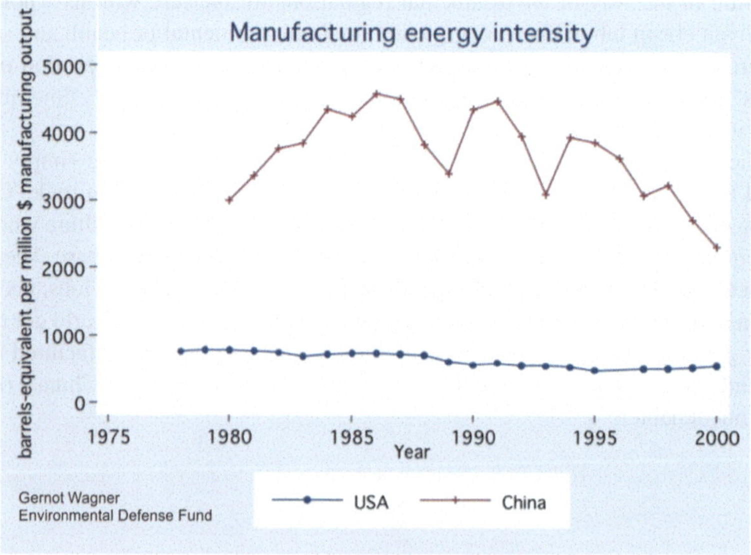

Fig. 7.11 Energy intensity of manufacturing in China vs. developed country (Wagner 2012)

Investment in a Transitioning World

It would be asking a lot of green energy to be able to build its own energy network and infrastructure while also repairing that left over from the oil era. Currently, many developed nations have aging infrastructure and large infrastructure deficits. These need to be upgraded and optimized for improved energy efficiency (Hub, Global Infrastructure 2019). Furthermore, the legacy costs of past energy and mining projects must be dealt with in addition to the energetic costs of upgrading existing infrastructure and creating the new renewable energy infrastructure. Clearly fossil fuel reserves must be invested in these areas as a transition strategy rather than depending on an embryonic green energy sector. As Jack Alpert of skil.org puts it, "The real test of renewable energy will be in the possibility of creating a renewable energy economy from renewable energy infrastructure, that is, can it both create and maintain itself?" The new system does not need the extra load of cleaning up after the fossil fuel era.

We should be investing fossil fuels now to more easily build the renewable energy infrastructure rather than squandering this one-time resource to simply continue to increase unsustainable consumption. And we should be investing consistently rather than at the whim of political opportunism and market fluctuations. As analyst Gail Tverberg of OurFiniteWorld.com noted, "Price must reflect real costs and market gyrations and distortions must be buffered out to maximize efficient and consistent investment." Real costs must be energetically correct and not distorted by the machinations of a money valuation system designed for purposes other than representing cost, real assets valuation, and national welfare.

Investment in renewable energy and in ways of reducing overall energy consumption will, somewhat paradoxically, be best accomplished by utilizing our remaining carbon budget, and fossil fuel optimized infrastructure already in place, to lay the foundations of a sustainable, renewable energy society.

Deal with Structural Issues, Not Symptoms

If a society's costs increase due to the higher costs of energy, a strictly financially based metric (management) system will most likely raise interest rates to combat the resultant inflation. This is exactly the wrong measure as it will make investment in renewables and conservation more difficult and less likely. Monetary policy focused governance will not provide an energy rational approach but rather hamper structural change. Reacting to real energy cost increases, monetarist policy will simply raise interest rates, negatively impacting the ever-greater investment levels needed in the energy and conservation sectors.

Increasing interest rates in the case where rising prices are caused not by the printing of too much money (printed promises exceeding the availability of real goods and services to purchase) but by a critical resource becoming more scarce and resulting in higher real costs. That is, the higher price would mean more resources would be allocated to maintain adequate supply. Raising interest rates would serve to restrict or choke off the required increase in resources being applied to the supply chain.

Money can be looked upon as a lien on energy as, in the end, no production will take place in the commercial economy without energy. The total value of monetized transactions map possibly go down in a more energy efficient economy, and policy makers cannot be distracted by monetary metrics sending distress signals when, in fact, progress is being made on the transition. Even if sustainability measures were certain to be a complete physical and social success, they may well appear as near economic collapse when represented by GDP metrics. And indeed they will be for growth-dependent sectors, but an easing in of policies and clearly stated goals will mitigate damage to individuals in the affected industries.

Money-valued goods and services will never exactly equal (accurately represent) the embedded energy required to create and maintain energy infrastructure. Project developers, whether small or large entity, or business or public domain, cannot possibly have the resources or expertise to perform thorough EROI analyses. Therefore, in order for the decisions of energy to be the most energy rational (energy is the society's bottom line), governments must incentivize renewable energy and conservation projects in such a way as to make the financial models within which the private and public financial entities must work, calibrate as closely as possible to the energy model reality.

Models, according to Rob Hoffman, must be "systemically coherent"—not designed to deliver one outcome but designed to make sure all parts work together regardless of which outcome is preferred. Clearly the final objective is to reach the intended goal, but the model has to be neutral, and the goal has to be reached with all of the parts of the economy operating in unison.

- Dollars are fungible, and in reality, there is far less substitutability than a dollar-based model would indicate. This does not guarantee success because the level of effort is left open as well as the timing, but it assures that any effort which is made is made in the right direction.
- A green energy infrastructure must be built before the shortage of fossil fuels becomes acute because it is not possible to build a green energy network using solely renewable energy.

Dollar metrics—fairy tales told by printed promises by which the printers profit by the telling.

Overly Good Corporate Citizens Do Not Finish

Similarly, businesses which might seek to be good or even leading corporate citizens could only do so to a certain extent if their competitors based all of their decisions on factors directly tied solely to the costs of doing business. Stores in the famous Canadian coffee chain, Tim Horton's, have business models in which 25–40% of total sales revenue goes toward wages. Typically their wage structure is minimum wage which was $11 an hour in Ontario. At that time a medium coffee was sold for $1.70. When the minimum wage went up to $15 an hour, the price of that same coffee should have increased to $1.85. But this cost increase would have been faced by all of Tim Horton's competitors, and the chain would not have suffered a competitive disadvantage.

However, if the minimum wage had not been raised and several socially conscious franchisees decided to raise wage rates to a living wage which would also be tax positive, say $22 an hour, the cost of the medium coffee would have jumped by a third to $2.25. Certainly those businesses would face lost sales as customers looked for a cheaper caffeine fix from the competing cheap labor business models. Unless national standards are in place to assure a level playing field for all companies, even the most socially responsible company will be unable to incorporate progressive programs and remain in business.

Geographical Context

Often we see capabilities of various technologies referenced without giving geographical context. What works well in some or many regions may not work well in all or specific regions. If a national policy were to promote solar installations, it should be in the context that these PV systems are only to be installed in areas where they represent the best option. Solar PV funding in Arizona might be a good idea but a bad one in New Jersey or Alaska. If New Jersey could be supplied by installations in Arizona through an enhanced grid, maybe that makes more energetic sense than adding low performing capacity in the east, even if dollar budget allocations call for a state-based division of funds.

In Canada, solar installations are better placed in southeastern Alberta than Northern Ontario or Nunavut no matter what politically derived funding formulas may have been laid out. Make hay where the sun shines (Fig. 7.12).

There are many caveats which apply to this graph based on a small, 300 facility random sample from the EIA database. As one would expect, capacity factor does drop the further north one goes but aridity levels allow a northern state like Montana to perform better than the more southerly state, Delaware with its high, east coast humidity and cloud cover. Central Ontario is almost on the same latitude as Montana but has fewer sunshine hours due to its proximity to the Great Lakes. Inuvik is near both the Arctic Circle and the ocean, giving it very low output.

As latitude climbs, so does the need for storage. Not only will the draw from storage grow larger in more harsh climates, the periods when it is needed will become longer. Low capacity factors and high demands for storage would seem to indicate that EROIs for full solar PV systems with storage would hit very low levels at fairly moderately northern latitudes.

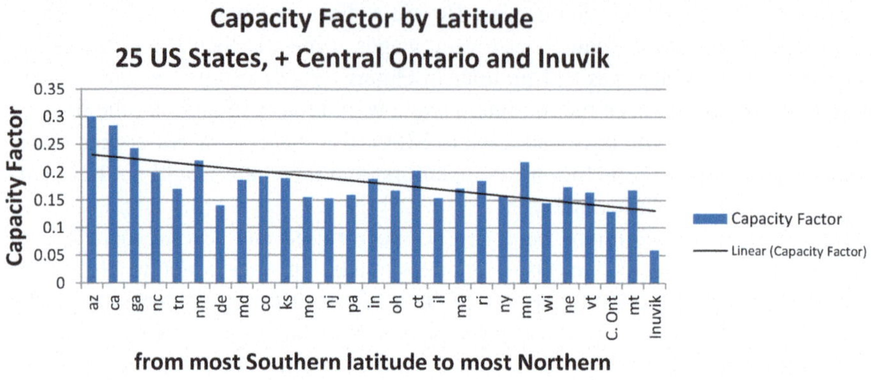

Fig. 7.12 Solar PV capacity factor in different locations of North America (EIA 2019)

The Profile of Energy EROIs Through Time

The time profile of the EROI Mountain closely follows the typical pattern of a resource depletion curve. Exploitation starts initially and then rapidly accelerates to harvest the richest fields of the resource first. These are depleted first but last well into the search for, and exploitation of, less dense resources, resulting in a long-term and fairly predictable decline. Innovations like fracking can claw back some of the decline for a short period of time, but in the end, either new types of energy have to be found or our pattern of life has to change dramatically. And it bears repeating that the above curve may be typical of any mined commodity, but without the commodity of energy, we will not have the ability to process any other commodities on a large-scale basis. A very late study, currently being finalized as we go to press from researchers in Spain, suggests that the world of renewable energy will be one of very low EROIs (Capellan-Perez et al. 2019) (Fig. 7.13).

As daunting as the "slide down the mountain" may appear, it has to be pointed out that the technology available to us now is much more advanced than it was when we began to scale the mountain. Although our consumption of energy could be classified as reckless in many areas, we are also able to do much more with it than ever before and do it much more efficiently. We have also developed the capability to harvest ever smaller and lower grade sources of energy. Possibly in the future we will be able to generate electrical energy from the heat in the air around us.

Whatever capability science has given us or will give us, we need to be able to look at the world around us and understand what it is capable of and how our needs can best be made to conform to that capability. This might be termed social intelligence or environmental awareness, but it must be developed and used to guide our national policies in a way that was never needed in the age of easy access to seemingly unlimited stores of energy. In fact, the skill of balancing our needs with nature's gifts has been built into human societies from our very beginnings, yet it has been lost through our development of, and obsession with, technology and the

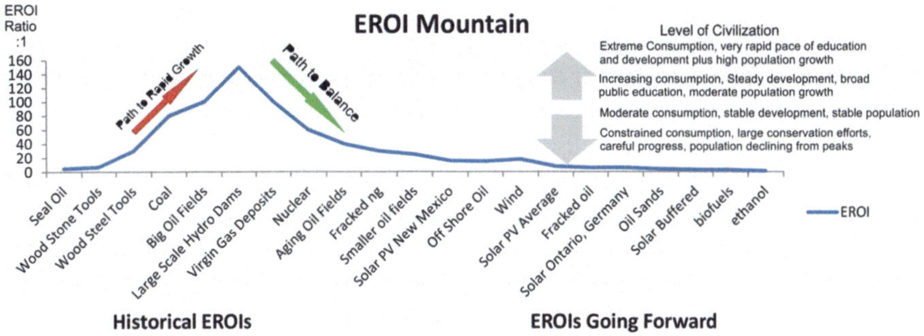

Fig. 7.13 The EROI Mountain illustrating how we have burned through the richest resources (author)

abundant energy supplies of the past 300 years. But technology can help us understand and rebuild that link in our highly quantified world as we return to life based on the earth's daily budget of energy rather than on drawing down its savings account.

We have just passed through the first measured century of human existence (Caplow et al. 2000). We are capable of assessing our situation relative to the world around us better than any generation before us, and this capability is now being included in the conversations of most nations. Certainly every society since the dawn of agriculture has been aware of critical metrics such as the number of people who could be supported by a given wheat field, but no society before us has had the wealth of statistical knowledge we currently do.

The world's first demographer, Thomas Malthus applied data analysis to express the reasons for the state of human affairs over two centuries ago, and since then, particularly since the intense use of fossil fuels spread uniform measurements throughout society, we have measured anything and everything. We are data rich, but without clear goals or understanding which are the correct metrics to apply in any given situation, we are effectively information poor. Industrial society humans are new to measuring our surroundings and building a coherent, long-term strategy.

With the information available to us, we should be able to differentiate between commercial intelligence and social intelligence. We can start asking questions such as, "Why would a country export a critical and finite resource on a massive level which, when used in small quantities can support a sophisticated society but when used in large quantities will destroy life on the planet as we know it."

In the era of fossil fuels and their ability to power endless growth, EROI primarily dealt with real cost and efficiency. If more energy was needed, EROI would indicate what the real costs to society were, but really, all we had to do was put in more effort and dig harder and drill deeper to access the stored energy stocks to keep energy in abundance. In the era of the flows of renewable energy, no matter how hard we "dig" or how deep we "drill," abundant energy will likely not be available at our current level of technology. Balancing the needs of a sustainable society with the flows of energy in the natural world will become a science in itself and the basic tool of Energy Returned on Energy Invested will play a critical role in forging a path toward a sustainable future.

The quality of our tools will determine our ability to navigate our way through the physical, social, and economic minefield that is the transition to renewable energy. A common goal and well-defined terms have to frame the national conversation which needs to take place and that conversation must be able to recognize the geographical and temporal limitations which will challenge a successful transition. And we have to understand where we sit on the learning curve of working toward a common goal, how successful we have been and how much better we will have to do to be able to climb over the threshold of sustainability.

References

Ahmed N (2017) Failing states, collapsing systems: biophysical triggers of political violence. Springer, Cham

Bauer C, Simons A (2015) Lifecycle assessment of hydrogen production

Breyer C (2016) Energy Return on Energy Invested (ERoEI) for photovoltaic solar systems in regions of moderate insolation: a comprehensive response

Capellan-Perez I et al (2019) Dynamic EROI and material requirements in scenarios of global transition to renewable energies

Caplow T, Hicks L, Wattenberg BJ (2000) The first measured century: an illustrated guide to trends in America. AEI, Washington, pp 1900–2000

Cleveland CJ (1984) Energy and economic activity

EcoFlight (2012). https://ecoflight.zenfolio.com/p147245797/h11742cb7#h11742cb7

EIA (2019) Annual Source and Disposition of Electricity for Non-Utility Generators, 2017 Final Revision https://www.eia.gov/electricity/data/eia923/

Ferroni F, Hopkirk RJ (2016) Energy Return on Energy Invested (ERoEI) for photovoltaic solar systems in regions of moderate insolation

Fizaine F, Court V (2017) Long-term estimates of the energy-return-on-investment (EROI) of coal, oil, and gas global productions

Hall CAS (1972) Migration and metabolism in a temperate stream ecosystem

Hall CAS (2009) What is the minimum EROI that a sustainable society must have?

Hall CAS (2016) Energy return on investment: a unifying principle for biology, economics and sustainability. Springer, Cham

Hall CAS, Cleveland CJ, Kaufmann R (1986) Energy and resource quality: the ecology of the economic process

Haney S (2017) Why the Canadian ethanol industry could be set to grow. https://www.realagriculture.com/2017/07/why-the-canadian-ethanol-industry-could-be-set-to-grow/

Hub, Global Infrastructure (2019). https://outlook.gihub.org/

Hyweb (2019) hyweb.de

King C W (2015) Comparing world economic and net energy metrics

Lambert J (2014) Energy, EROI and quality of life

Lambert JG, Hall CAS (2014) Energy, EROI and quality of life

MacKay D (2009) Sustainable energy—without the hot air. UIT, Cambridge

Palmer, Graham, (2017) A framework for incorporating EROI into electrical storage

Poisson A (2013) Time series EROI for Canadian oil and gas

Prieto PA, Hall CAS (2013) Spain's photovoltaic revolution

Rees W, Wackernagel M (1995) Our ecological footprint. New Society, Gabriola Island

Wagner G (2012) Energy content of world trade

Yan J, Feng L, Steblyanskaya A (2019) Creating an energy analysis concept for oil and gas companies: the case of the Yakutiya Company in Russia

Chapter 8
Public Policy Formation for Successful Change

Abstract The psychological and social ability to cope with different challenges and to adapt to changing conditions is a function of leadership, imbedded interests, culture, and perhaps the events that have taken place within the national living memory. There are many examples of governments rising to the challenge of both imminent and long-term threats and opportunities. There are perhaps more examples of governments which failed to identify and come to grips with problems which were in plain sight and for which directions forward, if not clear solutions, were easily identifiable. Nations dealing successfully with change generally have clear goals which are based in egalitarianism and broadly understood and supported by the population. Nations with highly stratified societies with detached elites, whose welfare and power stem from policies that exploit the people they are charged with leading, find it difficult to change. The effectiveness of government seems to depend less on the level of democracy than on the ability of leaders to leave their ideology behind and both speak to and listen to the people they are leading.

Different forms of government have different strengths, and the question of which form of government is best suited to handle the transition to renewable energy is an oft discussed issue in the environmental and energy communities. All forms of government have posted successes and failures under different conditions.

Can the rapid fire decision-making capability of well-funded private enterprise working in their own self-interest better respond to the challenges of transition to provide a better social outcome or can a top-down decision-making process, delivered through a bureaucracy of civil servants accomplish a smoother transition? Certainly both types of decision-making processes have delivered impressive results in the past.

The communist party of the Soviet Union moved Russia from a Czarist state of near serfdom in 1917 to become the first nation to put a satellite in space a mere 40 years later. It followed that up with the first human in space in 1961 when Yuri Gagarin did the first lap of the planet in less than an hour. The Soviets accomplished this while fighting a civil war through the first half of the 1920s and enduring famines in the 1930s. The Soviets defeated the invading German army in the 1940s with a loss of some 27 million of its own citizens along with the destruction of thousands

J. E. Meyer, *The Renewable Energy Transition*, Lecture Notes in Energy 71,
https://doi.org/10.1007/978-3-030-29115-0_8

of its towns and cities. Clearly the Soviet regime could accomplish a great deal in its early years.

In the first half of the twentieth century, the more-or-less free-enterprise west, on the other hand, also made progress in great leaps and bounds in the areas of building social infrastructure while creating mountains of consumer goods for its citizens, fighting wars, and recovering from market collapses.

Are large structural changes more suited to top-down decision-making while creation of smaller, more complex goods maybe better handled by smaller entities with distributed decision making?

We Will Bury You

Nikita Khrushchev's famous and most often misquoted statement of "We will bury you" (CIA 1962) was a reference to his expectation that the Soviet Union would out-produce "capitalist" democracies and "bury" the west in consumer goods while proving the democratic form of governance was a failure. The real bottleneck of Soviet production was the labyrinth of bureaucracy necessary for central planning to control every facet of every production process. It might have worked for tanks and dams and railroads, but it did not work for the innovation, myriad parts, and new processes which were needed to produce thousands of different consumer goods.

The great Soviet hope to clear this Gordian knot of red tape was the advent of the computer. It was thought the processing power of these new machines would finally streamline the production of the wealth of goods the Soviets were hoping to deliver to their long-suffering population; hence Khrushchev's boast that Communism could out-produce the western democracies and provide more for each citizen. "Let us better compete at who is going to produce more per capita." It did not work out the way Khrushchev predicted, as the freedom to design and flexibility to innovate by individual stakeholders in the west prevailed over any hoped for efficiencies and focus accruing to a centrally planned and managed economy.

The Soviet Union ended up producing a trickle of often crudely designed and generally inferior consumer goods. An example of this is the Lada automobile which was a wretched implementation of an outdated Fiat design. After placing their order, in the time it took for a Soviet customer to receive their new Lada, car companies in the west had gone through two new model cycles. Design and innovation languished and quality declined as a disconnected central authority sought to control all aspects of production. The Polish "furniture by weight" syndrome was just one manifestation of a system that could neither handle complexity nor foster innovation or efficiency. The Soviets did accomplish a great deal in the areas of improved housing, railroads, jet airliners, and basic consumer goods but simply could not match the dexterity and nimbleness of the west's productive infrastructure.

While Soviet manufacturing capability languished, and the Soviet state itself actually disintegrated into 15 different states 2 months after the fall of the Berlin Wall in November 1991, western countries were amassing great expertise in every aspect of product design and production to service the entire spectrum of industrial and consumer demand. Clearly free enterprise and individual responsibility had triumphed over state control as far as filling consumer demand and, it must be mentioned, creating some of that demand to start with. The net environmental impact of the different systems is an open question.

China: A Variation on an Iron Clad Theme

But about the time the Soviet Union was devolving into the Russian Federation with a few extra players left over, another experiment was getting under way. China was seemingly mired in chaos and a choking bureaucracy for decades after the defeat of Japan in the Second World War and the ascendance by Mao Zedong's Communists at mid-century. The central government delivered many improvements for its largely peasant population, but it also delivered some devastating blunders. Between 1958 and 1962, the Great Leap Forward triggered a famine in which tens of millions of Chinese died (New York Times 2010). Weather and population levels were certainly factors but much of the blame lies with policies ruthlessly enforced by the central government.

China was technologically backward in the early 1980s, but they had established social order and basic educational and health infrastructure. They then set out to deliver a better material quality of life for their citizens in the form of "a moderately sophisticated" lifestyle. What was the fastest way to do that? How to acquire technology and build manufacturing capacity quickly? The best way, of course, was to have the technologically rich nations give you their expertise along with the money necessary to build your factories.

The means by which China acquired a great deal of technology quickly and became an industrial and technological powerhouse were as follows:

1. Invest in people via education and science by
   (a) Sending students abroad
   (b) Hiring foreign professors
   (c) Attending academic conferences
2. Acquire technology from foreign corporations
   (a) Buy rights to foreign technology
   (b) Buy innovative companies
   (c) And, of course, utilize the backdoor method of industrial espionage
   (d) But most importantly, make products for foreign companies who will gladly hand over their expertise in manufacturing to allow a better product to be made even more cheaply. This would involve working hand-in-hand with the manufacturing experts as well as the design and engineering teams of corporations. No on-the-job-learning experience could be better than working directly with the most advanced companies on the planet to make their product and to gradually participate in the design of it.

(e) Become a technology leader and control leading edge development now that the west has forfeited the base of technology, namely manufacturing.

The following anecdote is an example of how much stock was placed in the program of learning through export manufacturing and how organized that learning process was. An Ottawa technology company had five different circuit boards they wanted a Chinese company to make for them. The Chinese manufacturing company looked at the five different parts and the volumes needed for each one and gave the Ottawa company a quote for the entire job. In the meantime, the Ottawa company, which did have its own domestic manufacturing plant, decided to make the most sophisticated part in its own factory.

They asked the Chinese company to re-quote on the four remaining circuit boards with the same quantities. Much to their surprise, the quote came back higher than the first despite the lower number of parts. When asked how this could be, the answer came back that "We already know how to make the other 4 boards." In other words, the Chinese had a program in place that subsidized very low prices on parts they wanted to learn how to design and produce.

Leveraging their immense labor pool with exchange rates they themselves set and using their vast stores of cheap coal, China was able to build both its domestic social and industrial infrastructure while building up huge reserves of foreign currency to extend its influence across the globe. Control over foreign supplies of vital commodities was necessary for the stability of China's growing economy and political system. Consequently, in their national interest, China exploited foreign resources and institutions to effectively expand its direct control over:

- Food
  - Farms in Africa
- Energy
  - Energy projects around the world
- Transportation
  - Harbor in Europe, trains in Africa, the modern Silk Road in Central Asia
- Minerals
  - Ownership of rare earth mining rights
- Leading industries
  - Nine of ten of the largest solar panel producers in the world are Chinese owned and managed.

The above summary of Chinese strategy as it appears to have unfolded may be seen as plot to some. In the free-trade west, planning often does not extend beyond the next deal with its attendant press releases. A well-thought out strategy might indeed appear to be a conspiracy, but, through history, clear goals and careful planning have been a common underpinning of most successful nations and empires.

What are China's goals and strategies? They are nicely laid out in 5-year plans. As much as the great Leap Forward was a disaster, the goal to industrialize has been a huge success. The goals of these two plans may have been equally clear but the means by which they were carried out were diametrically opposed. The Great Leap Forward was directed and micro-managed by the central government which exerted extreme pressure on the lower ranks to produce without giving them the decision-making capability to allow them to succeed. Millions starved.

Thirty years later, the drive to industrialize and create a modern nation with "a moderately sophisticated standard of living" has succeeded because the state partnered with independent decision makers in the form of private and state-owned corporations. The banking system supported this drive as did academia. It was a coordinated effort made possible by very low exchange rates, a huge wave of rural population entering the cities, and a single-minded focus by the government on the priority of industrialization regardless of the short-term human rights or environmental fallout.

The Chinese effort resembles that of the Norwegian management of its oil wealth as well as the German banking/industrial network which powered Germany to technological and manufacturing dominance in the late 1800s and early 1900s. In contrast, in the same period, the British banking system was focused on the short-term high rates of return possible by funding projects in its resource and labor-rich Empire and left British industry, the original industrial revolution, to meander into decline through the loss of manufacturing capacity, science, and labor force skills. The British Empire crumbled as all empires had done before and Britain was unable to re-establish its industrial base which foundered under extreme labor-management strife and lack of capital.

Now that the Chinese have their industrial base firmly in place and their technology developed to a very high level, they can begin to focus on other social and environmental issues. Their goals and priorities were outlined by Robert Lawrence Kuhn, the author of "How China's Leaders Think: The Inside Story of China's Reform and What This Means for the Future." There are four main components of the Chinese Dream:

- Strong China
  - Economically
  - Politically
  - Diplomatically
  - Scientifically
  - Militarily
- Civilized China
  - Equity and fairness
  - Rich culture
  - High morals
- Harmonious China

  - Amity among social classes
- Beautiful China
  - Healthy environment
  - Low pollution

Stability of the nation can be assured only by the maintenance of the commitment to law-based governance of China and by upholding the leadership of the Communist Party of China. Through control of the media, social networks, and the Internet, the Chinese government is well positioned to both explain its policies to the people and keep dissent in check. If push comes to shove, they are also easily able to identify dissidents and their relatives and contacts.

A "moderately sophisticated society" with an average income of $20,000 annually means a standard of living similar to that of Greece, Portugal, and Spain (World Data 2018). Given that much of China enjoys moderate temperatures, this level of annual income would provide a fairly high quality of life as it does for the Latin Mediterranean countries.

- Norway –$76,000
- Canada –$43,000
- Spain –$27,000
- China –$20,000
- Portugal –$20,000
- Greece –$18,000

The much higher levels of consumption dictated by life in a harsh northern climate means higher levels of paid work but not necessarily higher qualities of life.

Victims No Longer

China has been invaded and humiliated repeatedly over the past 1000 years. It has been torn by internal strife, floods, and famine. The Chinese government has succeeded in bringing order and progress to the huge Chinese population while also building China's strength, global reach, and influence. China has reaped the immense benefits from free trade agreements that they set out to achieve.

In the future, China will be focused on environmental health and transitioning to renewable energy. China has already begun the process of cutting their consumption of fossil fuels (mainly coal), but it will be a real challenge as coal has fueled Chinese export industries. Even if the Chinese population declines (from 1.4 billion now) as rapidly as predicted, sustaining 1 billion people on renewable energy solely drawn from installations on China's landscape will require huge investments and a truly committed populace. It is a very open question as to whether they will have the energy surplus to remain the world's factory.

Surely a market of 600 million to 1 billion will be enough to drive technology without the need for massive foreign inputs or access to foreign markets. If China chooses to retain its energy and material production for its own population, there is little they need from the world at this point beyond some food, raw materials, and possibly some energy. They now have access to a good deal of that with their expansion of interests into Africa, and Central and South Asia among other regions. China chose to both drive and participate in globalization because doing so was vital for the achievement of its national goals.

China knew it was paying a high environmental price for its rapid economic growth and now that the per capita income targets have nearly been achieved, the country has charted a course of "stable" growth which may, in fact, translate to stability of per capita income rather than growth in the size of the economy. As stated in China's Renewable Energy Outlook 2017, "the energy consumption and supply must adhere to the ecological boundaries for sustainable development." The concept of an "ecological civilization" has been designated as a leading development strategy by the Chinese government.

If China decides that free trade is no longer in its national interest, it will cease to export manufactured goods on the scale that it has. That means importer nations will be forced to produce their own finished goods with their own energy, which, in many cases, will require a substantial portion of their renewable energy budget in the future. This is one possible effect of the transition to a renewable energy world that few energy analysts or modelers take into account, the additional domestic energy required to enable domestic manufacturing of all goods the nation consumes.

And not only will the nations affected have to supply their own power, it will likely be renewable energy, priced according to real domestic costs rather than energy from Chinese coal priced through controlled exchange rates. Additionally, products will be made using domestic labor paid at the going domestic labor rate and in accordance with all domestic labor and environmental laws. This will likely push up real costs for virtually all goods dramatically with the increased prices of such vital goods such as solar PV panels perhaps being the most acutely felt. To paraphrase Jack Alpert, "a nation's ability to reproduce its renewable energy network will be its ultimate test."

Cuba: The Worst of Both Worlds?

Cuba has been late to the renewable party despite being perhaps the one country that would benefit most from domestically generated energy. Since Cuba suffered the loss of its oil subsidy with the collapse of the Soviet Union in 1991, it has been forced to undergo a crash conversion to organic farming. Yet that emergency did not inspire them to begin an equally motivated conversion to renewable energy.

The loss of their annual oil subsidy of 13 million tons from the Soviet Union created the urgent need to convert to organic food production. The loss of this oil subsidy was the equivalent of taking 40 kWh/day out of the energy budget of each Cuban. Initially staggered by the loss of fuel and fertilizer, the agricultural sector responded extremely well over a period of 5 or 6 years by bringing food production back up to where it had been before the cuts and also producing a greater variety of higher quality, healthier foods (Ho 2008).

With the encouragement of the government, over 8000 city farms were created. Urban seed houses were established to provide gardening tools, compost, and biofertilizers at low cost. The government also developed new biological products and organic gardening techniques for implementation by the now highly diversified agricultural sector. These measures were well within the normal operating procedures of the Communist government, but they were not enough. To truly deal with the food shortfall created by the loss of a significant portion of their fossil fuel budget, the Cuban government was forced to step back from industrialized and centrally controlled agriculture.

Cuba opened up to a more self-reliant agriculture where higher prices were paid to farmers, who were allowed to develop smaller production units and operate smaller scale, organic, and semi-organic farming systems. Key to success was the opening of farmer's markets and legalizing direct sales from farmers to consumers. Deregulation of prices combined with high demand for fresh produce in the cities allowed urban farmers to make two to three times as much as the rural professionals. Central decision-making in all aspects of production and sales gave way to the central government developing standards and supporting initiatives and letting individuals make decisions about how best to produce the final product and sell it.

Cuba is almost completely dependent on other countries for its fossil fuel and yet is one of the most inefficient consumers of that fuel. It uses oil to power its electric plants, a process which is both inefficient and expensive. Cuba has abundant solar PV potential and significant wind potential. Given its dependence on foreign sources for fossil fuels and its desperate foreign currency shortage, one would expect the Council of Ministers to have designed solar power into their economy as soon as it was remotely practical.

But there were some powerful interests opposed to any such initiative. For example, tourist hotels are large consumers of electricity on the big Caribbean island, and electricity costs represent a large expense for these businesses. Air conditioning and hot water for showers require a great deal of electricity. Solar hot water heaters, a technology that has been well developed for decades, and which would function 365 days a year in Cuba, could take a large chunk out of the hot water costs. But these have not been installed. Why? Because, according to well-placed sources, it would cause a significant drop in the revenues of the electricity utilities and they are a powerful group.

In 2013, a large solar PV farm was finally completed although years behind in its timing. Just 4% of Cuba's electricity came from renewables in 2014, but the goal is to increase that to 24% by 2030. By comparison, in 2014, the Dominican Republic boasted a figure of 14% while Germany, hardly the ideal location for solar, or even wind, was at 25%.

In the end, renewables now do seem to be a priority for Cuba but instead of becoming a world leader, as they were positioned to be, Cubans, because of the initial strong opposition of vested interests, are just trying to make up ground from the back of the pack. Had the Cuban government implemented the types of initiatives for the energy sector that it did for the agricultural sector, perhaps by now Cuba would not only be the first "solar nation" but be manufacturing and exporting solar panels on a large scale. It must be pointed out that Cuba has not had the room to maneuver that China has. The US sanctions limit Cuba's options, and the world did not need Cuban cigars as much as they wanted cheap Chinese manufactured goods. Once the Cubans are given an open field, they are very well positioned to run to the head of the pack in renewable energy generation.

North Korea

North Korea had the same food and energy dependence on the Soviet Union as did Cuba and suffered the same shock when the Soviet Union collapsed. But the North Koreans did not respond as well as the Cubans. Instead of engaging the creativity and initiative of their people, the elites drew inward to defend their privilege and a million of their subjects died of starvation (Chisholm 2018) (Fig. 8.1).

Fig. 8.1 North Korean generals with elaborate service awards (Coles 2019)

The North Koreans made nuclear bombs and a lot of medals. Actually a huge number of medals seem to have been created for soldiers who have never seen combat, and therefore, they probably represent more of a loyalty rewards program. This seems to be a pattern when elites cannot adjust to changing realities, and they make their continued grip on power the national priority rather than addressing fundamental issues. The Cahokian elites also closed themselves off from the rest of the population and accumulated more wealth as their society declined, probably hastening the collapse of North America's most advanced pre-Columbian society. The Cubans mounted a social cohesive effort to address the fundamental problem and grew more fruits and vegetables. As a result, Cubans are generally quite healthy and well-educated now, whereas North Korea boasts nuclear capable missiles, a much feared hacking capability and a huge standing army, but little positive is heard about the welfare of its people.

Communism Mutates and Evolves

The Russian Federation has grown out of the failure of the first generation of Communism, and Cuba is slowly evolving from that same kind of tight central control regime. China has clearly charted its own path and has to be considered an outstanding success on the basis of its social, economic, and technological progress. Those are three different outcomes of rigid Communist governments which grew from violent revolution. But Norway and Alberta started from close to the same point of democracy, standards of living, beliefs, and social norms. Yet, even with peaceful and stable governments, both their policies and their outcomes have been very different.

Norway and Alberta, the Inevitable Comparison

The oil and gas fortunes of Norway and Alberta unfolded at about the same time but very differently for many different reasons. In contrast to the Gulf nations, these two jurisdictions had a great deal of oil and natural gas, similar populations (Alberta now four million and Norway five million) but were western and democratic and, to differing degrees, "free enterprise." As events unfolded, Norwegian policies turned into an oft cited high water mark of socially responsible resource development while the Alberta experience has been made the poster boy for the most environmentally destructive development on earth, some of which it deserves.

In terms of raw resources, both Alberta and Norway had very large reserves of both oil and natural gas, but the logistics involved in developing these reserves and getting them to market were very different. Compared to conventional oil fields, the oil sands involves an inherently more complex and expensive process and is possibly the most costly large scale oil recovery process in operation in the world at this time.

But in looking at how governments develop policy and how they handle change, the difficulties the oil sands presents may be a moot point. That is because Alberta reserves of both natural gas and conventional oil were larger than their Norwegian counterparts. Norwegian North Sea fields may well have cost less to bring in than those in Alberta, but costs were much more comparable than were those of the oil sands.

All fossil fuels are not created equal. Natural gas, conventional oil, and oil sands bitumen are very different in the way they generate royalties and liabilities. Natural gas is an all-star with conventional oil being a distinct positive while the oil sands look to be a troubled investment that may never pay off. Aside from the current data available, when the world begins to take climate change seriously, and the number 1 (and currently only export customer, the USA) looks at its sources for carbon emissions, the very carbon emission intensive oil from the oil sands may well appear as the turkey with the longest neck in the flock just before Thanksgiving.

Wherever in the North Sea Norwegian oil is produced, its first point of arrival in Norway is along a coast which makes shipping vastly lower cost than the pipelines and freight cars which haul Alberta's crude. It also allows the Norwegians' direct access to more markets than the landlocked, one export customer (USA) market to which Albertans are tied. The Norwegians may have received higher net prices for their gas and oil, but Alberta has produced a great deal more including oil sands production (Figs. 8.2 and 8.3).

Since 1975, Alberta has produced in the range of 4 trillion m^3 of natural gas, 18 billion barrels of conventional crude and 12 billion barrels of bitumen with cumulative current value approaching $2 trillion. Government royalties totaled close to $200 billion. Peter Lougheed, the former premier of Alberta responsible for creating the Alberta heritage Fund, would be appalled at the thought of this huge bounty being squandered to a degree that has seen the provinces debt grow to nearly $60 billion. His may well have that number quadrupled as a surplus and invested in the Alberta Heritage Fund if his original vision had been honored.

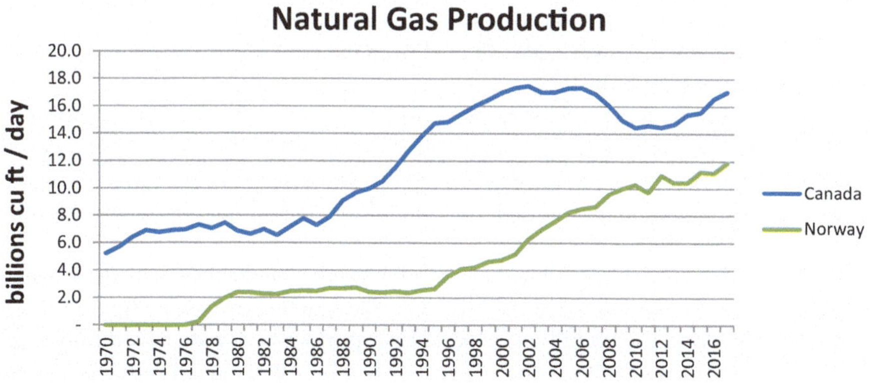

Fig. 8.2 Natural gas production history of Alberta and Norway (BP 2018)

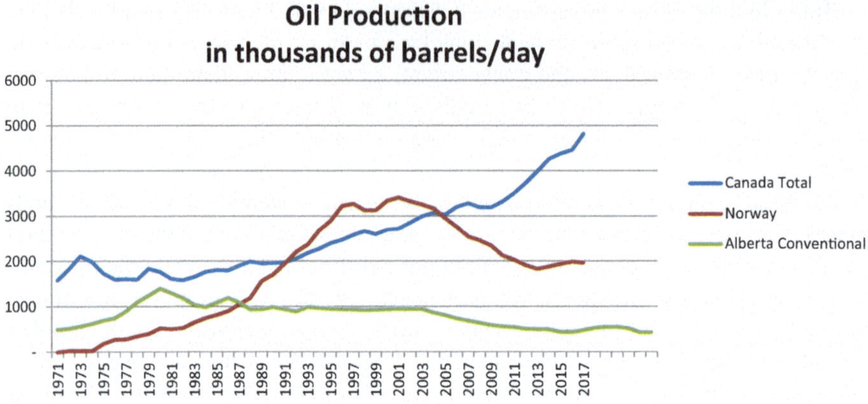

Fig. 8.3 Total oil production history of Canada, Alberta Conventional, and Norway (BP 2018; Hughes 2018)

The provincial government and the oil industry publicly set the legacy costs of oil field cleanup liabilities at $58 billion. Global News has reported that internal documents have come to light showing Alberta Energy Regulator Vice-president Robert Wadsworth's estimate in February 2018 that the cleanup cost of the Alberta oil patch may approach $260 billion, far higher than any liability amount made public by government and industry officials. He said that a "flawed system" of industrial oversight is to blame for the problem—but that ultimately—taxpayers would be on the hook to cover a portion of the costs.

From 1997 to 2017, Alberta produced a total of 24 billion barrels of oil that returned an average royalty of $1.20 per barrel. In the same period Norway produced 20 billion barrels, with no specific royalty, yet its fund results loudly speak for itself in Fig. 8.5. The amount invested by the Norwegian government is unclear as are the total amounts of federal and provincial subsidies directly paid to oil companies. The International Institute for Sustainable Development estimates Canadian government subsidies at the level of about $3.3 billion annually (IISD 2019). These amounts would have to be added to the legacy costs to determine the total out-of-pocket government expenses and deducted from the total royalty revenues to arrive at the net taxpayer profit or loss.

The oil sands, with their much higher costs, are a different story with an average of $1.20 royalties per barrel and total royalties of $41 billion. Remediation cost estimates from the Alberta Energy Board of $58 billion show a deficit of $17 billion currently, and with estimates as high as $260 billion, the oil sands may never turn a public profit (Fig. 8.4).

Legacy costs, which are today the cost of restoring the land to near its original health, are not clearly defined at the moment. The graph above uses a figure for oil sands remediation of $200 billion. Alberta Energy Board estimate is $58 billion to be paid by the oil companies within 70 years when they will almost certainly not be in the oil business. Still, at $58 billion, the oil sands in 2019 remains a money loser for the provincial government and the average Albertan even without including provincial and federal subsidies.

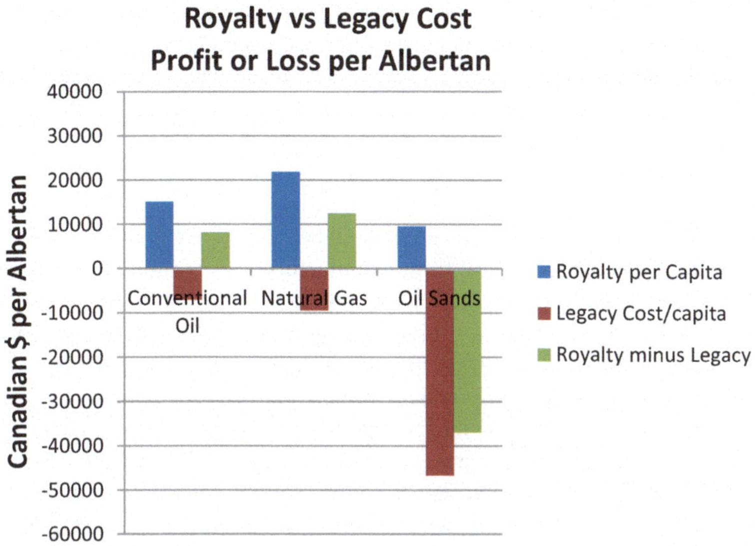

Fig. 8.4 Profit and loss broken down to per individual Albertan for the three sectors of the energy industry (Alberta Government 2018)

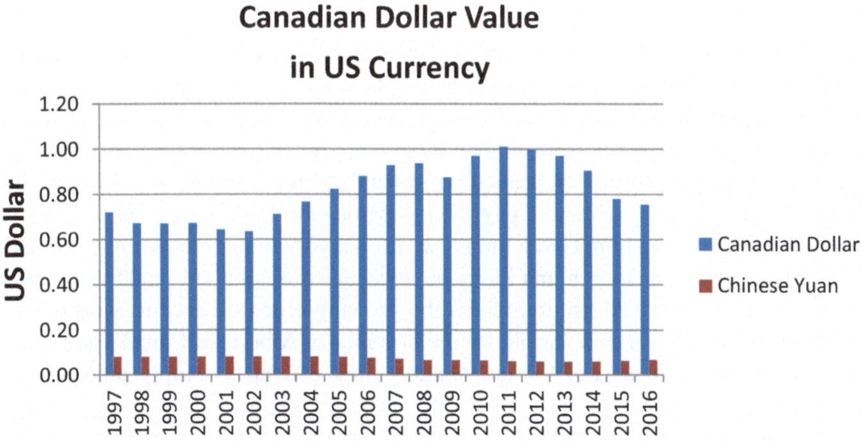

Fig. 8.5 Volatility of Canadian dollar which is effectively a petro-currency (BOC 2019)

In the Norwegian model, total system performance in terms of environmental standards is laid out at the beginning of the contract. Has Norway accumulated any legacy costs? There appear to be no records indicating that they have. A report detailing the total taxpayer dollars invested in the three fossil fuel sectors in Alberta and the royalties which have been received as well as the estimated legacy costs needs to be done very badly.

There are many differences in the quality of the resources of Norway and Alberta as well as in their cost of development but the social perspective in which development was conducted is perhaps the biggest difference. The Norwegians feel it is their oil and that any development should not hurt the other sectors of the economy but should benefit all. Alberta effectively turned over the control of development to private corporations without regard for the impact on other sectors or the rest of the country.

Turning the Canadian dollar into an inflated petro currency damaged many other sectors of the Canadian economy from manufacturing to farming and tourism via "Dutch Disease." Dutch disease is the term applied to the situation where one sector does well on the export front, resulting in a large net inflow of foreign currency causing the domestic currency to go up in its exchange rate relative to other currencies. This effectively makes all products in the country more expensive relative to their foreign competitors despite the fact that neither the products nor the processes have changed. With oil and natural gas comprising over 20% of total exports in many years, the Canadian dollar and the fate of many businesses outside of the resource sector rode inverse paths on the wave of commodity pricing (Fig. 8.5).

The graph above represents a fluctuation in the range of 60% in the prices Canadian exporters must charge to keep the cost of their products at the same level in foreign markets. Another aspect of the "resource curse" is the inability of many sectors to plan or invest given the volatility of the domestic currency.

Legacy costs are something governments deal with by kicking the can down the road if at all possible. The immediate issue for Alberta is cutting the high cost of getting its oil to market and the limits placed on sales by really having only one market into which to sell. Long-term debt may be building but, right now, thousands of high paying jobs are on the line. Consequently, government policy is driven more by panic and political opportunism than by good judgment or a long-term vision as subsidies in various forms are heaped on to help a struggling industry. Clearly the conventional oil and gas sectors pay their way, despite high legacy costs. The oil sands are another question. Given their low EROIs, very high greenhouse gas emissions and currently unknown, but likely huge, legacy costs, the oil sands are, to this point, a money losing adventure. Given the growing climate concerns worldwide, the oil sands may never turn a profit for the province of Alberta or Canada, and large investments made to access foreign markets may end up as empty pipelines to nowhere.

Government subsidies, through tax breaks, infrastructure construction, incentives and research, and lately even buying companies and rail cars, and offering to build pipelines are a separate question with estimates spread from hundreds of millions to billions of dollars annually. What is certain is that government laid the base for the oil sands with decades of research in developing the means by which the oil sands could be exploited and by mapping their extent. In other words, there have been huge government investments into the oil sands over decades and there remain huge, as yet unpaid, environmental costs. These appear to outstrip the large revenues to date (Fig. 8.6).

Fig. 8.6 Norway's sovereign "The Fund" vs. Alberta Heritage Fund performance comparison (The Fund 2019)

To sum it up: "For the last 10 years, when nothing went into the Alberta fund, and we put a lot of money aside, the profit went out of Canada," says Rolf Wiborg, a petroleum engineer who recently retired from Norway's public service (Ormiston and CBC 2015).

The Chinese Model

If the Norwegians have made the most of their opportunities with regard to their fossil fuel endowment, the Chinese have done just as good a job in building up their manufacturing and technological capabilities. And the formula does not seem to be radically different. National interests were spelled out in long-term visions for both the country and the people, and the state became active partners in the largest commercial entities.

In China, there were no power structures opposed to modernization. A very powerful central government laid out very clear goals, with specific 5-year plans. The government exerts tight control of the media and has very little tolerance for companies or individuals who begin to think they were more important than the interests the Chinese government feel it is necessary to serve.

Is It the System or Who Is in Control?

Does it matter if the system of government is one party or democratic or if the weighting of private/state enterprise is socialist or "free enterprise"?

It seems the ability of a government to develop clear goals and implement them is not restricted to one form of governance. What does matter is whether the government controls policy or whether interest groups control the government. That consideration, which impacts goals and strategies, seems to be more important than the system of government itself.

Communist China is a world leader in solar power and EVs. Communist Cuba, with much greater need and higher potential to become a solar powered nation, is a renewable backwater but now trying to improve. Communist Soviet Union suffered complete structural failure, and it is only in the past several years that the Russian Federation has committed to renewable energy. Of course, Russia has massive fossil fuel reserves, and renewables are a more difficult solution for this northern nation.

The United States holds itself to be the leader of the democratic free world and is one of the most technologically advanced nations on earth. Its progress on renewables is significant yet spotty. Nationally, it is lagging the most advanced nations.

Norway is a cohesive and democratic nation that protects its advanced social programs and human infrastructure with high tariff walls. It has shown itself capable of extracting the most beneficial arrangements for its people and society in the field of resource development.

Canada self-promotes on the basis of its advanced social programs developed in the shade of the American libertarian giant. It is a multi-party democracy yet heavily controlled by several extremely powerful interest groups. It appears to have no clear goals or coherent national strategy focused on citizen or environmental health. This makes it susceptible to strong influence by groups, both ex-national and corporate which do have very clear visions of their own interest.

Obstacles to Transition

It is in the interest of everyone on the planet to maintain the health of the planet's bio-systems. Yet there are interests who are forcefully opposed to implementing the changes necessary to save the very systems all human life depends on. In his book "Human Impact on Ancient Environments," Professor Charles Redman makes the point that when societies begin to establish trading networks with other societies, the common interests of the people diverge. When he examined the patterns of social development in several ancient Eastern Mediterranean cultures, Redman found they lived quite sustainably for centuries. It was only after they began to trade with other regions that their society experienced depletion of their resource base and civil upheaval.

The lesson he learned was that once a small group began to benefit from the trade in vital commodities, say timber, they promoted increased exploitation of the resource. It would have been in the long-term interests of the society as a whole to limit the harvest of resources to what the resource base could sustain. However, the traders had become sufficiently rich and influential that they were able to affect government policy to the extent that their short-term profit was given a higher priority than the long-term viability of their society.

Is successful transition a matter of the political system or of the power various interest groups hold no matter what the political system?

Financial and Political Interests

In the deal making, monetized political structure of today, it is even difficult to find a definition of what the interests of society are let alone design and implement policies to support them. We are geared to want more, and there are powerful interests whose business model can only exist when their market grows with more people consuming more goods year upon year forever. Like any forward looking interest group, they try to assure their survival and prosperity by locking in government policies which assures the strength and vitality of their market.

Fish cannot live out of water, and those who comprise the growth lobby cannot survive in a stable market where people may have as much as they could realistically want and are not demanding anymore. Stability and sustainability may look like progress to most people but it is, in very real terms, Armageddon for some of the most powerful interests in many western nations.

Any profitable company in any industry is in business to ensure its continued prosperity and even further growth. Name the current industry:

- Big coal
- Big oil
- Big media
- Big fast food
- Big corn
- Big finance
- Big real estate speculation

The list is long, and all want favorable treatment. Their interests might not coincide with the common good.

If the list of power wielding lobbies read as follows though, perhaps positive environmental and social action would be more certain.

- Big solar
- Big electric bicycle
- Big home insulation
- Big wind
- Big local produce
- Big heat pump

Unfortunately, these business models will not make billionaires out of their owners as easily as the consumer/growth economy has enriched its business leaders.

Oil Deep State

Kevin Taft has been around all sides of the Alberta oil industry and has served as an elected member of the provincial legislature. In his 2018 book "Oil's Deep State: How the petroleum industry undermines democracy and stops action on global warming," Taft details the means by which the oil industry bends public policy to suit its own interest. As he lays out the processes of manipulation, he reinforces the findings of Naomi Oreskes, the author of Merchants of Doubt, who documented the story of how a loose-knit group of high-level scientists and scientific advisers, with deep connections in politics and industry, ran effective campaigns to mislead the public and deny well-established scientific knowledge over four decades. Her examples included the issues of acid rain, cigarettes, fossil fuels, and sugar.

In the end, bad policy can survive only in an atmosphere of confusion and opaqueness which these "paid-for-results" consultants have delivered in many sectors. "Doubt is our product" wrote one tobacco executive and the same tools work for other major issues being controlled by powerful interests from coal smoke, to CFCs to limitless growth.

In Taft's book, he illuminates the tactics and techniques of the industry, its insiders, and lobbyists with active control of individual bureaucrats, academics, and activists are working together to pursue the oil industry's agenda. Taft sees the loss of Canadian democratic institutions to corporate influence and control, which is exactly the template of the resource curse Norway succeeded in avoiding and which the leaders of China are so determined to resist. In China's case, a more descriptive term might be "determined to seek out and destroy" interests working against the harmony of society.

But the oil lobby (deep state) effectively also includes consumers who constitute an embedded demand for high energy use and who are resistant to a change to conserve energy while paying more for it. High energy use, and mostly care-free energy use, has been built into our culture of expectations for over two generations whether in the areas of cars, travel, material goods or house size or the location of that housing.

Our current world was built by very cheap fossil fuels with the assumption that abundant fossil fuels would always be available and very affordable. Cheap, efficient natural gas warms our baby's bath water. The pro-fossil fuel campaign ads tell us so, and they are right. But profligate fossil fuel energy use will come to an end this century, and we need to build that uneasy realization into our planning.

In terms of metaphors, if humans had been building armed forces or a sports team, the teams we have built to this point have been designed for offense. And from the economic evidence, we would have to say we have been winning. But earth's systems play the long game, and the transition to renewables requires that we become almost 100% focused on defence. We have to begin to understand how "the other side," the natural world, works and learn how to live with the physical cards we have been dealt.

The Population Growth Deep State

Corporations know where their money comes from. Rather than just compete in a market as it exists, it is much more profitable to grow into an expanding and profitable market. Lobbyists try and facilitate this. Defensively, interest groups try to prevent restrictions on their business or cost increases levied in the form of environmental or labor standards. This becomes especially critical if they have foreign competitors who do not have to bear the expense of domestic standards.

Corporations want their markets to grow, and they want those markets to be profitable. They want to carry as few "external" costs as possible. Favorable government policies can assure all of these critical factors for many key industries. High-tariff walls would assure a healthy environment for manufacturers and farmers as well as virtually all sectors which create real product. The lower flows of currency in and out of the country would hurt some financial institutions and speculators. Retailers would be faced with fewer opportunities to carry import brands. Low-tariff walls favor traders, large distributors, retailers, and currency speculators.

High rates of immigration increase the demand for additional housing units. In Canada, 80% of this demand comes from immigration which increases to 100% of demand in many of the larger cities as there is a net outflow of Canadians. This creates a boom market for developers for the houses, condos, and commercial buildings they build, and it is equally lucrative for lending institutions. In 1969, Canadian banks held $14.2 billion in mortgages. In 2019, that had risen to $1.5 trillion, an increase of 10,800% in current dollars and 700% in constant dollars.

Part of this increase was due to the increased number of people buying house but a substantial part was due to housing inflation. Both were driven by the huge demand from many more people coming into the country than leaving. And the banks were ideally placed to exploit a market in which both the unit sales were increasing and the valuation of those units was inflating. This is great business if you are a lending institution or a speculator. But for the citizens paying for it, it is not a positive trend. In Canada per capita debt levels have quintupled in the past five decades in constant dollars.

Developers profit from this high demand for new homes and buildings by first building and selling them. They also profit from building and leasing them, particularly as valuations inflate, and they enjoy asset growth every year prices increase.

It costs money to influence government policy, and it can be spent by donating to politicians and by advertising in media outlets which influence politicians. The former was the subject of a 2009 study by Professor Robert MacDermid, "Funding City Politics: Municipal campaign funding and property development in the Greater Toronto Area."

There are a considerable number of sources who detail the spending by lobbyists in the United States. There are none in Canada, and Prof. MacDermid's study is the only one in the country to quantify the political donations of a specific industry. One explanation for this lack of oversight could be that Canadians are much more honest than people in other nations. But there could be other explanations as well.

Prof. MacDermid found in his study that "The development industry is by far the most important financier of the majority of winning candidates' campaigns in all municipalities with the exception of the (City of) Toronto and Ajax." In seven of the ten municipalities included in the study, corporate contributions made up more than 50% of all funding. In one case it was 77%. More than one third of the 132 of the winning candidates received over 75% of their funding from corporate sources, while three fourths of corporate contributions went to incumbents. In the case of Vaughn Township, "elected councillors frequently vote on development proposals submitted by those who financed their campaigns."

The study concludes that "Developers want councils that are favourable to rapid development and to their own development proposals and they spend accordingly. Unsustainable urban sprawl, high transportation costs, environmental deregulation and a weak sense of community that undermines political organization and representation are all traceable to pro-development councils and the provincial regulatory framework for urban development."

Large donations by banks and developers to federal politicians are designed to produce the same kinds of results experienced at the municipal level. In addition, media coverage (debt institutions and developers are large print media advertising buyers) is used to support the candidates likely to support wholesale growth. One hand washes the other.

This template with suitable variations can be applied across many industries powerful enough to afford buying access into the political decision-making process. Kevin Taft quotes the political historian Francis Fukuyama description of political donations as follows: "In a gift exchange the receiver incurs a moral obligation to the other party and is inclined at another time or place to return the favour .... A human being in any culture who receives a gift from another member of the community will feel a moral obligation to reciprocate." Political donations by individuals or corporations with clear interests that could be affected by the elected member are investments not a celebration of support for the institution of democracy.

Our nation's economic model has morphed from nation building to simple market growth and is now a valuation model in which the most wealthy and powerful acquire their wealth, not by producing it, but by molding government policy to produce the conditions under which the valuations of their assets inflate. The current model is not a sustainable one contrary to the models of hunter-gatherers, agrarian or even manufacturing tied to the output of real goods and in direct contact with the resource base may have been. Rather, it is a real estate, trading, and finance model where endless growth of the money supply is the only option, and there is no representation of the resource base or most of the individuals in the economy. Those entities have become merely market size, consumers and cheap labor suppliers.

At this point, it is apparent that it is the quick reaction, free-enterprise model which has been unable to change and, in fact, actively resistant to change, quite the opposite of what people would expect and more in line with the plodding failure epitomized by the early centrally planned economies. Open markets react instantly, whereas rigged or controlled markets are highly resistant to change. But markets cannot react to national interests or economic forces beyond their costing inputs.

Markets have to seek profit, and to date, western governments have been unable to establish the market conditions in which firms can profit from sustainability and the transition to renewable energy.

Decision Maker Background

The process of national learning is often painful. Geographer and historian Jared Diamond detailed the sectoral or national collapses in a number of societies in his best-selling book "Collapse." Some nations learned from their mistakes, and we able to recover while others simply could not halt the slide into collapse.

By the late 1860s, Sweden had depleted much of its forests to create charcoal to supply its iron ore production. The famines that struck Sweden at the same time resulted in the death or out-migration of one in seven Swedes, many to Wisconsin and Minnesota and the northwest USA. Sweden implemented policies to rebuild their forests sustainably, and learning from those disasters is built into its public policy. The same process took place in Japan in the late 1600s when forests were depleting and soil erosion and floods were becoming commonplace. The Japanese rulers implemented a number of measures which involved all levels of Japanese society in restorative efforts and means to establish sustainable forestry practices.

There is little in Canada's national consciousness to caution us against over-exploitation of our resource bases. Canada has had the depression of the 1930s and the cod fisheries crash of the early 1990s, but no famine induced population collapse remotely on the level of "older" nations. Our national mentality is one of a young nation growing and strengthening forever. The pain and education of a genuinely debilitating national crisis has never occurred here to the European majority, and we have no collective memory of the miserable conditions which drove the first pulses of colonists from Europe to the eastern shores of North America.

Certainly, The Science Council of Canada has often expressed policy concerns based on population growth and consumption rates in the context of resource limits and environmental health. However, their reports have been swept away by the constant bombardment of the corporate media for the need for growth as both a national goal and cure-all. Science has not been able to replace the common sense that comes with experience in Canadian policy since the Science Council was disbanded in 1993 as the growth lobby cemented its control of national policy. Science fell to its lowest point when federal scientists were forced to speak to other scientists through spokesmen appointed by then Prime Minister Stephen Harper's office. When they attended conferences, they were accompanied by these same chaperones. This was state control at its most extreme, but in whose interests did it take place?

The monitors are gone but science has gained no more influence in the policy formation process of the current Liberal government than it had in that of the previous conservative one. Hence our grow-forever mentality and the iron grip on national policy were held by the growth lobby ant its handmaiden, commercial economics.

Canada

The current leaders in Canada typically come from middle-class to privileged backgrounds and have had little exposure to the shortages or hardships their grandparents or great grandparents might have experienced. Few have had jobs which involve the production of real wealth but typically worked in law, education, or finance.

Canada is a new nation whose culture has been dominated by the concept of limitless frontiers and endless growth. Like the other new nations of Australia and the USA, whose entire histories have been taken place during rapid development and expansion, the business and government culture is based on the assumption that more will always be available. In the New World cultures, great forests have been cut down, swaths of farmland paved over and countless millions of tons of minerals extracted from the ground to support an ever growing population and increasing consumption levels without any significant penalty being exacted by natural system failures. Mature countries, which have gone through the cycles of boom and bust, are typically more cautious and forward looking.

Of course, there have been episodes of regional collapses. The crash of the North Atlantic cod fishery in Newfoundland, the dustbowl of the 1930s in the US Midwest and Canadian Prairies, and the die-off of the American chestnut in the Appalachians all drove regional out-migration but none of these have resulted in large national population declines or threats to the elites or system of government.

If, instead of development decisions being made by the descendants of European colonists, they were made by indigenous peoples whose presence on the land, approaching 15,000 years, had fused a working relationship with nature into their culture, the decisions made would certainly have been different. It is an open question as to whether the conventional oil and natural gas resources would have been developed or not, but if they had, it is likely it would have been done in a much more limited and conservative fashion with careful attention paid to environmental and social impacts. What is more certain is that the mining of the oil sands would not have taken place at all.

But the new Albertans are not at peace with the land, they are there to take what is possible and then, after several generations, move on. Since Alberta represents very close to the end of the chain of "endless" resources on this planet, "move on to where?" might be a relevant question, but it is something this culture has yet to be confronted with. The Canadian development culture has a 400 year history which emerged when the desperate refugees from a depleted Europe landed on the shores of a huge continent whose resources appeared to be infinite. As their technology evolved, and their disposable energy supplies grew, the illusion of endless development became more entrenched, and it is that culture that made the decisions on the means by which the Alberta energy resources were to be exploited.

The culture driving Alberta fossil fuel policies is an exploit-and-move-on culture. The exception to this, a thoughtful and moderate voice, was that of Alberta Premiere Peter Lougheed who established the Alberta Heritage Fund and believed the oil and gas industries should always be just a small part of a broad based Alberta economy. In his youth, Lougheed had witnessed the human and social toll the boom and bust oil took in Oklahoma and wanted to build in safeguards against such consequences in his province.

But as the power of the oil industry grew, so did its influence in the political system, and it managed to narrow the focus of government policy and shorten its time horizon.

The people making policy in Canada has had little personal connection with the resource cycles which affected the lives of their ancestors nor do they have any sense of boundaries their grandparents learned in the Great Depression of the 1930s. Certainly their grandparents, even if just as privileged, would not have shared their fascination with simple GDP growth. There is almost no scientific or engineering education represented in the Canadian political demographic as this would have been scrubbed out by the media and special interest filters of the electoral process. Consequently, policy making in Canada like that of the other "new" nations is informed by the optimism engendered by the rapid growth of the past and not by a studied examination of the real options which lie ahead.

Norway: How Did They Get It Right?

Norwegians have been on their land for almost as long as the First Nations have roamed the Canadian landscape. They have hundreds of generations of cultural learning that has taught them to live within the limits of nature's sustainable bounty. This process would have undoubtedly involved exceeding those limits and learning what the consequences of doing so were. By the time it became apparent in the late 1960s there was a great deal of oil in the North Sea, Norway had well developed fishing, timber, manufacturing, mining, and tourism industries. It was more than just a resource development nation, and it was a socially progressive nation with a broad-based economy and coherent national policies. Oil development was not going to be dealt with in isolation.

The Norwegians were afraid they would bring the "resource curse" down up on themselves and were trying to understand the implications of a flood of wealth from a new industry. They knew full well this flood of wealth would eventually dry up and that it had the potential to disrupt both their economy and society as such rapid wealth infusions had done to so many other countries. In the end, they laid out an economically rational and socially coherent plan. How did they manage to do that?

Given their culture and their apprehension, the Norwegians were very willing to listen to expert advice. They urgently needed someone with an overview of the problem, and one spring day, the right someone simply fell into their lap. Farouk al Kasim, an ex-pat Iraqi oil geologist with a few hours to spare in Oslo, walked into the offices of the Ministry of Industry. He ended up spending 3 months examining test data and concluded that Norway was sitting on "a doozy" of an oil field.

As he put it to journalist Lisa Margonelli, "Norway didn't know anything about oil." But leaders knew about the resource curse; they knew they did not want a tsunami of fast money and corporate influence to wash over their neat country. "It scared them" (Margonelli 2014). Al Kasim knew firsthand the control foreign firms and governments could extend over government policy through his work in the oil industry in his native Iraq. But getting rid of foreign influence was not really an answer. "If you simply replace international oil company monopolies with state owned monopolies, it's not an improvement," because such large flows of money concentrates wealth and power in the hands of a small minority. The plan he and a co-worker drew up called for the creation of a strong regulator in the form of the Norwegian Petroleum Directorate as well as Statoil, a national oil corporation.

The mandate of these two entities was to develop local expertise, provide jobs for Norwegians and make sure national interests were served by minimizing environmental impacts while maximizing profits for Norway. Norway would put up 50% of the capital and have a high degree of ownership yet the plan would foster competition. At first, oil companies balked at this kind of arrangement. But Norwegian negotiators remained firm, and after a year, the oil companies returned to work within the framework laid down by Al Kasim and Industry Ministry bureaucrats. This arrangement, driven by both government cooperation with the companies and competition between the companies, resulted in very low environmental damage and saw 95% of profits stay in Norway.

It allowed technology to be applied in concert with social objectives. Social context is vital in the application of technology, and as al Kasim put it, "One of the great myths of our time is that tech solves everything. It's not just that science hardly ever provides the right answer at the right time, he says, but that technology without a social vision is just stuff." In the end, Farouk al Kasim worked as the director of resource management for the Norwegian Petroleum Directorate for 18 years and was knighted by Norway's king in 2012.

After 40 years of this arrangement, Norway has clearly succeeded in avoiding the resource curse by creating many high paying jobs, extensive infrastructure, a highly educated population, and progressive social and environmental policies. For instance, there are over 900 road tunnels in Norway with a combined length of 750 km, the longest of which is 24 km (Transport Norway 2019). Tunneling, although extremely expensive, is technically possible in the stable rock of Norway's mountainous terrain, and it features lifespans of extreme length. It speaks to the long term vision behind these very expensive investments.

Norway has no public debt while the May 2019 Alberta provincial debt was \$58.3 billion or \$13,500 per Albertan. Adding that to the Canadian federal debt of \$686 billion or \$18,500 per Canadian produces a total debt of \$32,000 per resident of Alberta (Debtclock 2019). Municipal debt is not included nor is the public infrastructure deficit which sees Fort McMurray serviced by only a single highway despite having been the only way into the center of Canada's largest oil development. Fort Mac's municipal debt stood at \$385 million in 2017 or \$5700 for each of its 67,000 residents.

China

China has a system of meritocracy in which individuals with ability (and the appropriate political outlook) are given progressively more responsibility. China has a long history of this process as the civil service exam which could be taken by anyone who could read and write, regardless of their social class. This exam process was said to have originated in the Yao Dynasty circa 2300 BC and was copied to some degree by the British, French, German, and American governments in the nineteenth century. The effect was to gentrify policy making and remove, or at least substantially counter, the influence of powerful interests.

The current crop of Chinese leaders are well educated, some having PhDs in science and mostly have had experience running industrial or regional organizations before being appointed to the Central Committee. Currently there are seven leaders in the Committee. Their life experience and that of their families differ dramatically from those of current western leaders. They were all alive and possibly old enough to be aware of the trauma of the Great Leap Forward (1958–1962) and its accompanying famine and violent repression in which tens of millions of Chinese died. Their parents and grandparents would have lived most of their lives in a country awash in conflict and war with outside powers.

China has thousands of years of recorded history. Every Chinese learns about the famines, floods, natural calamities, and invasions which have been formative in their culture and worldview. Chinese leaders know there are limits to what the land can provide and the problems too many people present, because these limits have been made clear time and again through their history and they understand what social and political stresses occur when those limits are exceeded. Famines bring death and political upheaval, something which is an anathema to the holders of the doctrine that social harmony and loyalty to the party are the foundation of the new China. An understanding of the critical nature of resource limits is baked into Chinese culture.

Xi Jinping is the most powerful leader China has had in decades, to the point where his "Xi Jingping Thoughts" have been incorporated into the constitution of the Communist Party. When he was 13, student militants ransacked the family home killing one of his sisters. His father was imprisoned 2 years later and Xi was sent to work in a small, remote village. After several months, he ran away to Beijing but was arrested and sent to a work camp to dig ditches. From there he gradually worked his way up through posts of increasing responsibility and finally became Chairman.

Central to Xi's concept of the "Chinese Dream" is the glorification of self-made individuals who have helped themselves as well as the nation as a whole. It allows and promotes individual achievement and therefore differing levels of reward. Collective egalitarianism is still the focus of Chinese policy, but individuals are urged to pursue their own interests and be self-supporting. This could be viewed as a form of "free productive enterprise" in which all activities with a positive social outcome are permitted. "Productive" being the operative word, since it implies a net real wealth gain and activities aligning with the interests of the Chinese people and with the priorities of the Communist Party.

China's government may be too new, national goals too clear, and the control exercised by the Central Committee too strong for the development of any special interests powerful enough to sway public policy away from the priorities referenced above to have developed dominance. And the Chinese leadership has been very aggressive in bringing those with growing influence back into the party fold if they are deemed to have strayed from party ideals or if, as the Romans put it, have become "ambitious." This could be read as being too big for their britches or too eager to put their interests before the common good.

China does not lack for strong industries like manufacturing, power generation, or growth-based interests like real estate and banking/investment, but to date, these groups have not seemed to have been able to sway government policy in their favor to an untoward degree. Attempts must have been made though, and the fact that several of the current Central Committee members have made their reputations on fighting corruption perhaps indicates the presence of constant pressure from powerful interests to bias decision-making away from the core principles of the Party. It would appear the Party leadership is vigilant against the rise of power of groups whose interests differ from national interests as determined by the Party.

The concept of "ecological civilization" (Marinelli 2018) has now been designated as a leading development strategy by the Chinese government. The phrases "biophysical economics," "sustainable society," and "ecological civilization" can all co-habit the same sentence, whereas "infinite growth" simply does not fit into any discussion using physical or social metrics. It will be interesting to see if the GDP-centric media in the west describe the tapering off of GDP growth in China. Will the alarm be sounded describing the unraveling of the Chinese economy or will it be heralded as a positive step toward sustainability?

USA

The development history and the origins of the political class of the USA bear a lot of similarity to those of Canada. There are many differences in social attitudes and social policy but, in the end, the American government policy is geared to provide high growth to match the GDP performance of the past, and resource exploitation is still a major part of the economy as are construction and banking.

The American politics are famous for the influence of its lobby groups through direct political donations as well as public advocacy through think tanks and media organizations. However, most of these activities take place in public view, and most of the money flows are documented.

The most powerful American lobbies do exert a great deal of control over the US policy, but, for the most part, issues are argued strongly and in public forums by all sides. This contrasts to Canada, where major issues like growth and immigration levels are not discussed in public. Canada may, in fact, be the outlier of the developed world in this area. The business-as-usual interests may be in control of the US government policy, but the status quo is regularly challenged in the mainstream media.

Yet somehow, the fundamental problems of the pursuit of population and economic growth are left largely unmentioned.

Below are excerpts of an article by historian Joseph Ellis on the CNN website in which he addressed the problems of the country and the challenges which lay ahead with the partial context of the transition to renewable energy and the government structure needed to achieve it. Ellis presents the Green New Deal in light of the ideals of the American founding fathers (Ellis 2019).

There are pressing challenges facing American society that have thus far eluded solution. They are:

- *reforming our extremely expensive and inefficient health care system;*
- *reducing our currently unprecedented levels of economic inequality;*
- *rebuilding our aging infrastructure;*
- *and recovering our global leadership role to combat the existential threat posed by climate change.*

If we fail to address and resolve these problems, all talk of America as a "city on a hill" needs to cease.

The great advantage of the Green New Deal initiative is that it forces and focuses a much-needed debate about the role of government in defining our agenda as a people and a nation.

Socialism is a political theory based on the principle of government ownership of the means of production; in short, the abolition of private property.

What they (Green New Deal advocates) are proposing is a collective response to our common problems.

The pursuit of a collective good is the very essence of the Founding Fathers' vision for America.

It is impossible to become elected to high office in the United States without the backing of powerful and well-funded lobby groups.

Powerful Interest Groups

The form of government may lend itself to one type of approach of national policy formation, but how effectively the policy can be implemented is dependent on the ability of policy makers to maintain the democratic integrity of the process in the face of the strength and influence of the interest groups within the country. Every nation has powerful interest groups, and these have normally been formed during the surges in development catapulting one sector to prominence. The coal industry in the USA was followed by the oil industry and then by the industrial might of weapon suppliers, and the military itself after World War 2 finally elicited President Eisenhower's famous reference below to the Military-Industrial Complex.

The American President Dwight Eisenhower's 1961 speech deals with one specific power group, but there are others whose interest is always maintaining business-as-usual in a manner which produces a favorable environment for their business models (Eisenhower 1961).

In the councils of government, we must guard against the acquisition of unwarranted influence, whether sought or unsought, by the military-industrial complex. The potential for the disastrous rise of misplaced power exists and will persist. We must never let the weight of

> this combination endanger our liberties or democratic processes. We should take nothing for granted. Only an alert and knowledgeable citizenry can compel the proper meshing of the huge industrial and military machinery of defense with our peaceful methods and goals, so that security and liberty may prosper together.

"He who pays the piper calls the tune." Political donors influence policy decisions, which is why they make donations. Advertisers influence media coverage because media corporations want to provide a healthy market for their advertisers and are reluctant to bite the hand that feeds them. This is particularly so as the media have become "corporatized." In the past, the national conversation to halt advertising for cigarettes and alcohol was conducted in the pages and over the airwaves of media companies. And despite the very large and negative impacts the cessation of advertising would have on the media companies themselves, they facilitated this discussion which ultimately cost them hundreds of millions of dollars annually in lost sales revenue.

The mantra of the committed journalists in charge of these media outlets could be summed up by a quote by Arthur Miller, one of the most famous American playwrights of the twentieth century. "A good newspaper, I suppose, is a nation talking to itself" (Miller 1965). Few would confuse the politicized and commercialized news coverage of today with the intent of that comment from what seems like another age.

Who are the powerful interests which dominate today's policy making and do their interests align with those of the general populations of their nations? Possibly the groups listed as highly influential in public policy development offer unhealthy and unsustainable social and environmental choices but modification of their business models to support the long-term public priorities might not be possible.

Generally speaking narrow interest groups have interests which diverge from the public interest as follows (Table 8.1).

Table 8.1 Conflicting interests (author)

	Special interest group priorities	Public interest
Economic measurement	GDP growth—continuously higher dollar flows	Higher or stable incomes per capita. Size of economy is unimportant
Environmental requirement	Environmental considerations are externalities—costs which do nothing to enhance the bottom line	Environment is critical to health, quality of life, and future prospects of generations to come
Resource priority	Exploit as quickly as possible	Maintain high ratio of resources per capita for the long term
Time reference	Quarterly report, annual bonus, maximum 5 years view	Three-generation view for most families, nation needs 7+ generation view
National goals	Growth, strong markets for their sectors, business as usual	Quality of life, stability, progress for the individual, the family, the community, and the nation
Transparency	Corruption can only take place in an environment with low levels of information and high levels of confusion	Full information and clear national goals are critical to transparency and public policy in the national interest

Table 8.2 Time horizons for different groups (author)

Interest	Short term view	Long term view
Commissioned sales rep	Today	End of the month
Small retail shop	This week	End of the year
Med-sized manufacturing company	This month	5–10 years
Electrical utility	1 year	20–40 years
Nation	1 year	7 generations (2 centuries)

Different interests not only use different metrics as barometers of success, they also have completely different time frames upon which they build their plans and expectations. Despite the natural tendency of various interest groups to try to impose their priorities and their timelines on government decision-making, policy makers must be cognizant of the short- and long-term priorities that drive the national interest. Public infrastructure, resource depletion, and the energy transition have long wave dynamics and need to be dealt with accordingly (Table 8.2).

To sum up, the interests dependent on growth need to be bigger now and always, while the individual and public interests are in making things better. For those interests, the size of the economy is irrelevant, and in a resource-stretched world, further growth is a distinct negative for most people.

Way of Life Interests: The Soft Lobby

But Canadians, along with people everywhere, are not clamoring to use less energy, have smaller homes and drive smaller cars, or to walk more and fly less. They may see the necessity of saving the planet and understand it might involve changes on their part, but they are not lining up to demand the government cut their consumption levels. People almost universally understand that supplies of fossil fuels are finite and even that the end of the fossil fuel era might be within sight, but they have not parked their cars by the side of the road and started bicycling 40 km to work every day.

To maintain the core of our lifestyles, education levels, and health, fairly high levels of energy consumption is not a luxury but a necessity. Not only do we enjoy our way of life, one that is more luxurious and convenient than any in the history of humanity, but to even maintain a portion of this quality of life in many regions of the world, will still require very large amounts of energy.

People can be expected to resist changes as it affects their lifestyle in many areas, a few of which are listed below:

- Mobility
- Outdoor sports requiring transport
- Living comfortably in a cold climate
- Travel
- Luxury goods

- Large homes
- The anytime, anywhere lifestyle where activities can be engaged in as quickly as they are conceived. Currently, a cross country trip in an EV requires forethought that is never needed with an internal combustion car.

Change will not come easily, and therefore the longer change takes, the more smoothly it will go. Since Mother Nature will not be changing her schedule to accommodate us, the amount of time we have depends on how early a start we make.

Structural Time Drags

In terms of reducing carbon emissions and avoiding acute resource depletion shocks, change cannot come quickly enough. But deciding what has to be changed quickly and what can be changed quickly has to be arrived at by estimating the rate of carbon reduction payoff against the cost of modifying or making obsolete, existing infrastructure. Getting the most out of our current layers of infrastructure can help assure asset upgrading does not become large-scale asset destruction for the majority of people either privately or in their business lives. Depending on the lifespans of these various machines, complex systems, and structures we have already built, there is a lot of time drag involved in changeover whether it be electrification, modification, elimination, or replacement.

Changing people's habits and expectations is impossible to quantify but changing their habits and estimating the order in which to best change infrastructure is a task suited to educated guessing and very sophisticated models. Below are a number of infrastructures and the roughly estimated time it would take to replace, modify, or eliminate them as a means to reduce emissions and energy use (Table 8.3).

The above table is an outline of the issues involved for bringing the various forms of investment into compliance with the transition to renewable energy or sustainability requirements. It does not mean the scrapping and rebuilding of infrastructure. In the case of agriculture, at this point, fossil fuels will still be required—see the page on John Howe's work in the next chapter. Also, any commercial, emergency, or military air transport will have to use fossil fuels as battery densities technology (kWh of energy to kilogram of weight) is far short of that required to make long distance, high speed, high carrying capacity travel practical.

The amount of energy in L or kg of jet fuel is approximately 10 kWh, and the efficiency of a jet engine is 70%. A Boing 747 burns 10,000 kg of fuel an hour and can carry 238,604 L of fuel (238,604 kg) which would give it a range of about 14,500 km. These figures would mean that the energetic content of the fuel is 2.4 million kWh. The 75 kWh battery pack of a Tesla Model 3 weighs 478 kg for a weight per kWh of 6.4 kg. Batteries of equivalent energy content to the jet fuel would increase the weight of the "fuel" on the 747 from 238,000 kg to 15 million kg. It might be easier to put wings on a super tanker and try and get it airborne.

Table 8.3 Infrastructure conversion time estimates indicate potential for stranded assets (author)

Existing infrastructure	Life span	Time to "transition"	Action
Private automotive fleet	17 years	25 years	Convert passenger and light vehicles to 100% EV with extensive charging network
Housing stock	110 years	50 years	Upgrade all residences to be energy neutral or energy positive
Heating and cooling equipment	20 years	15 years	Replace all inefficient heating and cooling systems
Recreational vehicles	20 years	10 years	Lower usage of energy intensive, discretionary vehicles and electrification of some
Commercial buildings	80 years	25 years	Upgrade to energy positive
Mining equipment	30 years	20 years	Does the technology exist to replace fossil fuels in mining applications
Agricultural equipment	30 years	0	Technology (batteries) does not exist to electrify farm tractors at this time
Air transport	25 years	0	Technology (batteries) does not exist to electrify commercial aircraft at this time
Travel	n/a	15 years	Air travel drastically reduced, high-speed train and electrified ships for long distances
Where we live	n/a	35 years	Stop sprawl and consolidate in tighter communities
Public transportation networks	50 years	35 years	Extensive coverage of major routes in and between communities
Food distribution systems	n/a	30 years?	Imported food cannot be replaced any faster than local food production can be increased
Trading networks	n/a	20 years	Reduction in overall consumption and increase in domestic manufacture

Even taking into account the higher efficiency of electric motors versa the combustion-based jet turbines, it will be many years before battery technology allows practical commercial flight to take place (Flight Deck Friend 2018).

Assuming there will be no technological miracle fix just in the nick of time, we will have to both work with the tools we have and plan on using them for the foreseeable future.

Socialism or Capitalism: Does It Matter?

Human systems evolve to meet needs or to consolidate power or represent interests. Possibly the starting point matters less than how well the system evolves to meet its goals. Systems of government which do not allow clear goals to be developed or which use only the goals of the past or of narrow interests will ultimately be replaced.

Whatever form of government can manage the transition to renewable energy will be one that continues to exist. There is no restriction on the number of successful methods, but if a form of government is unable to provide, food and heat and some transport for its population, it will be replaced by a form that can. If one form of government is much more successful at providing the necessities than another, in this well-connected world, it is certain that eventually variations of the more successful form of government will be fairly universally adopted. To paraphrase Khrushchev, "Let us better compete at who can provide the best life for its citizens in the new world of renewable energy."

It is apparent that the form of government is not necessarily a restriction on the ability of a nation to solve its problems. Rather, the determining factors are the power of vested interests, the clarity of national goals, and the openness of the national conversation.

References

Alberta Government (2018). https://www.alberta.ca/royalty-history.aspx#toc-0

Bank of Canada (2019). https://www.bankofcanada.ca/terms/

BP (2018) Statistical Review of World Energy. https://www.bp.com/en/global/corporate/energy-economics/statistical-review-of-world-energy.html

Chisholm P (2018) The Food Insecurity of North Korea

CIA (1962). https://www.cia.gov/library/readingroom/docs/CIA-RDP73B00296R000200040087-1.pdf

Coles P (2019) Photographer

Debtclock (2019). www.debtclock.ca

Eisenhower D (1961). http://avalon.law.yale.edu/20th_century/eisenhower001.asp

Ellis J (2019). https://www.cnn.com/2019/04/18/opinions/green-new-deal-not-socialism-ellis/index.html

FlightDeckFriend(2018).https://www.flightdeckfriend.com/ask-a-captain/how-much-fuel-does-a-jumbo-jet-burn

Ho, MW (2008) Organic Cuba without fossil fuels

Hughes D (2018) Canada's Energy Outlook. Current realities and implications for a carbon constrained future. https://energyoutlook.ca/

IISD (2019). https://www.iisd.org/faq/unpacking-canadas-fossil-fuel-subsidies/

Margonelli L (2014) How Farouk al Kasim Saved Norway from Its Oil

Marinelli M (2018) How to Build a 'Beautiful China' in the Anthropocene. The Political Discourse and the Intellectual Debate on Ecological Civilization

Miller A (1965). www.goodreads.com/quotes/28419-a-good-newspaper-i-suppose-is-a-nation-talking-to

New York Times (2010). https://www.nytimes.com/2010/12/16/opinion/16iht-eddikotter16.html

Ormiston S, CBC (2015) Norway's sovereign wealth fund holds lessons for Canada

The Fund (2019). https://www.nbim.no/

Transport Norway (2019). https://www.vegvesen.no/?cid=1092405927948&pagename=vegvesen%2FSVVartikkel%2FSVVartikkelMedRelatert&c=SVVartikkel

World Data (2018). www.worlddata.info

Chapter 9
The Transition from the Ground Up

Abstract Transition means a change from one state to another over time. We are currently in a state of high fossil fuel dependency but will arrive at or be forced into a state of low fossil fuel dependency by the end of the century. How we arrive at the final state of renewable energy dominance is a matter of determination and time. If we choose to start now to aggressively change our pattern and level of consumption and invest heavily in renewable energy for the next 30 years, it is quite possible we will enjoy a relatively successful transition to renewable energy. If we wait until fossil fuel supplies begin to destabilize or for climate to become dangerously erratic, a smooth transition and a soft landing will likely be impossible. The goal of transition must be to preserve social stability and the highest possible levels of planetary health, not to simply keep growing economically. The more time we have to learn and adapt, the more successful the transition. Fossil fuels are energy dense, while renewable energy is diffuse and requires great expanses of infrastructure. Once past determining the amount of renewable energy we need, how much physical infrastructure, spread over what area will be needed to produce it? How much fossil fuel will be required to create that investment?

For any nation to successfully transition to renewable energy, the national consciousness will have to be very clear on the need to undertake the transformation. Nations changed slowly in the thousands of years before fossil fuels, but since then, particularly in the New World, re-invention has been almost continuous. Canada has moved through the stages of dominant renewable resource sectors from fish, to furs, to timber, to farming. These were succeeded by the industrial-based power structures of railroads and manufacturing. Then the non-renewable resources of mining and energy ascended to prominence.

All through this process, real estate and banking were building in size and influence; until now, they are the dominant power group across the nation, rivaling fossil fuels even in the province of Alberta.

All of these groups have been based on commercial enterprise, and their rise has coincided with the growth of the nation and, generally speaking, with the welfare of the citizenry. Change was led by these commercial interests because their power was based on the increased benefit they brought to society. Government could serve as

J. E. Meyer, *The Renewable Energy Transition*, Lecture Notes in Energy 71,
https://doi.org/10.1007/978-3-030-29115-0_9

overseer but, with the possible exceptions of the railroads, did not have to function as the source of inspiration and drive. The transition to renewable energy will be the first shift in power that is not done in lockstep with the rise of specific commercial interests. In this, it is very similar to a war. Critical national interests are determined to be under direct threat, and the nation as a whole turns its attention to dealing with the threat no matter the level of personal sacrifice or commercial displacement.

However, the time frame is completely different from that of a military war involving clear and immediate threats which might impose devastating impacts on the populace if not prosecuted successfully. Fossil fuel depletion and destructive new weather patterns do not wear uniforms, and their threat is both nebulous and possibly temporally distant. The end result of not making a successful transition to renewable energy will be every bit as devastating as losing a major war, but the negative impacts will be different from war damage and take place over a much greater time span.

The negative effect of this is to cloud national perception and delay the implementation of necessary changes, while the positive aspect is that we have the time to change without threat of immediate destruction. The de-carbonization of our society can only take place through nationally coordinated action over many decades. Commercial interests simply do not have the breadth of interests, the scope or the planning horizon to be able to successfully lead a process of replacing nearly ideally cheap and abundant fuels with more expensive fuels that are less abundant and, in many ways, more difficult to use.

Time Compression from Transition to Crisis

Does this over-dramatize the issue? Terms like war, threat, danger, and destruction really seem out of place in our buttoned down digital world in which a satisfying snack is never more than moments away and we can change the temperature of our surroundings in minutes. We are in control. World War II lasted for 6 years and devastated Europe and China and Southeast Asia and the Pacific. That is where the war was fought. Many other countries from other regions participated but did not suffer damage directly. If the impacts of fossil fuel depletion and a global temperature rise of 4 C took place over 6 years, it is fairly certain that the devastation experienced across the entire planet would be far worse than that the Europeans and Asians experienced 70 years ago.

What would be the impacts of the loss of 80% of our energy supply over 6 years? Global food production would drop dramatically. Trading networks would collapse as would the financial system. Very few would have the ability to heat or cool their homes. Transport would be restricted largely to electric train, bicycle, and foot. All of this would happen with a very rapid increase in global temperatures as a backdrop, quite possibly resulting in increased energy consumption. More frequent and more powerful weather systems and ocean levels several meters higher would rearrange a good deal of our society's infrastructure. Droughts, forest fires, and floods would be almost constant threats.

The outcome would be negative for virtually everyone. Some climate models indicate that climate could change dramatically over a period of a very few years. The results would be extremely destructive as we simply are not prepared for that kind of event. However, even the most pessimistic models indicate we have at least decades of oil and natural gas left (Mohr 2015). We do have decades to invest our fossil fuel stores in building the renewable energy structure that will allow us to mitigate the worst of effects of climate change while eliminating our need for fossil fuels themselves in a tapering off withdrawal. But this would involve diverting energy from consumption to investment, a slight twist on the commercial economics term of "guns or butter" where a nation has to choose between two options when spending its finite resources. And the investment will be enormous.

Our Clock Ticks in Real Time

The scientists of the IPCC give the global community until 2030 to be well into the implementation of fossil fuel use reduction and replacement. Failure to meet that deadline will leave us open to the vagaries of a climate system becoming increasingly unpredictable, dynamic, and destructive. In that time frame, the proposed Green New Deal could not be put in place even if governments placed their nations on a full war footing. But, given determined efforts to reduce carbon emissions, certainly major strides can be made and, a solid foundation for weaning ourselves off fossil fuel dependence can be built.

And "dependence" is the operative word with many parallels to its use in addiction recovery. The difference between fossil fuels and someone's drug of choice is that the drug user can go "cold turkey" and recover, whereas a society stopping fossil fuel consumption "cold turkey" would produce social chaos. For example, our agricultural systems are very dependent on fertilizer and mechanization. "Going back to the land" simply is not an option that will feed more than a fraction of the world's current population. We cannot survive without fossil fuels at the moment, but we have to make sufficient changes to allow us to put them largely behind us.

Addressing any dependence problem involves the same process. Here are the five steps to ending dependence on either drugs or fossil fuels or "transitioning to renewables" as developed by CRC Health, a California addiction recovery clinic.

- Awareness and Early Acknowledgment
  - This first stage is marked by a growing awareness that there is a problem. Few experiences are as essential to an addict as the moment when he or she shifts from denial to a willingness to make a change.
- Consideration
  - The second stage of the recovery process involves a shift from awareness to action

- Exploring Recovery
  - Moving past denial, the addict is motivated to overcome his or her addiction and begins taking small steps such as exploring the concepts of moderation and abstinence.
- Early Recovery
  - On the positive side, addicts in early recovery have not only stopped using the substance to which they were addicted, but they have also begun learning how to remain drug-free for the long term.
- Active Recovery and Maintenance

They have also learned that **they will need to continue to work hard for the rest of their lives. Recovery** is **a complete transformation of mind, body, and spirit**.

For any nation, the sooner all levels of government, in all regions, arrive at the conclusion that ending fossil fuel dependence is their top long-term priority, the sooner the nation, as a whole, can make progress. Going green and sustainable means a new life with new visions of prosperity for everyone. The fossil-fueled past cannot be relived on a renewable energy budget using depleted resource bases to divide up among a growing population.

Developing a strategy will be easier for some nations than for others. For those countries with a recent history of nation-building programs, assembling the consensus to embark on the transition to renewables will not involve any disruptive change to the national conversation or the way policy is determined. For nations focused on market-building benefitting specific interests, the process of developing transition policy may well be very difficult because it will involve the loss of influence for some powerful voices. Beyond influence, it will certainly involve their experiencing economic decline.

Nations have been led by surges in sectors of the economy which may have lifted all boats. This catapults corporations and financial interests into positions of power and influence for as long as their sectors remained healthy and dominant. These groups base their decisions on their own interests which may have coincided with the common good for a period of time. But national interests involve time frames, cost issues, and moral judgments which simply lie outside the operating sphere of personal and corporate decision-making. The responsibilities of a government are far different than those of a business leader, and governments must assure that decisions made within their purview are made by government in the national interest and not by other, more narrowly focused parties.

In Canada's history, the fur trade gave way to fisheries, timber, and farming as the basis for power and influence. In 2018, the fur industry makes up less than one tenth of 1% of the commercial economy. Things change. Railways built Canada and concentrated power in the hands of a few but have now stepped back into the role of an industry just fulfilling its transportation job. Currently real estate, banking, fossil fuel energy, and construction wield the greatest political power. In a sustainable society, where simple commercial growth has been left behind, these sectors will recede back into a normalized supporting role as new leading sectors provide the impetus for progress. However, unlike the relatively smooth transitions which took place over Canada's history, the displacement of the current power groups will not be taking place by the force of an even larger economic wave; it will be taking place because the nation feels it must change direction. Are the existing power elites likely to loosen their grip on the reins of market-making gracefully?

Build Solid National Structures

National priorities and standards have to be set by national governments, but systems have to be built from the ground up regionally and sector by interdependent sector. Assembling the parts so that they grow together requires the understanding that they all will not transition at the same speed. Blanket standards for fossil fuel reduction may even prove to be counterproductive. Hence a flexible push like higher fuel prices and carbon taxes can maintain the pressure to change while allowing flexibility. A coordinated program of incentives and hard standards could be applied to sectors with no concrete obstacles to the process of change, whereas sectors facing the lack of viable technological alternatives could be given a different set of targets.

This requires a comprehensive energy map to identify critical systems and their energy flows. How much and what kind of energy does this sector use? How easy will the transition to electricity be? How critical is this sector to the functioning of the economy and general welfare? The table below sets out the amounts of energy different sectors use and how easily they might be converted to electrical energy from renewable sources (Table 9.1).

Sectors requiring very large energy inputs, particularly those with mobile machinery like farming, mining, and air transport will be much more difficult to electrify than many other sectors which are essentially stationary users of energy in buildings. Education, health, services, accommodation, and residential pose few problems to the electrification process since they are already electrified to some degree, and their fossil fuel infrastructure just has to be gradually replaced over time. Passenger transport is very easy to electrify as the technology to do so exists today. Conversely, the limitations of battery technology will keep viable commercial electrified airplanes on the ground for the foreseeable future.

The food chain is made up of many steps, but the first step is the actual production of crops and farm tractors will remain fossil fuel based for decades due to their high energy consumption and need to react quickly to unpredictable crop and weather conditions demanding periods of intense use. Forestry and fisheries have somewhat the same usage profiles and need powerful machines with a great deal of storage to carry out their work at the right time and safely. For the foreseeable future, this means fossil fuels.

Table 9.1 Ease of conversion to electric of various economic sectors (author)

Sector	Ease of electrification	Demand of total energy	Technology available?
Agriculture	2	Low	No
Forestry, fishing, mining	3	Medium	No
Energy	1	High	No
Utilities	7	Low	Yes
Construction	4	Low	No
Manufacturing	6	Medium high	Yes
Residential	8	High	Yes
Services	8	Low	Yes
Wholesale retail	9	Low	Yes
Transportation—ground	9	High	Yes
Transportation—air	1	Low	No
Transportation—water	5	Low	No
Finance, insurance, real estate	9	Low	Yes
Education	9	Medium low	Yes
Health social assistance	9	Medium low	Yes
Information technology	10	Low	Yes
Accommodation services	9	Medium low	Yes
Public administration	8	Medium	Yes

Underground mining may well lend itself to electrification, but surface mining, with its much more extensive work areas, particularly sites such as the oil sands, will take some leaps in technical development before fossil fuels are able to be banished from the jobsite. Large-scale construction projects like dams and road and railways have much the same profile of energy demand as surface mines.

Of course, if carbon emissions become a paramount consideration and rapid electrification does take place, and oil sands development will likely cease. In that case, overall energy consumption in Canada would drop by over 10% and greenhouse gas emissions by closer to 15%.

Energy in Food Production

What is involved in reducing a sector's fossil fuel consumption? Food is the top priority for governments, but few in the west have had to address availability issues for almost a century. In Canada, this was last a concern during the Great Depression of the 1930s. In a far more localized world, the steps in food production and delivery will be altered significantly from what they are now, but here is a good overview of the food chain as it currently exists in most western countries.

From Lester Brown, the environmental activist and author, we have the following condensed summary (Brown 2008).

- "Tractors use gasoline or diesel fuel. Irrigation pumps use diesel fuel, natural gas, or coal-fired electricity. Fertilizer production is also energy-intensive. Natural gas is used to synthesize the basic ammonia building block in nitrogen fertilizers. The mining, manufacture, and international transport of phosphates and potash all depend on oil."
- Fertilizer accounts for 20% of the US farm energy use.
- Irrigation, more necessary in an increasingly drought prone world, requires 19% of the energy budget of an American farmer.
- 14% of system energy is required to move food from the farmer to the end consumer.
- 16% of food system energy use is devoted to canning, freezing, and drying food, from frozen orange juice concentrate to canned peas.
- The shipment of fresh fruits and vegetables over vast distances by air is extremely energy intensive.
- Food miles represents the distance that food travels with fresh grapes in winter typically coming by plane, almost 8000 km from Chile, almost 5000 miles. Fresh produce moves out of California by refrigerated trucks.
- Packaging accounts for 7% of food system energy use.

According to Brown, the most energy-intensive segment of the food chain appears to be the kitchen. Counter-intuitively, more energy is used to refrigerate and prepare food in the home than is used to produce it in the first place. The big energy user in the food system is the kitchen refrigerator, not the farm tractor. While oil dominates the production end of the food system, electricity dominates the consumption end. Pimentel (Pimenetel and Pimentel 1979) might disagree, but clearly the energy devoted to food storage and preparation in the home is significant.

Once the clinical analysis of the energy flows of the food system during and post oil era, one has to ask how healthy and how appealing is the resulting diet? Can local substitutes be grown to take the place of a rich variety of far flung food sources? Can the many expensive and energy intensive foods which have become de rigor be replaced locally? Given the importance of food in our culture, adaptation to changing conditions will certainly be rapid with small, specialized local growers jumping into opening market niches.

Being able to lay out the lifecycle of all industries will underwrite good transitional planning. National strategies are essential, but the work to identify workable initiatives across all sectors will come from small entities and individuals. The residential sector is a large energy consumer and efforts from individuals such as those below are helping to make the transition real.

Retired But Useful

If the world is able to step off the fossil fuel high-speed escalator without breaking a leg, it will be due to the efforts of people like John Howe (1935–2018) and Art Hunter. A great number of individuals are working on the issues John and Art have been addressing for a good part of their lives, but the efforts of these two are well worth highlighting both for their accomplishments and for the unique perspectives they offer.

They have demonstrated a high level of personal initiative and social responsibility in addressing national needs and have used their lifetime of education and experience to step up and put their shoulders to the wheel of the transition to renewable energy.

John brought a common sense practicality to the picture of life after oil—he tried to connect the dots of our current life with the ones we will be living in 100 years. He suggested by rationing now and cutting our consumption now, we can ease back into a sustainable life and he put his thoughts into the useful metrics of per capita resources.

Art is bringing science to domestic and neighborhood energy flows. His purpose was not to build a clean sheet, energy-positive house but to take an existing house and design and document a complex system that would monitor and optimize the energy flows within it. The learning that would fall out of this effort could then be condensed and packaged into a form that would be applicable to every house. The solar energy and geothermal systems he has installed in his house are integrated and closely monitored. All good decisions start with good questions that result in good data. Art is supplying the data and the framework for designing software to integrate and optimize energy flows in a home and in a home's connection with the grid. For both Howe and Hunter, their projects were about getting good numbers and producing a framework upon which others could build.

John Howe

In his own words, "I am a retired-engineer, farmer, speaker, and author who has focused on every aspect of this subject, especially searching for quantitative facts and possible solutions. Initially, I thought I could help by building and testing solar-powered cars and tractors as examples of successors for the petroleum-fueled machines we take for granted today, thereby choosing www.solarcarandtractor.com. The website now attempts to educate the main-stream public about all issues related to our urgent need for action."

Howe's mainstream work history began designing distribution transformers for GE and ended with his work at Head Ski Company designing skis and tennis racquets. Howe put head skis on the map in the 1960s with the new process of aluminum/epoxy laminates and took them into the tennis racket market with a radical composite design, one of his 12 patents. He then turned his attention to larger issues and fought long and hard to build public awareness of the risks of our dependence on fossil fuels. He was a critical thinker who strove to make the world a better and more sustainable place.

Howe was determined to bring the issue home to people and policy makers and found expressing the oil availability issue in per capita terms and the oil age in

Fig. 9.1 Rudimentary solar powered vehicles from the ground up (Howe 2014)

human lifetimes i.e. two lifetimes. The time remaining for us to deal with the two big problems the inevitable end of oil and increasing population was a critical consideration to him. He pointed out the recklessness of increasing our dependency on a resource which is limited and brought the question of recyclability of batteries to the fore. Also, in a renewable world, how local is local? What distances are practical and energetically viable? Howe covers this many other considerations involved in the gradual removal of fossil fuels from our lives in his publication "The End of Fossil Energy and per Capita Oil" (Fig. 9.1).

Energy budgets can become quite complex but here are a few statistics that would be of use to anyone maintaining a rural property or farm post-oil. Howe's home featured a 4 kW capacity solar PV array that he rated at a 14% capacity factor. This yielded an average 400 kWh/month "energy budget" which could be allocated as follows:

- 200 kWh for personal transportation in a battery electric vehicle (bev) similar to the MG Midget. This would be enough for up to 1600 miles (2560 km) of travel or 55 miles/day. (John gets 7 miles/kWh at 30 miles/h on level ground with his MG. Commercially available EVs use about 0.2 kWh/km at 100 km/h. Lower speeds would reduce this but probably not to the 0.08 kWh John has experienced with his MG Midget.)
  - During farming season, the 200 kWh provides up to 20 h of tractor power (at 10 kW or 14 hp). This is enough power and energy to plow and harrow about 5 acres of land for planting.
- The 200 kWh remaining could energize the following household items:
  - 85 kWh for hot water heating (preferably augmented by thermal-solar), enough for one shower or bath every 2 days.

- 15 kWh for full-time of a high-efficiency, full-size (12 ft^3) refrigerator.
- 45 kWh for 30 h (1 h/day) of cooking on a hot plate.
- 12 kWh for 4 h/day using the TV, home entertainment, and computer.
- 8 kWh for about 80 h with the washing machine.
- 12 kWh for 720 h of lights (2000 h with LEDs).
- 12 kWh for typical water pump usage.
- 11 kWh miscellaneous power tools.

  Say 5 h of use with a 2 hp electric chainsaw.

The total household monthly budget was 400 kWh for everything including minimal transportation. A 1.5 kW array on a good sunny day would provide enough energy to plow 11,000 ft^2/day or 0.25 acre or one tenth of a hectare. Basically John Howe energy quantifies many of the daily tasks required for farm and rural life. 400 kWh a month, even if allocated completely to heat would not put much of a dent in a Maine winter, so it is safe to assume that his two chimneys experience a great deal of wood fire smoke passing through them over the course of a year (Fig. 9.2) (http://www.solarcarandtractor.com/pieces/Ch5.pdf).

Howe knew what area of land his 15 hp tractor could plow with the 1200 lb stack of lead-acid batteries he built into it and how long that battery stack would take to charge with a given array of solar panels.

His legacy is the awareness he has engendered in thousands of people around the world through his publications (downloadable from his website) and by his leadership. John Howe's hard numbers and broad perspective will be used to inform practical decision-making at all levels well into the future. The issue of battery recycling

Fig. 9.2 A man plowing his field. Sustainably, how much can a lead-acid battery tractor plow? (Howe 2014)

is reviewed in detail in a 2019 study with the development of 10 principles for making batteries sustainable and efficient as possible. The paper is referenced by Design News in their coverage of the issue (Design News 2019; Arbabzadeh et al. 2019).

Art Hunter: As Close as We Come to a Rocket Scientist in Canada

Dr. Art Hunter was a member of the Telesat Canada's spacecraft design team for Anik A and project manager for the Canadarm project at the National Research Council of Canada. Hunter's expertise spans IT to Tunnel Boring Machines to "Condensed Matter Nuclear Science" (look it up). His major "hobby" now is the conversion of his standard suburban bungalow into the "Fossil Energy Freedom Project" which he expects will generate, store, and direct sufficient electrical power in a microgrid, to provide heating, cooling, and urban transportation without use of ANY fossil fuels during all seasons.

Whereas virtually every sustainable living project's first objective is to immediately reduce energy use, the first stage of Hunter's project is to quantify and understand the thermal and electrical energy flows in his house and electrical and geothermal system systems. Once the patterns are clear, he will go about developing an AI system to manage the flows with the aim of reducing the overall energy demand.

Hunter is quick to point out that 82% of his energy use is thermal which is pulled from the ground source heat pump. Unlike the typical geothermal system which has a one-way flow of heat into the house, Hunter's system stores excess house heat back in the ground. The process works as heat dumping (air conditioning) the summer months as well. The overall efficiency of the geothermal side of the project seems to be extremely high, but several years of data will be needed to determine how high. And Hunter's sophisticated data gathering capability will undoubtedly determine *exactly* how high.

The electrical side of the project sees power being generated by the solar panels and either used in the house as electricity, deposited in the grid on a net metering program or turned into heat and stored in the ground. The geothermal system and the electrical net metering work in a similar manner. Excess energy is sent out of the house and pulled back in when needed. The difference is that when the geothermal heat is pulled back in, it is free, whereas using the electrical credits built up with the local hydro utility results in a charge equal to 50% of the regular price. The complex and well-maintained utility infrastructure does not come free and using it as storage incurs a cost. Ignoring the initial cost of the geothermal "infrastructure" in his side yard, thermal energy storage is free. Hunter lives near Ottawa, and Ottawa is cold. His results have been very encouraging.

But there are both energy and financial costs to making the kind of investments both Howe and Hunter have built into their lives. Clearly, these are built-from-scratch systems and therefore extremely expensive, but, even at that, there has been

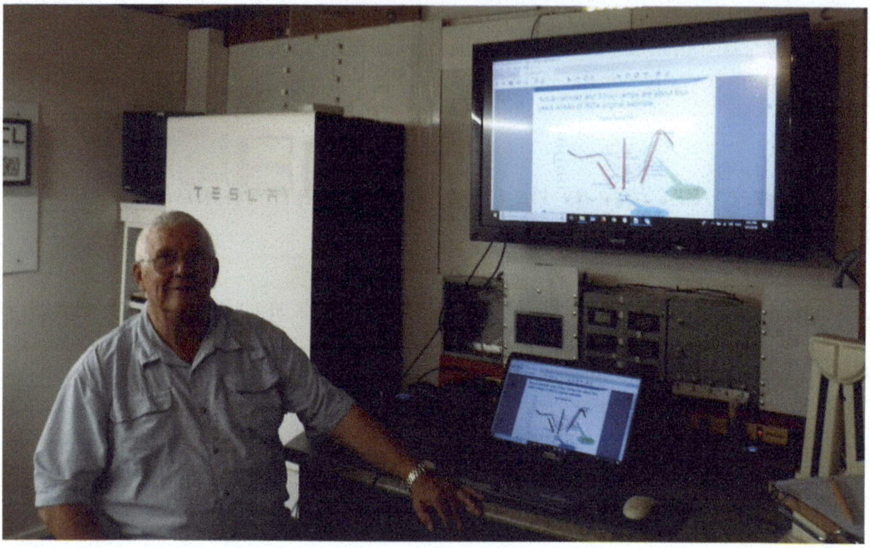

Fig. 9.3 A man and his data collection system. There are three Tesla PowerWalls totaling 39 kWh of storage behind Art (author)

Fig. 9.4 Solar panels on a conventional bungalow roof with Mitsubishi EV in the driveway (author)

no effort to modify the structure of the houses themselves to optimize their energy collection and minimize their energy losses. Multiply these efforts by 14 million residences in Canada plus every building in which citizens work, shop, and gather. How many hours of labor, kWh of energy, and money will it take to make renewable energy viable in Canada and other nations? How many years will the transformation of our living and working spaces take? (Figs. 9.3 and 9.4).

Arrays of 15 kW worth of solar panels are installed on the southeast and southwest facing roof surfaces, and these panels power the house and the Mitsubishi EV sitting in the driveway. Under the grass, beside the house, the real star of the show, a geothermal system, stores excess heat from the house and pulls the heat back in as needed (Fig. 9.5).

Fig. 9.5 Geothermal glycol circulation plumbing (author)
Not your average basement plumbing. Complex systems are required, but they must be simple to operate or made to be completely operationally invisible.

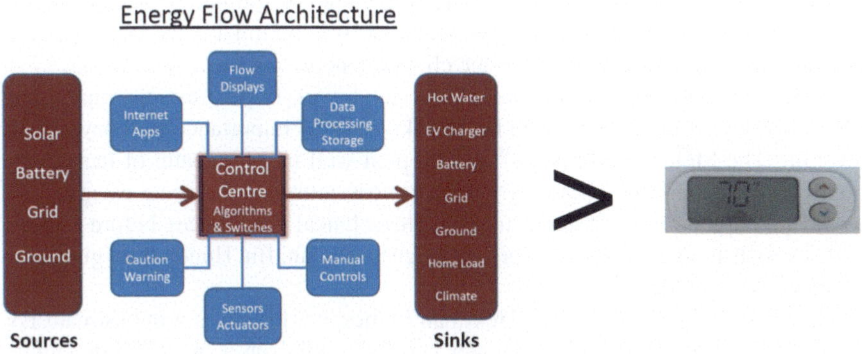

Fig. 9.6 Sophisticated heating system needs to be made accessible by all (author)

Most horizontal ground tubing is a series of trenches in which glycol filled tubing is laid out flat. Svec coils (Svec 1991) resemble a tube continuously wound around a mandrel or a stretched out coil spring. These provide a much higher working volume and surface area that dramatically increases the heat transfer between the coils. Otto Svec was one of Art Hunter's fellow researchers at Canada's National Research Centre (Fig. 9.6).

Hunter's objective is to quantify the flows of energy in the systems installed in and around his house, optimize their interaction, and allow a consumer interface to be designed that will take all of that complexity, move it into the background, leaving the consumer with very simple decision-making: On or Off, Warmer or Cooler. Added layers of complexity will always be available.

Hunter's system currently includes the following technical components and it is growing.

- Three PCs plus two in the pipeline
- Six raspberries (the Raspberry is a basic computer board "breadboard" for hobbyists)
- Ten geothermal sensors and a further 21 on other systems
- Many actuators and much wiring

Very few individuals want another complex piece of technology in their lives, nor do they wish to be faced with a steep learning curve as part of the price they pay for doing the right thing and minimizing their energy usage. Living in a house with a complex and efficient energy management system needs to be no more difficult than driving a car or getting into an autonomous vehicle.

Upgrades and refinement of the system is ongoing almost daily, and once Hunter has several years of data to work with, he will be able to provide valuable insights into the day-to-day practicalities and efficiencies of the systems he has built into his house. This can help inform government priorities and perhaps serve as a starting point for educational programs for young people in the fields of sustainability, architecture, engineering, and energy management.

Ideally, this kind of system and monitoring would be ultimately made available to students across the country to develop their skills in optimizing the latest technology for application in their own unique climatic/energy regions.

If Howe's work seems like a back-to-the-land work-in-progress, the quantification of energy required to perform basic tasks takes its importance well beyond the level of "The Mother Earth News" and drops it well into the realm of facts every policy maker should be aware of. Hunter's work falls into the same category but would perhaps be better filtered in through technical committees before landing with a very firm thud in front of a broad policy politician. But Hunter does give tours of his working model.

Hunter's system is more Star Trek than homespun, his car is a mainstream EV (Mitsubishi i-MiEV), and his computer network could run a small utility. Hunter can tell you how far his EV will run on a kilo Watt hour of energy, just as Howe can, but Howe could tell you how much energy is needed to cut a cord of wood with an electric chain saw and how many days of charge from a certain size of solar panel array it would take to energize the farm machinery needed to feed ten people. The costs of these working experiments have been very large for both Howe and Hunter, but the feeling of "filling up" your car for free or heating the house with ground heat is priceless.

Energy budgets per capita should be front and center in national policy planning and Howe's work gives that metric focus as does his work on local recycling, transport, and manufacturing infrastructure.

Hunter's work will provide results which are immediately actionable. But he has also had to deal with electrical providers whose mission in life is not, to this point, driving the renewable energy agenda forward. Even at this early stage, it is clear that a call for integrated planning with all levels of power providers and consumers is badly needed. Currently, new systems looking to be tied into the grid are bound up in red tape rather than being expedited. They are treated like foreign bodies found in an organism's bloodstream. Innovation and initiative need to be nurtured at every stage of the process and at every level of management in the utilities and governments involved.

This begs the question of how private companies can be expected to actively work toward reducing their own sales and work with small entities which provide new challenges and little profit? How will they deal with "storing" the energy from many small systems when the number of them rises from a handful to tens of thousands? What will be a reasonable "storage charge" if the utilities have to make large investments to support this new service they are providing?

The data from the systems of both Hunter and Howe are readily available, but the same cannot be said for the vast majority of systems in Canada. Right now the real-world outputs of private solar PV and wind systems are not openly available to researchers. Real-world capacity figures need to be used if planning is to be effective and researchers and policy makers need as much data about the differences in each location as possible. Data from every grid-connected system over a certain size is collected in the USA by their Energy Information Agency and is available online. Canada has no requirement that energy producers publish data on the energy they produce. Spanish researcher Pedro Prieto brought reality to the table in his EROI studies of solar PV systems with a number of years under their belt. We need to have the same historical data to make the most of our renewable energy investments.

Small, Northern, Unconnected

Canada has scores of communities from coast to coast beyond the viable reach of the electrical distribution network, since high voltage power lines (240 kV) can cost close to $1.5 million/km to erect. These communities are powered by stand-alone electrical systems, typically driven off a nearby waterfall or by a diesel generator. The difficulty of delivering the fuel results in very high electricity costs which are heavily subsidized by the government. The scale of the current costs and those involved in any type of transition to renewables is well illustrated by the case of Colville Lake in the Northwest Territories.

Colville Lake has 160 inhabitants and prior to 2013, electrical power was provided by diesel generators. The annual fuel usage was close to 1000 L per resident. The proposal to upgrade the diesel generators and add solar PV generating capacity estimated the total cost of the project to be $7.9 million ($49,000 per resident) with the solar portion accounting for $3.2 million. This broke down to perhaps $1.47 million for the panels and $1.73 million for the storage. The lifespan of the storage batteries is an open question as is their replacement cost.

This expenditure on renewable generation and storage would cut down the demand for fossil fuel by 20%. The cost of the renewable section was $20,000 per resident for providing 20% of their electricity consumption or 700 kWh. Total electricity consumption was 3500 kWh annually, and this would have been supplemented by oil and wood heat.

Colville Lake is an extreme example of the challenges in delivering power of any kind to small communities without grid connection and depending on seasonal transportation access only for material which cannot be flown in. Its northern location demands a high level of energy use, yet offers low and highly seasonal energy returns on solar PV installations.

Your Town: Latitude Matters (Again)

For larger centers, the options are better. Grid connection and easier access to more energy sources year-round assures dependable energy supply. But latitude still plays a huge part in the costs of assuring that energy supply. Using rough estimates, below we have two tables examining the size and cost of solar PV systems for six different cities and towns in different regions at different latitudes.

A single family residence in Igloolik might optimistically have a 40 kWh annual average daily heating budget and need 120 days of storage during the winter since the solar array would produce almost no energy. It would have 120 days of full production to cover current daily requirements and charge up the batteries in the summer for winter drawdown. During the remaining 125 days in the fall and spring periods, the energy produced would match energy demand.

The variables of maximum production and demand change by latitude so for the most southerly reference city of Guadalajara, Mexico, there are 365 production days and zero storage days. Wilmington, North Carolina would have 320 maximum production days and need only 4 days of storage to assure reliable delivery of its 12 kWh daily budget.

The panel arrays have to be large enough to allow for daily consumption plus filling the storage batteries in the period of the year when there is strong sun. At the extremes again, Igloolik requires a large 63.4 kW array (254 × 250 W panels) at a cost of $127,000. In Guadalajara, a single family dwelling could be provided with the heat it requires (none) and the household power needed with a 0.7 kW array (3 × 250 W panels) at a cost of $1300. Storage, if utilizing Tesla PowerWalls at a cost of $850 per kWh, would cost $4.1 million for that single family in Igloolik and $2125 for Guadalajara.

Storage

Clearly, for basically all of Canada, a dependence on solar PV energy with battery storage for the complete annual energy supply and storage is not economically feasible, nor would it likely be energetically sustainable. Monetary metrics were used

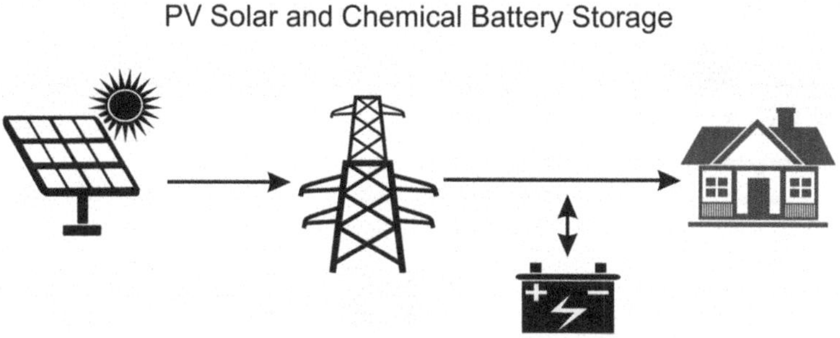

Fig. 9.7 Solar PV array and electric battery storage required in different locations (author)

Table 9.2 Array and storage costs for different locations

	Annual budget (kWh)	Storage required days	Storage required annual (kWh)	Size of array required (kW)	Storage	Array	Total
Igloolik	14,600	120	4800	63.4	$4,080,000	$126,736	$4,206,736
Calgary	10,950	90	2700	27.2	$2,295,000	$54,315	$2,349,315
Victoria	7300	60	1200	17.3	$1,020,000	$34,564	$1,054,564
St. John's	10,950	90	2700	28.5	$2,295,000	$57,031	$2,352,031
Wilmington	4380	4	48	3.2	$40,800	$6337	$47,137
Guadalajara	1825	0.5	2.5	0.7	$2125	$1302	$3427

for these examples since dollar costs were readily available. The author makes no claim that these, or any, dollar figures represent energetic reality. To properly represent the standard of living possible at different latitudes and in different regions, an overall index of viability would be useful. It could be based on EROI of the source at that point on the planet, divided by the ratio of the demand at that point vs a fixed reference point, say Wilmington, NC and further divided by the storage required in weeks. It certainly should be part of government assessments, which it is not now.

Solar PV system with chemical battery (Tesla PowerWall) for one single residence (Fig. 9.7) (Table 9.2).

The costs change dramatically if hydrogen storage is used. The efficiency of the electrical batteries was assumed to be 100% while that of the hydrogen process was set at 35% round trip. That 65% energy loss in the hydrogen storage process has to be made up with larger arrays. Hydrogen storage costs are extremely nebulous, particularly over a range of capacities so cost was arbitrarily capped at $50,000 with no minimum set for this very immature technology. A cost of $10 per kWh was used or 1/85th that of electric battery storage.

Costs of the system using hydrogen storage vs the chemical battery installation are dramatically lower and bring forward the point that, despite its hugely less efficient process, hydrogen storage for stationary applications might be economically feasible. That, however, does not mean it would be energetically feasible. For more

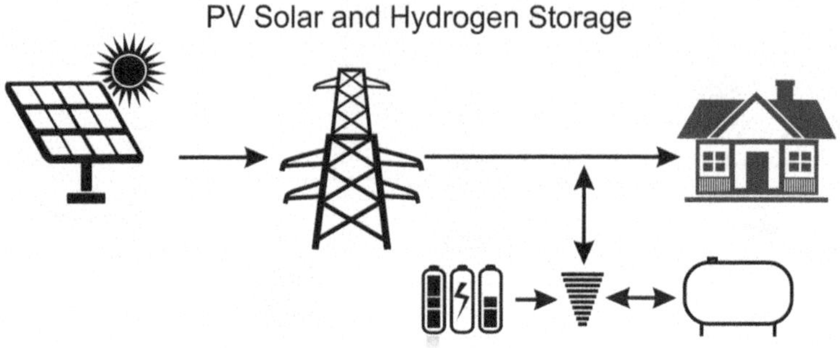

Fig. 9.8 Solar PV array and hydrogen storage required in different locations for one single residence (author)

Table 9.3 Solar PV and hydrogen storage in different locations for one single family residence

	Annual budget (kWh)	Storage required days	Storage required annual (kWh)	Size of array required (kW)	Storage ($1 k/10 kW)	Array	Total
Igloolik	14,600	120	4800	102.1	$50,000	$204,117	$254,117
Calgary	10,950	90	2700	39.6	$50,000	$79,188	$129,188
Victoria	7300	60	1200	22.6	$50,000	$45,116	$95,116
St. John's	10,950	90	2700	41.6	$50,000	$83,147	$133,147
Wilmington	4380	4	48	3.2	$4800	$6466	$11,266
Guadalajara	1825	0.5	2.5	0.7	$250	$1305	$1555

northerly locations, it might require a doubling of the size of the renewable energy generation capacity, whether it be PV, wind or hydro. Given the losses involved in hydrogen storage, EROIs could be expected to drop dramatically for the system as a whole. Further, hydrogen storage is tricky and hydrogen fuel cells do use consumable parts. However, electrical batteries must also be replaced at some point possibly within 10–30 years.

What is the difference between "economically feasible" and "energetically feasible"? Economically feasible means it is possible to allocate the resources necessary, using the purchasing power of money, to build the system. "Energetically feasible" means that the system will deliver more (and hopefully much more) energy over its life than went into building it. A system can be economically feasible at a point in time, but if it is not energetically feasible, that is energetically positive over its life, it is a failure, and it reveals the financial reasoning that went into its approval did not represent physical reality (Fig. 9.8) (Table 9.3).

In the above examples, capacity factors ranged from 8% in Igloolik to 32% in Guadalajara. Ten percent was used for humid and cloudy St. John's, Newfoundland and 12% for much drier Calgary even though it is further north.

A further consideration is the location of very large arrays. Currently a 280 W solar panel measures 1.0 m × 1.7 m for an area of 1.7 m^2 or 17 ft^2. The 41.6 kW,

148 panel array needed in the St. John's hydrogen example would occupy 250 m^2 or 2500 ft^2, much larger than most residential roofs even if they were bungalows with the entire surface one plane and oriented south.

Reports on the potential to mount solar PV arrays on urban roofs have been done for at least one region in Ontario using crude footprint estimates and, in the USA, using the much more advanced and reliable LIDAR technology (Gagnon 2016). Solar energy has a number of positive characteristics which make it attractive for use in Canada as well as all countries.

- Solar panels, whether photovoltaic or hot water, can be placed on most roof surfaces.
- They make power at the point at which it is consumed.
- They can be integrated into a heat pump system creating a cogeneration system.
- Mounted on roofs, they have no environmental displacement footprint.
- They create user awareness of energy flows.

However, at Canadian latitudes and with the humidity, cloud and snow issues which exist for every region of the country, solar has the following negatives:

- It has low capacity factors at northern latitudes, i.e., throughout Canada.
- It is highly seasonal.
- If used in industrial ground based arrays, it displaces crops or other vegetation.
- It is every expensive on an energy per dollar basis and has quite low EROI, the exact figures for which have yet to be determined on a regional basis.

Consequently, wind is the better choice for industrial-sized power generation in Canada as its EROI is several times higher and it is able to generate energy at a far lower dollar cost. Wind output is highly variable but does not have the inherent seasonal or nighttime limits that solar PV does. Storage requirements would be much less with perhaps 2–3 weeks being safe in regions with cold winters.

Major towns and cities have the luxury of being connected to electrical distribution networks which span large areas and connect diverse power generation facilities to a wide range of customers. Given the large numbers of hydroelectric dams, it is likely there is some significant degree of storage capacity built into the grids and, as long as the discussion is within the realm of the 85% fossil fuel free parameters, there would also be gas plants on tap to cover renewable shortfalls.

And, as time goes on, the battery storage capacity of EVs plugged into the grid can also be seen as a source of stability. Once electrified, the automotive fleet of private passenger vehicles in the Ontario region, currently 6.8 million vehicles, would provide a buffer of 270 million kWh assuming each vehicle had at least a 40 kWh battery which is pretty much the starting point for the 2019 crop of EVs. This battery asset would provide the equivalent of almost 20 kWh per person of backup. In reality, large chunks of the theoretical maximum EV/grid storage capacity would be removed by practical considerations, but it would remain as a very significant buffer for short-term energy fluctuations. And it would likely eliminate curtailment losses.

Home thermal storage and battery storage can also moderate short-term demand peaks but will not be able to serve as useful buffers against months long seasonal demand peaks at northern latitudes. But possibly geothermal storage can and that prospect will be further addressed in Chaps. 10 and 12.

District Heating

There are many local opportunities to implement cogeneration strategies employing district networks for heat distribution. Gas plants, incinerators, and some types of industrial processes produce substantial quantities of heat which can be captured and distributed by a hot water network among dozens or hundreds of nearby houses and commercial buildings. In Sweden, garbage that cannot be recycled is saved for the winter months and burned, with the end energy output breakdown being 10% electricity and 90% heat. In regions with suitable geology, there are opportunities for industrial sized versions of Art Hunter's geothermal storage with deep drilled piping storing very large amounts of energy hundreds of feet underground for use months later by nearby buildings. Heat sources can range from solar hot water systems all year round or the heat "effluent" from air conditioners in the summer, whereby the geothermal system would both make cooling in the summer more efficient and store the heat for later use.

District heating is like an electrical grid except it connects a heat source to many customers using hot water in a closed loop system. Little discussed, but widely used, especially in Europe, district heating systems are estimated to save up to 40% of energy costs for those connected to it. Of course, the energetic cost of the source dictates the ultimate cost, but if the system uses waste heat, or a strong geothermal source, then costs can be kept quite low.

The earliest known commercial use of geothermal energy in a district heating system was in Chaudes-Aigues Cantal, France in the 1300s (Bloomquist 2003). Seventy-five percent of district heating is provided by geothermal sources, and the first systems drew their heat from hot springs which must offer a constant flow of water of at least 50 C to make such projects viable. Bloomquist estimates that such systems are in use in 12 countries with a total heat annual output of 12.4 billion kWh or 7.3 million barrels of oil equivalent, roughly the amount of oil Canada uses in a week. Clearly, district heating may have strong potential, but it is currently providing a vanishingly small portion of world energy demand.

According to Enwave, a Canadian company specializing in district heating and cooling, there are over 5000 district heating systems in the United States suppling 8% of the heat used in commercial buildings (Enwave 2019). Canada's first district energy system was built in London, Ontario in 1880 to serve its university, hospital, and government complexes. The University of Toronto launched a district heating system in 1911, while Canada's first commercial district heating system was established in 1924 in Winnipeg's commercial core. There are now over 150 district heating systems operating in Canada, but their impact on total energy consumption remains minute. If, as expected, natural gas generators do become more widely used in Canada, then their heat would likely be used in in district heating systems, in the more densely populated cores of urban centers.

A variant of this is the district cooling system in use in the central core of Toronto which uses the coldest water, around 4 C, from Lake Ontario to cool scores of large buildings near the waterfront before the water is then fed into the municipal water system. This system saves 90% of the air conditioning costs of conventional systems although one would have to ask about the effect on the health of Lake Ontario if such systems were to remove significant quantities of its coldest water (Acciona 2019).

Renewable Energy Technology Development Communities

There are a considerable number of building and energy research projects being conducted across Canada by private individuals, corporations, universities, and government agencies. These stand-alone technologies will eventually have to be knitted into working communities. And if the community as a whole is to function well, while minimizing energy consumption and maximizing energy production, then the best way to learn is to develop advanced technology communities from the ground up. These "Greenvilles" or energy "Transition Towns" would allow planners to get a feel for the complexity of integrating the needs of a community with the dynamics and limitations of renewable energy.

Such a project is underway in the UAE (Sustainable City 2019). Clearly, the circumstances are very much different for this country compared to Canada and other northern nations, but despite their status as a rich petro-state, the United Arab Emirates face the end of the oil era just as everyone else does. In fact the lifespan of UAE oil is 68 years (BP, 2018) while its natural gas is projected to last for 97 years at current production rates.

"The Sustainable City" is a solar-powered development slated to accommodate 2500 people, and it represents the forward thinking of the leadership which sees the inevitability of the transition to renewable energy. The climate and energy requirements in the United Arab Emirates could hardly be more different from any conditions found in Canada, but the point of the Solar City is to create the best living conditions possible for its residents for as far into the future as planners can see. It cannot be viewed as a solution, but is rather, a learning center. These are desperately required in all of the major climatic regions in Canada to facilitate learning and to allow policy to be rolled out with some confidence that it is practical. These also serve as incubators for innovation and enterprise (Fig. 9.9).

The sun-advantaged UAE makes the most of its solar potential in this new community, but a northern latitude version would have to have similar solar PV coverage in addition to large storage capacity and substantial wind and/or hydroelectric generating capacity (Fig. 9.10).

Fig. 9.9 Rooftop utilization in Sustainable City, UAE (Sustainable City 2019)

Fig. 9.10 Energy-based design of both community and buildings in Sustainable City, UAE (Sustainable City, https://www.thesustainablecity.ae/, 2019)

Energy Diamond Innovation Centre

Over its anticipated 50-year lifespan, the Energy Diamond Innovation Centre itself will produce 140% of its energy requirement, thus offsetting emissions during construction, operation, and decommissioning. This building will be off-grid and provide an amphitheater for 700 people, as well as a spacious atrium for exhibitions. But to be truly sustainable, the lifespans of buildings has to extend into the hundreds of years.

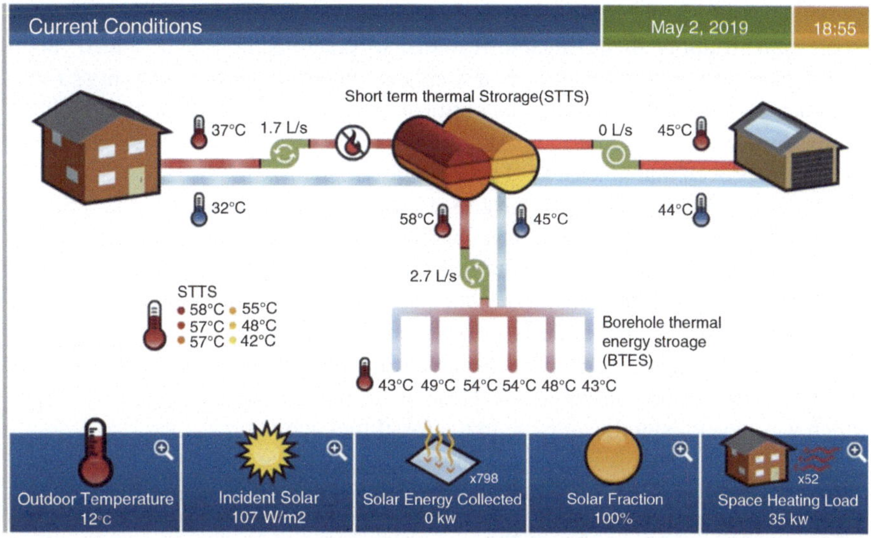

Fig. 9.11 Drake Landing monitoring software (Nguyen 2017)

The Sustainable City, on the outskirts of Dubai, qualifies as a renewable mega-project given its 1100 residences and 500,000 m^2 coverage. Clearly, projects of even several score homes and community facilities would require the involvement of all levels of government in addition to the private sector.

The Okotoks, Alberta community of Drake Landing utilizes solar PV and hot water panels extensively as well as borehole thermal storage (DLSC 2007). All 52 houses are serviced by direct-buried, pre-insulated piping. An array of solar hot water panels generate 1.5 MW of thermal power during a typical summer day and supply heat to the district heating system. Surplus heat is stored in a collection of 144 holes extending down 37 m below the surface. The temperature of the earth will reach 80 °C by the end of each summer. With over a decade of data, this project can serve to inform the design of similar projects across the country (Fig. 9.11).

The Province

Provinces have to look at the province-wide availability of energy and the sources available outside their borders. They have to integrate grid expansion planning with their overall internal strategy. It is at the provincial level that the policy decisions on whether to generate locally or import or possibly restrict exports will begin to come into play. Does Quebec have enough hydroelectric capacity to sustain itself in the post fossil fuel era with enough left over to export to other Canadian provinces or US states?

Will clean energy be in short supply, making it a strategic commodity? Will provinces want to apply their valuable resource to creating high value-added, high-paying manufacturing jobs rather than simply selling off the raw commodity? Those priorities pervaded the thinking of political leaders in Canada's nation-building phase. Remember that Ontario restricted natural gas exports in the late 1800s until it could be proven that at least a 30-year supply existed.

Provincial planning horizons have to stretch out almost as far into the future as do those of the nation. Nuclear plants, hydro dams, and large distribution networks require decades to build and decommission. Population size has to be front and center as does industrial demand. For the past 300 years, planners have strategized in an environment of continuously greater demand fed by increasingly more abundant and cheaper energy. That era is at an end, and the mindset of planners will have to embrace the energy budgets of the future as well as the added wrinkle of storage which was never a significant issue in the fossil fuel age.

Changes in the energy mix of provinces mean a change in their industrial structure which then calls into play many levels of social support. The phase-out of coal plants from the Alberta electricity grid means that the workers would undergo a shift into other industries and perhaps other communities. The suite of programs involved in Rachel Notley's NDP government supporting them through this transition included the following.

- Extended unemployment benefits
- Re-location expenses
- Tuition assistance
- Transition employment counseling
- Community adjustment funding
- Creative employment program development working with employers and various government agencies

In terms of nuts and bolts, policy has to reach down to make all systems work together. EVs need charging networks where the owners live, work, and shop. Some of the colder cities in Canada have engine block heater outlets in parking lots for shopping malls and offices. Upgrading these networks to provide a higher rate of charge would be quite simple. EVs do not require block heaters, so there would be no stranded infrastructure loss in this case (Fig. 9.12). Charging networks for condos, apartment buildings and parking lots will have to be made viable through coordinated changes in legislation and subsidy programs.

Provincial policy has to recognize and address the complex issues presented by the conflicts involved in the landlord/tenant/energy relationship. Well-to-do and high-functioning individuals may find it easy to control their energy costs as they both own their property, are capable of making all energy investment decisions, and can operate their household systems competently. The exact reverse situation exists for many people living in rental accommodation. One in three Canadians rent rather than own their living spaces. A rational response to fuel cost increases is more difficult with consumption and investment responsibilities split between renter and property owner.

Fig. 9.12 Plugged in cars in the winter, block heaters now, batteries soon

This is compounded, in many cases, by low incomes of the renters and poor maintenance practices of both the landlords and renters. Community programs have to be developed to address this problem.

Clearly, energy transition policy is not simply a question of physics or economic cost. Its impacts reach deep into social structures, and provincial governments need to plan to actively manage the wide array of tools needed to both change the energy supply and deal with the impacts which will fall out of that change.

The Country

Ultimately, the success or failure of a nation to transition to renewables rests on the shoulders of the federal government. This level of government must set the overall strategy and put the larger mechanisms in place which will assure the transition takes place smoothly. The federal government must create the environment of price and program stability and implement programs which create scientific transparency and accessibility at every level. It has to see to it that technological incubation programs are able to support initiatives in all regions of the country. It also has to assure that major policy, in the areas of trading networks in energy, population, major climate policy, and international agreements are integrated coherently rather than being independent policies working at cross purposes.

First and foremost, the federal government must stimulate "A grownup conversation" to determine the countries' real options and make sure that all interests including science are at the table, rather than just the one with powerful voices and embedded commercial interests. If the transition does not go well, the system of government fails. If the transition is not implemented in a transparent and fair way, the transition fails. Energy policy cannot be separated from either social or industrial policy.

Policies succeed when everyone sits at the table, and all productive interests have an equal voice. Examples are many, from China on industrial development to Norway on resource development, to the way Germany and Alberta are working with the workers in the coal industry as coal is phased out of the economy. Relying on the existing social safety net to catch the damage caused to individuals and communities by large-scale energy system changes will result in the kind of chaos that resulted from the collapse of the North Atlantic cod fishery in Newfoundland. Plans have to be formed long before the damage becomes acute.

Conditions necessary for a smooth transition:

- Stable, predictable rise in pricing of fossil fuel energy with clear projections for the fuel and the licenses of the devices which use it.
- Stable investment environment, stable costs, returns, and interest rates.
- Buffer real product system from financial shocks as growth- and inflation-dependent sectors destabilizes or even declines.
- Consistent, long-term incentives and market forces to move away from fossil fuels and to conserve energy across the board.
- Expectations and context matter. "Are we going to do this?" A social contract between all levels of society and several generations.
- Honest communication of challenges and progress is critical. Responsible and technically competent full-context media coverage is critical.
- Build real metrics into the daily conversation. For instance currently, house valuations and sales are more important to the media than citizen well-being and energy policy.
- Resiliency, the ability to stay the course through inevitable setbacks, and unforeseen events. Change is not linear—apparent false starts in energy supply and climate change will embolden skeptics and laggards and give pause to decision makers. More on energetic and social inequality on Chap. 10.

Localism, resiliency, and broad-based economies are all in direct conflict with both the letter and the spirit of the current form of globalism with its free trade agreements. The federal government has to prevent previous fossil fuel era agreements from interfering with progress in the domestic transition to renewables. The change from maximum consumption and the sell-off of natural assets to a focus on conservation, efficiency, and resiliency will have profound effects on trade agreements and international finance.

Is the challenge of transition best met by the attributes of one form of government or another? Is it the decisive action of a strong leader and the unwavering will of a nation that will implement the replacement of fossil fuels by renewable energy? Perhaps long-term success depends more on the inclusiveness and integrity of the decision-making process involving the culture of the country, the experience of the leadership, and the power structure of the nation. It is difficult to point to an ideology which can best handle the challenges of transition. Perhaps, any system which is flexible enough to leave ideology at the curb in order to implement the practical, real-world policies will be effective enough to be successful.

The National Conversation

A national conversation would involve temporarily putting aside the expectations developed in the era of every growing supplies of cheap energy and looking beyond the economic structure that evolved from it. It will be necessary to look at the physical realities Canada faces rather than casting a longing eye in the rearview mirror to when natural resources seemed inexhaustible and the concepts of legacy costs, climate change, and depletion were unknown.

We can no longer expect technological miracles to deliver complete solutions at the last moment. Technology does not deliver solutions, it buys time. Norman Borlaug's Nobel prize acceptance speech, which reminded us that the Green Revolution, was a technological leap that was only buying us time, should be hauled out every time a politician cuts a ribbon on a new technological advance, whether it be a new wind generators or an EV and declares global warming or resource depletion "Mission Accomplished." The "Mission Accomplished" banner cannot be unfurled until both our demands on the earth and our numbers stop growing.

We have come to view the ready availability of large amounts of energy as business-as-usual, but over the span of 1 million years of human existence, or even over the past 12,000 years of the rise of civilization, 97% of the time, energy abundance was not usual. We are returning to an era where we will have to be energy parsimonious. But now we have the ability to be very efficient and to do a great many things with the energy we are able to harvest sustainably.

A productive conversation has to involve everyone being heard, not simply everyone listening to the speeches of the influential elites. Hard numbers have to be absorbed by all parties, and policy makers must develop the ability to distinguish between processes which produce real goods and those which merely transfer dollars from one pocket to another.

The new world of public policy will be further discussed in Chap. 10, but if the national government is not leading the transition effort, a fractured and ineffective result is most the likely outcome. The transition will have to have a firm base of national will and national context in order to provide the stability necessary for the long haul of change.

Not that this process needs to be non-stop pain, but various groups, legitimate contributors, will be pressed and dislocated from time to time, and the rest of the society will have to make sure they do not shoulder their burden alone. Simply put, businesses, individuals, smaller government organizations, and utilities will be unable and unwilling to enact policies which put them out of business. Finger pointing and sensational reporting will quickly bring rational discussion to its knees in a toxic and polarized public forum which would see knowledgeable voices marginalized. All parties are responsible for maintaining an informed, cause-and-effect conversation in which everyone can participate and which no one can dominate.

The conversation must be informed and egalitarian and so must the policies on the ground. The corrosive effects of inequality and the promotion of endless consumption are almost universally recognized. Fortunately, the lust for material goods is being widely challenged (BBC 2018). Pope Francis has called on people in developed countries to live a simpler and less materialistic life. The Pope said:

"Let us ask ourselves: Do I really need all these material objects and complicated recipes for living? Can I manage without all these unnecessary extras and live a life of greater simplicity?" "For many people, life's meaning is found in possessing, in having an excess of material objects. An insatiable greed marks all human history, even today, when, paradoxically, a few dine luxuriantly while all too many go without the daily bread needed to survive."

Highly divided societies eventually crumble under the weight of internal conflict. Wilkinson and Pickett, authors of "The Spirit Level", document (Wilkinson and Pickett 2011) the debilitating effects of inequality in modern societies. When a nation can face their problems together, the nation can endure but when one group unloads national problems disproportionately onto another group, national structures can crumble.

The national government has to develop illuminating metrics and detailed datasets to make planning and monitoring of progress effective. The process has to be well explained and open to all, and it has to be made resilient. Persistence and realistic expectations can tolerate some setbacks but perhaps unrealistic hopes for fast results cannot.

The transition will be driven by millions of large and small decisions by individuals, businesses, and government agencies. These have to be effective and energy rational. The only way of assuring energy-smart decision-making is by presenting the decision-makers with solid information, appropriate incentives, and accurately costed energy. No matter the jurisdiction, the national government must assure that even if uniform policies for such things as building codes and penetration of renewables are not possible, uniform positive results must be.

Asset Changeover

Regardless of the level of government, if an efficient transition is to take place, it must waste the least amount of existing assets and build up the most appropriate mix of energy efficient practices and technology.

- Stranded Assets
  - If the oil sands are deemed to be an uneconomic or unacceptable source of energy, then the oil within them and the facilities for processing it will be termed "stranded" since, although they are still functional, they will no longer be used.

- The millions of internal combustion engine vehicles we use daily will undoubtedly still be functional beyond the date they are mothballed either by rapidly rising fuel prices or climate change priorities.
- The huge houses we have built well away from urban centers and their transit networks will last far longer than the transition process. How will they be re-purposed?
- Fleets of aircraft with lifespans of many decades will be partially shelved.
- Millions of ICE recreational vehicles may not have access to gasoline.
- Huge and remote roads, when does the maintenance/use ratio become too onerous to continue their use?
- Oil pipelines and refineries will see vastly less volume.
- Suburban shopping malls (in decline even now due to online shopping).

Transitioning to a new energy regime will be disruptive to significant sectors of the economy, and these disruptions can be transmitted through the rest of the economy and society by means of the finance sector and the labor markets and distribution systems which service these sectors. Good planning is critical to the process, and an important part of a good plan is timing. The longer people and businesses have to adjust, the better they will be able to adapt and the greater the use they will be able to wring out of current assets before they are set aside.

In Homer-Dixon's "The Upside of Down," he relates the story an immense monument being abandoned when half complete. Workers simply laid down their tools and left the jobsite. Whatever change caused the cessation of work, if the leaders had seen the change coming, surely it would have been better to apply the effort of monument making toward mitigating the impacts of the impending crisis. But if the society insists on carrying on in a rigid manner with no forward-looking leadership and no flexibility in the decision-making structure, then change will arrive in the form of a destructive shock.

In the worst case scenario similar to the monument episode, if there were no gradual scaling up of the transition process, business-as-usual would continue until Sunday night, and a new regime would begin Monday morning. On Sunday afternoon, everyone is having a great time driving their brand new internal combustion engined trucks and 400 hp sports sedans around the neighborhood of big homes, with garages and driveways stuffed with ATVs, snowmobiles, and boats. Then on Monday morning, gasoline is no longer available, and the vehicle inventory becomes inoperable while the houses go unheated. Yet, the payments for all of these shiny new stranded assets will need to continue. Abruptness has large negative impacts.

Adequate foresight would allow future obsolescence to be taken into account, and the use of assets with no future to be wound down while the acquisition of assets that will work in the new energy regime is ramped up. The more slowly and deliberately this process takes place, the more successful it will be for all levels of society. It will take decades to build up the renewable energy infrastructure to produce sufficient energy to keep Canada and other countries functioning. Likewise, the current fossil fuel-based infrastructure has decades of life left in it, and we should replace the oldest and least efficient parts of it first.

Electric Vehicles (EVs): Deserve Special Mention

The automobile is a big part of our lives, and in the future, electric automobiles or EVs will fairly seamlessly take the place of gas-powered cars. The ranks of EV detractors are melting away in the face of these improvements in the driving experience.

- The responsiveness of the electron-fueled drive system will become obvious to any driver. Tesla is famous for the "ludicrous" mode in its Model S which allows very rapid acceleration. But aside from the power the car can put on the road, there is the rapidity with which it can deliver the power. Electrons move virtually instantly making an internal combustion engine feels comparatively sluggish.
- Despite the power that a big V-8 can produce, it takes time for an open throttle to allow the airflow and fuel delivery systems to make their way into the combustion chambers. Then it takes time for the valve settings and spark timing advance to adjust to the new demands and further time for the transmission to spool up and select the right gear. Then what follows is a strong push in the back as mass of the driver and passengers compress the cushioning in their seat backs.
- Until the arrival of EVs, this process was taken to be instantaneous. But now, by comparison, it seems lethargic. This can be described as the difference between a strong shove in the back and being slapped forward. Electrons simply move faster than complex mechanical systems.
- Lack of vibration, noise, and jerkiness.
- Regeneration allows for mostly one-pedal driving and extended range.
- Tire and wind noise are more obvious due to the lack of engine and mechanical noise.

Electrifying ground transport is one of the easiest things a society can do to wean itself off fossil fuels. Transport burns gasoline and diesel fuel and is our largest sources of carbon emissions. When combined with the energy to produce the gasoline and diesel fuel, it produces over one third of Canada's greenhouse gas emissions (Environment and Climate Change Canada 2019).

Tracking Clean Energy Progress

The IEA tracks clean energy progress over a range of energy technologies and sectors critical to the global transition. Below is a table listing the sectors and their progress to date toward the Paris 2 Degree target. A national scoreboard would certainly help in putting the efforts of each nation in perspective (IEA 2019).

Some technologies have made tremendous progress in 2017—particularly solar PV, LEDs, and EVs—but most are not on track. Energy efficiency improvements have slowed, and progress on key technologies like carbon capture and storage remains stalled (Fig. 9.13) (Table 9.4).

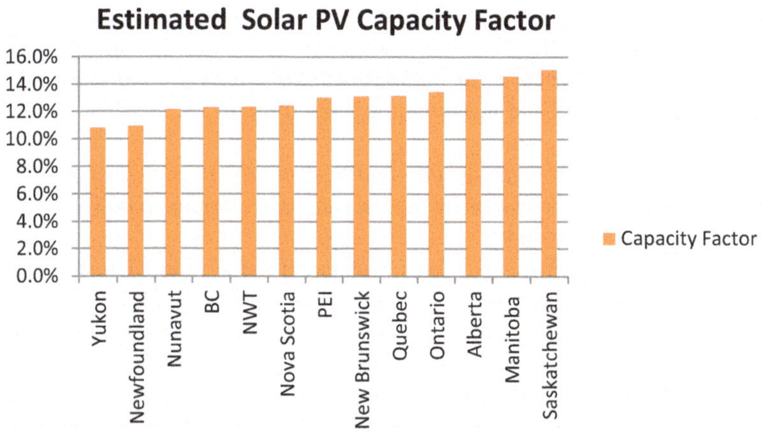

Fig. 9.13 Solar capacity factor by province—even the highest is about half that of the US South West (NREL 2019)

Table 9.4 Progress to date on energy efficiency by sector (IEA)

Power[a]	Buildings[a]	Transport[b]	Industry[b]	Energy[b]
Renewables[b]	Building envelope[c]	Electric vehicle[d]	Chemicals[b]	Energy storage[b]
Solar PV[a]	Heating[c]	Auto efficiency[c]	Iron and steel[b]	Smart grids[b]
Onshore wind[b]	Cooling[b]	Trucks and buses[b]	Cement[b]	Demand react[b]
Offshore wind[b]	Lighting[d]	Transport biofuel[c]	Pulp and paper[b]	Digitalization[b]
Hydropower[b]	Appliances[b]	Aviation[c]	Aluminium[b]	Hydrogen[b]
Bioenergy[b]	Data centers[d]	International ship[b]	CCUS in industry[c]	Renewable heat[c]
Geothermal[d]		Rail[b]		
Concentrated solar[d]				
Ocean[d]				
Nuclear power[b]				
Natural gas power[b]				
Coal power[b]				

[a]One to watch
[b]More effort needed
[c]Not on track
[d]On track

Energy-Optimized Processes

A successful transition to renewable energy would see compact communities with centrally located markets within walking distance of all residents in all-weather surrounded by farmland and forest (Tungare 2011). These communities would be linked primarily by electric rail systems, and when the need was found for personal transport, it would be rented. Buildings would be surfaced with energy harvesting devices from solar PV panels to hot air and hot water systems. Windows would have exterior blinds to control the heat/cooling flows through the windows which are, thermally speaking, every buildings' weakest link. Each building would have both thermal and electrical storage capacity and heat pumps to provide both heat and cooling. Any power not produced by the building itself would be delivered by wire rather than by pipeline or tank.

There would be a central recycling center and some manufacturing capacity. Transporting material long distances would simply not be feasible as John Howe has pointed out. Currently, Toronto, Ontario ships its garbage to Michigan and North America ships its toxic waste overseas. Plastic suitable for recycling is shipped from North America to China. These flows survive solely on the availability of cheap energy and the lack of commitment to embrace lifecycle consumption, which is to say the failure of consumers to pay the full price for the goods they consume, from the costs of production to the costs of disposal and neutralization. A district heating system might be powered by the biologically based waste that could not be recycled. For instance, Syracuse, NY, burns all its starch, including 20% plastic, to make enough electricity for 70,000 homes for periods in the winter. Materials that could not be recycled to an extremely high degree would only be used in applications where it would last scores or hundreds of years. Product quality will be higher, and there will be higher degrees of upgradability.

Private machinery inventories would be more oriented toward more "tool" than "toy" compared to today. The trends below would become widespread.

- Industrialized farming displaced by a wide variety of crops and breeds.
- Education systems include energy and biosystems from pre-school up.
- Conventional economics replaced by BioPhysical economics in the national policy conversation.
- Size of the commercial markets will be information found to be useful only by larger business enterprises and tax departments.
- Broad-based economy.

This vision may simply be a version of Europe moved 40 years into the future, a logical extension of their current course. For North Americans, it is a radical departure requiring fundamental shifts in the structure of the economy and in the lifestyles of their citizens. Suburban communities will condense as suburbs are drawn back into denser communities and those living in rural will be largely the people who work there, rather than those willing to commute long distances to work.

Without the growth to inflate housing prices, accommodation will be more affordable, although any good material or energy will likely be more expensive by today's standards. However, given the higher quality of goods and food, perhaps the value per dollar will end up being higher than the mass market goods and industrialized food we buy today.

Band aides and coats of paint will not lift society to the level of transformation needed to achieve prosperity in a renewable energy world. Commitment by all levels of government and by leading citizens will be required to bring about the deep structural changes necessary to arrive at a point of balance with earth's environment and energy flows. Canada, as a northern country, faces immense challenges in transitioning to variable, hard to store energy sources. There may be a solution from a new direction, but overall success cannot be built on out-performing technology. We will have to be firing on all cylinders across the policy spectrum.

References

Acciona (2019). http://www.acciona.ca/projects/construction/port-and-hydraulic-works/
Arbabzadeh M et al (2019) Green principles for responsible battery management in mobile applications
BBC (2018). https://www.bbc.com/news/world-europe-46676268
Bloomquist RG (2003) Geothermal space heating
BP (2018) Statistical Review of World Energy. https://www.bp.com/en/global/corporate/energy-economics/statistical-review-of-world-energy.html
Brown L (2008) Plan B 3.0: mobilizing to save civilization. WW Norton, New York
Design News (2019). https://www.designnews.com/electronics-test/10-green-principles-ev-sustainability/167432097460891/page/0/1?ADTRK=InformaMarkets&elq_mid=8952&elq_cid=4572813
DLSC (2007). https://www.dlsc.ca/district.htm
Environment and Climate Change Canada 2019). https://www.canada.ca/en/environment-climate-change/services/environmental-indicators/greenhouse-gas-emissions.html#transportation
Enwave (2019). http://enwavetoronto.com/district_heating_system.html
Gagnon P (2016) Rooftop solar photovoltaic technical potential in the United States: a detailed assessment
Howe J (2014). http://www.solarcarandtractor.com/pieces/Ch5.pdf
IEA (2019). https://www.iea.org/tcep/
Mohr SH (2015) Projection of world fossil fuels by country
Nguyen A (2017) Borehole thermal energy storage systems under the influence of groundwater flow and time varying surface temperature
NREL (2019). https://pvwatts.nrel.gov/pvwatts.php
Pimenetel M, Pimentel D (1979) Food, energy, and society. CRC, Boca Raton
Sustainable City (2019). https://www.thesustainablecity.ae/
Svec O (1991). https://nrc-publications.canada.ca/eng/view/accepted/?id=d9e93bc8-26c7-4a22-842e-0c5c178b0eca
Tungare A (2011) Le Corbusier's principles of city planning and their application in virtual environments
Wilkinson R, Pickett K (2011) The spirit level: why greater equality makes societies stronger. Bloomsbury, London

Chapter 10
Building a Renewable Energy Network—Canadian and Northern Options

Abstract The human energy network began with the shipment of coal from northern Europe to England during Roman times. It expanded during the Industrial Revolution as purpose built transportation began to deliver stored energy by railroad car and shipload to an expanding array of destinations. When oil came into use, both the extent and the depth of this network increased, and when electrical networks exploded across the globe, it became possible for the majority of the world's population to flick a switch or put a plug in the wall in any room of their house and access power that might have been generated thousands of kilometers away, mere milliseconds before. The development of energy networks has been a huge transition for human society, and these networks will continue to transform to adapt to the replacement of fossil fuel source by renewable electric sources. Canada's energy network will be more complex than those of most countries, due to its geographic span and variety of sources but mostly due to the very high levels of energy required per person and the very large amounts of seasonal storage that will be necessary.

Our energy network currently consists of scores of large energy generation sources connected by comprehensive and sophisticated transmission systems to millions of large and small energy consumers. In the renewable energy future, that system will change dramatically. The vast majority of the energy that moves any significant distance will be electric. There will still be large energy generators, but the majority of these will be intermittent. There will still be millions of consumers, but most of these will have the capability of generating energy of their own on an intermittent basis. They will likely also have storage capability, either in the form of electric batteries, heat batteries, or hydrogen tanks. The energy distribution network will have to be able to adjust to both the intermittent demands of these consumers and the intermittent energy generation of millions of small home and business solar and wind systems.

On the other side of the distribution system will be the major producers, and the distribution system will have to balance their reaction time, storage capacities, and output capability with the demands of the end users/part-time producers of energy. How much variation in the time of use of energy will consumers tolerate, and how

J. E. Meyer, *The Renewable Energy Transition*, Lecture Notes in Energy 71,
https://doi.org/10.1007/978-3-030-29115-0_10

well will they be able to tailor both their demands for and their production of energy to the requirements of the distribution system, often referred to as the "the grid"?

But first it is necessary to clarify our terms. Technically speaking "the grid" is the transmission system over 50,000 V of bulk lines moving electricity from generators to load center transmission stations. "The distribution system" is less than 50,000 V operated mainly by municipally owned companies (Richardson 2019). For the purposes of this book, the generic term "grid" is applied to the system of energy transmission to and from the end user, that is, the entire network outside of your home or business, regardless of voltage or extent or form of energy. In North America, a home in Labrador connected to the grid would theoretically be connected to power suppliers 5000 km away in southern California and vice versa.

Currently, most societies have at least two energy distribution systems. One is electric and the second is for oil and gasoline transport by either pipeline or tanker. The third might be the natural gas pipeline networks, and there might even be a fourth in the form of a hot water distribution in district heating systems. These will be heavily transformed in a renewable energy society. The most efficient means of transporting power is down a copper conductor, through an electrical network, and since most energy used by the future end consumer will be in electrical form, the electrical grid will have to be upgraded to handle larger flows.

Eventually, most of the oil pipeline grid will be either left to rust and leak in the ground, or it will be pulled up and the land remediated. For some critical uses, there will still be some oil consumption with some refinery capacity on-line, and these might be supplied by pipeline but, given the small volumes of oil involved, perhaps rail tankers would replace pipelines during oils last decades. Thus rail lines have an open future, whereas oil pipelines do not. It will certainly pay to invest in the maintenance of rail lines and perhaps even in extending the rail network. It may not pay to maintain oil pipelines, and it would be hard to justify building new ones except in exceptional circumstances. But oil will continue to be used at some level for many decades, and the most efficient means of transporting it in the necessary quantities will endure.

Weaning ourselves away from the fossil fuel network will be no easy task given the incredibly high level of energy and reliability the fossil fuel "grid" has delivered over the past several centuries. Not only has net energy availability increased dramatically but the consistency or reliability of energy availability has also risen to near 100% in most developed nations. That consistency is a function of the fuels themselves which are pulled from static, slow declining reserves and then transported with a very low level of difficulty to the end user.

Stability Challenges of the Renewable Energy Grid

Above are the outputs of four renewable energy sources. The top graph is the forecast output of the Cohora Bassa dam in Mozambique over 60 years allowing for climate change. The second graph is the annual river discharge in cubic kilometers for the Eastern Arctic. The third graph illustrates the power output of a solar PV

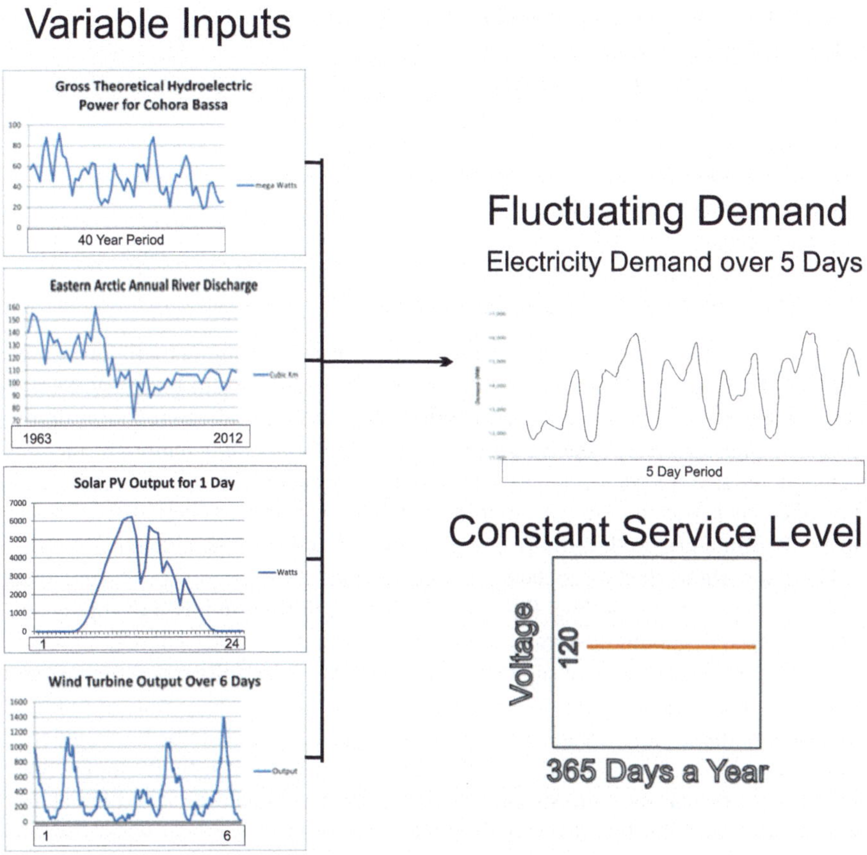

Fig. 10.1 Highly variable and unpredictable inputs and demand (author)

plant for 1 day in July, and the last graph gives the output of a wind turbine over 6 days. The grid must integrate these short-term and long-term variable inputs and match them to the fluctuating end user demand while maintaining high service levels. Service levels in Ontario require the grid operators to maintain voltage in the range of 110–125 V (Yamba et al. 2011; Dery and Stadnyk 2016) (Fig. 10.1).

The renewable energy grid will have a complex of energy sources feeding into the network, and it will have a complex of energy consumers on the other end, with the capacity to both produce and store energy intermittently. The end consumer will move beyond mere conservation via better insulation, LED lights, efficient appliances, etc. and build energy harvesting and storage capacity into their buildings, processes, and vehicles. The end consumer will be equipped to minimize their energy consumption and maximize their energy generation as well being able to exert control over the energy flows within the residence or business and the energy sub-systems. Eventually many energy sub-systems will be built into every home and building with a short list including:

– Solar photo voltaic panels on the roof

- Heat pump, air, and/or geothermal
- Electric battery storage
- Roll down shutters on the windows
- Thermal mass storage in the building
- Thermal mass storage in the ground (geothermal)
- Hydrogen storage
- Solar hot water
- Solar hot air
- Cogeneration with solar PV and hot air.
- Integrated master energy management system

In order for the grid to maintain a high service level, consumers will have to be able to both look after their own needs to a much greater extent that they do currently and modify their demands on the grid to fit what the grid can reliably supply. More complex systems, which require energy to produce and maintain, mean a change in the EROI which can only be accurately estimated through the application of detailed energy accounting. A complex system which supports high service levels must still ultimately produce more energy than required to make and maintain it.

There are three themes in energy management: total energy produced, total energy required, and smoothing the transition between different levels of supply and demand. A large amount of storage spread throughout the system will help to seamlessly integrate fluctuations in supply and demand. Also, the willingness of consumers to accept less than the energy they wanted at the moment they wanted to use it will also be critical in assuring a high functioning grid. It needs to be reiterated that the fossil fuel system renewables will be replacing had almost infinite storage built into it because fossil fuels are simply stored energy. They could be set on fire at any time or place and the heat from their combustion used to perform whatever task needed to be done.

All of these systems at the consumer end should lighten the load the large, industrial energy producers have to carry but even in being lighter, the variability of the demand and supply will make it more difficult to manage. A post-fossil fuel world can expect to use significantly less energy, and the capacity of the grid will have to be sized accordingly.

One Sub-System: Cogeneration from Solar PV and Heat Pump

There is an expanding galaxy of technologies which can be employed to increase the amount of energy that we obtain from fairly standard devices (Fig. 10.2).

Solar irradiance for Calgary is approximately 1 kWh/m^2/day in November, December, and January, 2 kWh in February, and 3 kWh in March (SEH 2019). If a 10 kWh solar PV system were installed on a roof in Calgary, it would occupy approximately 60 m^2. If the solar PV panels were 18% efficient, 82% of the sun's

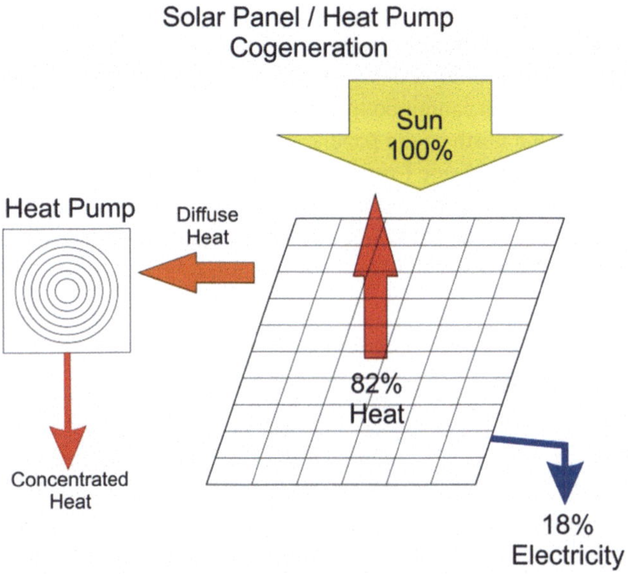

Fig. 10.2 Heat scavenging from solar PV panels increases system efficiency (author)

Table 10.1 Heat scavenging from solar PV panels increases system efficiency (author)

All figures in kWh/day	Total irradiance/ day	Electrical output at 18% efficiency	Waste heat generated on panel surface	Heat captured at 25% efficiency	Total energy harvested
November–January	60 kWh	10.8	51.2	12.8	23.6
February	120 kWh	21.6	98.4	24.6	46.2
March	180 kWh	32.4	147.6	36.9	69.3

energy striking the surface of the solar panel is lost as heat which escapes into the atmosphere. If the air trapped below the panel array was pulled into a heat pump, the energy it contains could be used to heat the house. If that system managed to capture 25% of the heat from the panels, then the total energy output would be as follows (Table 10.1).

In this example, the energy harvested from the solar array would almost double during the months in which heat was required, and the efficiency of the system would be increased dramatically. This would boost the EROI of the solar panels and allow them to be used at more northern latitudes than can currently be justified on the basis of electrical energy output alone.

Few recognize the amount of solar energy hitting the exterior surfaces of our buildings. Even if the skies are overcast and the increase in the temperatures of the building skin may be slight, there is a lot of surface area. If this energy can be absorbed into the heating system, large savings can be realized. The perforated SolarWall system preheats incoming ventilation air and saves between 20% and

50% of energy costs (Solar Wall 2019). It has to be noted that any solar-based system comes into conflict with trees, particularly tall trees. In a dense urban environment, very clever neighborhood design will be needed to accommodate both trees and solar panels. The two cannot be in close proximity with one another unless the trees are strictly to the north of the panels, which would place them to the south of next row of the buildings. Aside from being esthetically pleasing, trees also contribute to clean air and health and act as temperature moderators. Eliminating them from large areas of the urban landscape comes at a social cost that would have to be weighed against any energy benefits.

Other Sub-Systems

External roll down shutters can double the R-value of conventional windows in the winter and greatly reduce heat loading in the summer (Enviroblind 2019). Windows are the greatest source of heat loss and gain in a building since their R-value rarely goes over R5 while the walls of new buildings in Canada must be at least R30. In the summer, preventing direct sunshine from coming in the windows can save a great deal of air conditioning expense. In Canada, simply keeping the windows mainly on the south side of a building can save a great deal of energy three seasons of the year.

Thick concrete pads in buildings sitting on foam insulation can absorb a great deal of heat. A 15 cm slab in a 70 m^2 building can store 4.7 kWh for every increase in temperature of 1 °C. The thermal mass of concrete walls and floors gives a building a large thermal battery capacity if the heating system is designed to exploit their potential. Concrete is probably the cheapest way to add thermal mass to a building. Its use has to be designed, but at $200 for a cubic meter, with a mass of 2000 kg, very few other materials can approach its cost of just over 10 cents/kg.

Solar hot air and solar hot water systems can easily be installed to bring energy directly inside the envelope very efficiently. Vacuum tube solar hot water systems boast efficiency levels of 35%, a very attractive energy return for the buildings with high heat demand and good sun exposure.

In-home and in-room thermal storage can be accomplished by employing high temperature ceramic brick units which draw electricity during low demand, low cost periods, typically overnight and on weekends and store heat in ceramic bricks. The commercially available units store from 13 to 40 kWh and take up little space. While 13 kWh is enough to get a household through the morning rush, 40 kWh is more than enough to heat a modest house for a day and certainly for long enough to last until the next low-demand period begins. If not installed in the basement, make sure the floor can handle the 300 kg weight of the largest unit (Steffes 2019).

Electric battery systems are widely available and can be scaled to almost any size and are used in much the same way as the thermal storage, that is, excess power is shunted into them for use during demand peaks. Tesla markets PowerWalls with a 13 kWh capacity, and the Nissan Leaf has the ability to be used as a home energy backup should the grid fail. It would be reasonable to expect all EVs to have this capability in the future.

District Geothermal Subdivision

The Drake Landing development on Okotoks, Alberta consists of 52 homes linked to a geothermal storage "heat battery" which derives its heat from the banks of solar hot water panels. With 10 years of operating results to verify its original design estimates, the development has reduced the external source energy consumed by the average house by over 90%. The houses were constructed to a high initial standard (R2000) which meant they would use about 30% less energy than the average home constructed in the same period.

Energy harvested by the hot water arrays in the summer is stored underground for use during the winter. This process has proven to be, on average, 45% efficient over the last 8 years of its life and has proven to be extremely reliable with only two of its 798 solar collector panels requiring replacement. These essentially conventional 140 m^2 (1400 ft^2) homes still have an average daily energy budget of 35 kWh but, although very high by global standards, it is significantly lower than that of the average Canadian house.

In Canada, space and domestic water heating account for more than 80% of greenhouse gas emissions in the residential sector; hence, the emissions from these houses will have been cut by 72%. Certainly further cuts would be easily achieved if, rather than conventional designs, the houses had been constructed to take advantage of passive solar heat. Possibly heat pumps could have been integrated into the system a la Art Hunter, and these might have eliminated the need for any natural gas backup. Additionally, the house roofs would easily accommodate 6 kW solar PV arrays which, at a capacity factor of 10%, would be more than enough to power a fleet of EVs, 1 per house, 30,000 km/year.

But the point of the development was to prove that solar hot water collection and geothermal storage are technologies ready to be used now and using conventional, well-made homes drives that point home. Drake Landing stands as a huge step forward from conventional subdivisions. Future developments can further optimize the designs to progress beyond this solid base (Fig. 10.3).

The large arrays of solar panels on the garage roofs are hot water. Note the lack of trees and the ample space on the house roofs for additional photovoltaic arrays.

Grid Implications

The more storage available throughout the system, the lower the curtailment losses will be. Curtailment is the dumping of power that cannot be handled by the grid when, for instance, strong winds during a warm night create high wind turbine

Fig. 10.3 Drake Landing solar hot water panels on garages and solar PV on home (Mesquita et al. 2017)

output, but there is not enough demand to absorb all the energy being created. Curtailment can be expected to drop over time as more storage and more progressive time-of-use regimes are developed. Cheap electricity would be welcomed at almost any time by thermal storage systems which could always use another degree or two of battery temperature. If those batteries match the Drake Landing system efficiency level of 45%, then prices down into the range of several cents a kilo Watt hour would be extremely attractive. And geothermal storage can soak up a great deal of energy very quickly. So can a large fleet of parked EVs. Another great advantage of geothermal storage is that it is simple. Simple to build. Simple to maintain. Simple to recycle, if that term even applies here.

But clearly, the challenge to respond to changing demand and generation levels is a problem that will be front and center for utilities going forward. They will be looking for fast response energy sources, and the two major ones will be hydro and natural gas. Both of these can ramp up in as little as 5 min. For smaller scale fluctuations, electric batteries and hydrogen could also fill in very quickly (Table 10.2).

Buffering Bonus

The willingness of consumers to gradually ramp up their demand rather than serving up a forest of small spikes for the utility to handle would certainly be welcomed. Perhaps, as monitoring becomes more sophisticated, a "Buffering Bonus" could be

Table 10.2 Time to bring on line (author)

Source	Response time
Solar	On its own schedule
Wind	On its own schedule
Wave	On its own schedule
Tidal	Like clockwork
Geothermal	variable
Electrical battery	Nearly instantaneous
Natural gas	Under 5 min
Hydro	Under 5 min
Hydrogen	Seconds
Coal	Hours
Nuclear	Hours or days, complete shutdown is extremely impractical

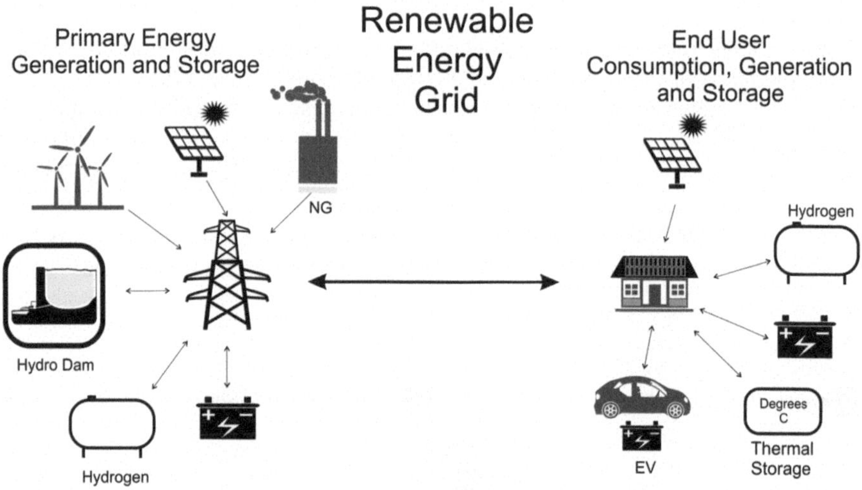

Fig. 10.4 The complex grid in the renewable energy era (author)

remitted monthly to customers with more moderate demand profiles. It would certainly make consumers more aware of the grids issues and give service providers more leeway in minimizing the size and expense of their storage capabilities. In the end, the size and complexity of the energy system dictates the costs consumers will eventually bear, so initiatives by individual consumers to reduce both the overall load they place on the system and the dynamic profile of their demand should be rewarded (Fig. 10.4).

All of these examples serve to make the point that where once "the grid" was a fairly simple line drawn between source and end user, it is evolving into a complex and dynamic network that surrounds us with many different kinds of energy flows and storage systems.

"The Copper Plate"

Energy is not always produced where and when it is needed. The graph below indicates there is a considerable difference between the intended purchase (1 day ahead) of energy and its actual delivery through the European grid (Fig. 10.5).

In the graph above, the physical flows were what ended up taking place while the scheduled flows were what was planned. The copper plate would be infinitely simple. Our electrical transmission systems come with the unpredictability that complexity inevitably imparts.

The ideal grid system is often referred to as "the copper plate," where all electricity production, consumption, and storage would be perfectly connected together with no restriction of flow from one point to any another. But rather than living on a thick plate of copper, our populated landscape is criss-crossed with a web of copper wires with different capacities and carrying different voltages. Demand comes and goes in various areas, and building and maintaining transmission lines are very expensive.

Necessarily, transmission lines are tailored to the size of flow anticipated. A network composed of small producers and consumers might have a fine web of light lines while a network with large hydro dams, nuclear plants and aluminum, and cement factories would include very high voltage and high capacity lines. And these lines have to have sufficient capacity to carry peak loads, not just average loads. Peaks are associated with time-of-use questions and storage capacity at the consumer end which can greatly reduce the loads the grid has to be designed for. However, for large consumers like aluminum plants, storage is not practical.

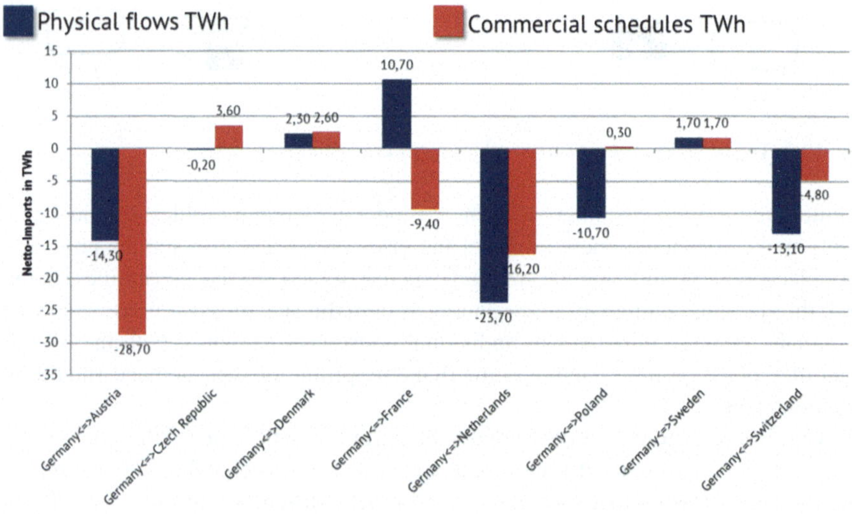

Fig. 10.5 Energy distribution plans change over a 24 h period (EnergyBrainPool 2019)

Promising Technologies

Building enough capacity into the distribution system to allow for any eventuality is simply not economically or energetically possible, nor is building enough solar and wind "overbuild" capacity to cover all possible demand peaks. The means by which service levels can be maintained and investment in grid infrastructure minimized can come in the form of storage technologies and energy demand minimization strategies. On the supply side, storage plays a very significant part in how successfully we transition to renewables. If we cannot store enough energy to minimize disruptions during the long winter high demand period, the consequences in northern climates will be severe. Unpredictable and frequent fluctuations in the level of power available will ring through society and the commercial economy with deleterious effects. A complex energy system and the complex society that depends on it require stability and predictability in order to maintain their high level of function.

We have the ability to store energy in two categories. One is thermal storage and the other is electricity. Hydrogen and pumped storage are processes used for electrical storage, something chemical batteries do more directly. The resulting high-grade electrical energy can be used for any purpose. Thermal storage, on the other hand, can only be used for applications which require low-grade energy such as heating buildings. And it has to be used very locally since heat suffers high rates of loss during transmission.

Geothermal Storage

Geothermal storage is where Canada's path to renewable energy diverges distinctly from that of more southern nations. Heat storage is of little use in the south (although cold storage may be), and any geothermal system will not need any more heat than what it can draw given local the ambient soil temperature. For many points south, the added energy and financial cost of installing infrastructure to allow two way energy transfer cannot be justified. Northern countries, however, can potentially benefit greatly from burying excess heat underground and retrieving it months later when it is needed.

Up to this point, storage has been treated fairly homogenously by researchers since a good deal of renewable research has been done in countries with strong solar PV potential and hence low building heating demand. But a look at the differences in the infrastructure required for chemical electric battery storage and geothermal storage shows a massive gap in both energetic and monetary expense between the two. The counter-intuitive concept of pushing heat into the ground and pulling it out months later has been proven to work, both in Canada and elsewhere.

The soil and rock formations have to be suitable which means there cannot be significant water circulation around the geo-heatmass wicking heat away. Solar insolance and wind energy maps exist for most areas of the populated globe, but there appear to be no maps of geothermal storage. To complete the picture of the complete spectrum of renewable energy options, we need surveys indicating potential of inhabited areas for geothermal storage. Subterranean energy storage will be most important for northern climates, but regions which experience extreme heat could also benefit from the use of geothermal cold storage.

Low-grade energy can be used to heat buildings as the application does not require very concentrated energy or even extremely hot water. The Okotoks hot water system operates at between 30 °C and 70 °C. This minimizes losses and allows enough heat to be delivered to the houses through its hot water grid.

Geothermal is a simple system, whose most complex mechanical part is a pump for water (glycol) or a fan for air. The total cost for the solar hot water and storage first-of-a-kind Drake Landing was $4.3 million or $82,000 per house. Developers feel that to be competitive, the number of homes serviced should be several times larger. But their comparison was with the cost of natural gas, which is by far the cheapest heating method. Financial reality changes dramatically in a renewable energy world.

In the fossil fuel world, the cost of a modest new home's natural gas heating system might be $15,000 including the natural gas boiler and hot water radiators. Then there would be the hookup cost to the natural gas pipeline running down the street. In a renewable energy world, the costs would escalate with the installation of geothermal piping either drilled down or trenched horizontally 3 m under the surface. The cost of a geothermal heat pump system starts at $25,000.

In the Drake Landing project, the thermal mass of the storage is effectively 20,000–30,000 m^3. When the temperature of this mass is raised to the 70 °C peak, it represents storage of over 1.5 million kWh of energy or 30,000 kWh of storage per house. If that amount of storage were provided by electrical batteries, say our Tesla PowerWall reference, 2300 of them would be required for a cost of over $2 million per house. The cost of chemical batteries such as lithium-ion is dropping but will probably never approach the cost level of geothermal storage for space heating now available for the aforementioned $80,000. Art Hunter's 3 PowerWalls on their own would heat his home for less than 1 day in the depths of a Canadian winter.

Aside from cost, the other advantage of geothermal storage is that it is self-charging to a fair degree. If the ambient ground temperature is 10 °C from a 5 m depth and below, imagine if it were heated during the summer with an array of solar hot water panels to 50 °C. Over the course of the winter, perhaps there is a heavy demand and all of the energy invested is withdrawn, leaving the heat mass (soil or rock) back at the 10 °C level. A heat pump can pull energy from a 10 °C environment with a coefficient of performance (COP) of over 4:1 meaning that a 1 kWh input creates a 4 kWh output. In the heat pump industry, COP is the metric for performance, and it equates directly to our core national metric of EROI.

In other words, if 1 kWh of energy is used to drive the heat pump in the example above, it will generate 4 kWh of heating for the home. This will tend to reduce the temperature of the underground thermal mass below 10 °C, but heat from the surrounding soil and rock will then slowly flow into it, reducing the temperature drop and, once demand declines, heat it back up to the 10 °C ambient. If electrical batteries could absorb electrons from the air around them, it would be a similar process. But they cannot, in fact, they lose charge over time, and once drained, chemical batteries will stay flat until they are recharged by sufficient electric current.

In effect, as a heat battery, geothermal storage never goes completely flat. One can overdraw one's thermal account by drawing the temperature down below the natural 10 °C level, but, over time, Mother Nature will bring the account back up to zero. Electric batteries do not work like that and neither does your bank account.

The Heating Bill

Northern countries use a lot of energy to heat buildings. In Canada, here are a few statistics for reference.

- Thermal heat and cooling account for two thirds of the energy used in the Commercial Sector in Canada.
- Combined with residential thermal uses, this represents 20.3% of total national energy consumption.
- Average household use of energy is 30,555 kWh annually or 84 kWh/day.
- Currently, air conditioning forms only 2.5% of residential energy use but this is likely to increase.

If all space heating went to the combination of electric and geothermal, the total reduction in fossil fuel consumption would be 20.3%. Two further steps, shutting down the oil sands (not conventional oil) and electrifying the passenger vehicle fleet, would reduce fossil fuel consumption by 10% and 15% respectively for a total combined reduction of 45%.

In the future, every building will get its external power from a wire rather than a pipeline or a tank. On the surface, it may seem to make more sense keeping natural gas heating to heat buildings directly with 95+% efficient boilers than to drive natural gas plants operating at 70% efficiency to generate electricity to provide heat. However, if the electricity is used to run a heat pump with a COP of 4:1, that 4 to 1 lever makes it a much better energy bargain than simply using the natural gas for heat. If 1000 Wh of natural gas was used as heat, it would deliver an impressive 950 Wh into the building. If used in a generating plant, it would produce 700 Wh of electricity. Applying that through a heat pump with a 4:1 coefficient of performance would deliver 2800 Wh into the building. Hence, it is likely the electrical grid will be the only source of energy for small consumers.

Hydrogen

Hydrogen can be created only by high-grade energy, either electrical or fossil fuel. However, it can also be converted back to electricity for use on-site or hundreds of kilometers away. Hydrogen requires higher technology than geothermal as its fuel cell stack facilitates a process combining hydrogen from the storage tanks and oxygen from the air to produce electricity and water. It also utilizes rare earth metals and requires careful recycling at the end of its useful life. The "stack" has an estimated service life of 2000–4000 h of short-term use in an automobile and up to 40,000 h in a stationary application running continuously (Valente and Iribarren 2019). There are 8760 h in a year. Some early adopters of hydrogen found they had to replace the stack three times a year, which negated most of the systems' potential benefits. But the technology is young.

The storage tanks must be very high quality to safely hold the very small molecules at a pressure of 700 bar (10,000 psi). For reference, a full 20 lb portable propane tank at room temperature would be under a pressure of 145 psi or 10 bar, which is 1/70th of the pressure a hydrogen tank must be built to withstand.

The current default storage system for electricity is pumped storage, but its potential capacity in most parts of the world is limited, whereas hydrogen storage can be installed anywhere and theoretically built to any scale. Electricity stored as hydrogen, of course, losses over two thirds of its energy in the process, whereas the energetic cost of pumped storage is less than one tenth.

According to energy expert Pedro Prieto, other problems of hydrogen storage include: "It reacts with all the metals trying to confine it at either 700 Bars+ or a −250 °C and forms metal hydrides and decomposes the metals (brittling) very fast." And "the rate of leaks through metallic deposits is very high" (Prieto 2019).

If chemical battery storage experiences a technological breakthrough allowing very high capacity in small packages at very low cost, then electrical storage will no longer be a bottleneck on the road to widespread adoption of renewable energy. But until then, the options for mass storage of high-grade energy are quite limited and, in the case of hydrogen, very inefficient and largely untested.

Storage in the Fossil Fuel System

Just for reference, the amount of storage in the current Canadian fossil fuel system is considerable and, technically speaking, has been achieved relatively effortlessly. Large natural gas holding tanks smooth out the seasonal fluctuations in demand while oil terminals hold only enough to maintain a smooth daily flow. Gasoline in distribution terminals, 11,928 service stations, and the tanks of vehicle fleets smooths the flow of energy delivery over a period of days and weeks.

The average fossil fuel (not including coal or material being processed at refineries) storage capacity converted into kWh works out to the following:

- Natural gas—500 billion ft^3 equaling 146 billion kWh (CGA 2019)
- Oil—460,000 bbl equaling 782 million kWh (Oil Sands Mag 2019)
- Gasoline—1.9 billion L = 19 billion kWh

The total storage amounts to 166 billion kWh or approximately 12 kWh per person per day for a year. If we attempted to simply replace fossil fuel storage with chemical battery electrical storage, we would require 12.8 billion Tesla PowerWalls costing $128 trillion in 2019. The total global GDP is currently $80 trillion.

Sound discouraging? Unless we discover a way to make small, fast reacting fusion plants, the renewable energy network of the future will not be a simple mirror image or carbon copy, if you will, of the current fossil fuel network. The energy will be made and used differently and at different times and, in many cases, electricity will do several times the productive work that the same energetic amount of fossil fuels will. The fossil fuel "grid" and storage capability have been built up over a period of several generations, but we have at most one and half or two generations to build up this new grid. However, a good part of the basic electrical grid already exists and much of the work that remains to be done will be in adding various storage and generation capacities at either end of it.

High-grade energy storage is expensive, but low-grade energy storage used for heating buildings may not have to be. And if we do indeed develop the ability to store energy geothermally over periods of months, with round trip efficiencies close to 50%, then the roadblock of energy storage becomes a much, much smaller problem. We certainly have the technology now to generate huge amounts of heat for 6 or 7 months a year in the most populated areas of Canada, Asia, and Europe. If we can get half of it back when we need it the most, then one of the largest technical obstacles on the path to a sustainable northern society will have been sidestepped.

Thermal vs. Electric Storage

The energy systems of all nations have, until now, been based on the stocks of very dense, very high-grade energy. Natural gas, oil, coal, and electricity can be produced in one location and efficiently transported to far distant consumers because their energy density is so high and any loss is minimal. Not so with low-grade thermal energy. It has to be used very close to where it is harvested. The energy available from geothermal sources varies greatly in density. In terms of locations featuring the very high temperatures needed to produce very high-grade energy (electricity), there are relatively few areas where this is practical. But in terms of being able to draw heat from the earth at a very small energetic cost, there are huge areas of the planet where this is possible. In effect, every building may be standing on its own thermal battery.

Access to thermal storage may be problematic in more densely populated areas, and in some areas with unfavorable geology, but many areas are well positioned to exploit this resource. Geothermal storage does not produce something that can be transmitted through the grid; it represents a load that will be taken off the grid. Our energy system should not be viewed as simply a transfer mechanism from producer to consumer but as a system which will allow the minimization of energy demand and of energy movement.

Just as we need better high-grade energy accounting to better determine the energy rationality of our decisions, we need to have a better understanding of low-grade energy systems, where we make it, where it is available naturally, and where we can store and use it. It is a question of efficiency once again. What was waste is now a vital resource just as we progressed from open fires in longhouses to high efficiency woodstoves to natural gas boilers—a progression from 5% efficiency to 50–95%.

Grid Technicalities in Ontario

Electricity is distributed in Ontario through a network of transmission lines owned and maintained by public and private utilities. To be effective, government policy to both conserve energy and increase production from renewable sources must penetrate this maze of overlapping responsibilities and interests. The coordinating body of the system is the Independent Electricity System Operator (IESO) which manages the power system in real-time plans for the province's future energy needs and enables conservation. Here are two descriptions of the levels involved in the distribution system from individuals who have worked on both sides of the system.

From Art Hunter: "Each of the 60 Ontario LDCs (Local Distribution Company) are privately owned. Ottawa Hydro is owned by the City of Ottawa, for example. They are the retailers in the grid and they own their own transmission and distribution (T&D) assets (transformers, telephone poles, 230/115 kV tower lines, etc.). They buy their energy from the wholesaler, the IESO which is owned in a major part (40%) by the Province of Ontario. The LDC buys wholesale energy from the IESO and sells it to retail customers. The IESO energy cost passed to the LDC includes T&D, generation, overheads and profit. The LDC then adds its own T&D and other costs to arrive at a price to the retail customer. Each LDC arrives at their own pricing so rate comparison from one side of the street to the other where there is an LDC boundary is often an issue locally."

There are varying mixes of generation plant and transmission networks in every province, making it difficult to apply countrywide standards. However, all have the same challenges of continuing to offer very high levels of consistency of energy delivery. They will have to maintain those service levels and transform themselves during an era of decreasing energy source reliability. Complex systems, no matter how miraculous their performance on day one of operation, are ultimately no better than the maintenance practices which support them.

Resilience is the ability to carry on in an acceptable manner despite encountering difficulties. The most resilient grid will:

- Be the most widely distributed generation
- Be the most overbuilt
- Have the most storage
- Be the best buffered
- Be well connected to other grids
- Clearly, resilience, like good insurance, is expensive

What Is a Microgrid?

Think of a microgrid as your friendly neighborhood energy co-op. They can be configured in any way the cooperative feels is best without having to defer to the choices of big-utility or big-government. Microgrids have yet to be embraced in Canada by either explicit government policy or larger service providers. In the USA, however, they are officially seen as a critical component in the effort to build independent generation capacity, intelligence, and storage into what used to simply be the consumer terminus of the power line.

According to the American Department of Energy, "Microgrids have been identified as a key component of the American Smart Grid for improving power reliability and quality, increasing system energy efficiency, and providing the possibility of grid-independence to individual end-user sites. The DOE defines the microgrid as "a group of interconnected loads and distributed energy resources within clearly defined electrical boundaries that acts as a single controllable entity with respect to the grid. A microgrid can connect and disconnect from the grid to enable it to operate in both grid-connected or island-mode. The benefits of microgrids include: Enabling grid modernization and integration of multiple SmartGrid technologies. Enhancing the integration of distributed and renewable energy sources that help to reduce peak load and reduce losses by locating generation near demand"" (Ton 2012).

Microgrids can shaving peak loads from the main grid and have the ability to operate independently for short periods of time. This gives the main grid more latitude in maintaining service levels while reducing costs. Microgrids will go hand-in-hand with district heating initiatives, and they will bring both a degree of grid independence and energy consciousness to the parties involved. Entire communities can benefit by taking initiatives best suited to their own unique circumstances rather than waiting for the trickle down benefits of macro-energy policy which may never arrive.

However, governments will have to be watchful that these positive initiatives do not lay the groundwork for greater levels of energy inequality caused by well-to-do areas heavily investing in energy infrastructure and allowing the overall energy network to decline. Microgrids should be supported as long as they promote the concept of energy-haves and have-mores and discouraged when this tends toward causing social polarization into energy-haves and have-nots.

The Energy Cliff

The "energy cliff," a concept derived by Euan Mearns of Scotland, is a term used to explain the consequences to society of sliding off our current plateau of very high EROIs onto the slope of progressively declining energy resource productivity. As we use up the richer deposits of stored energy and develop ever-lower quality renewable resources, such as smaller hydroelectric developments and more northern solar

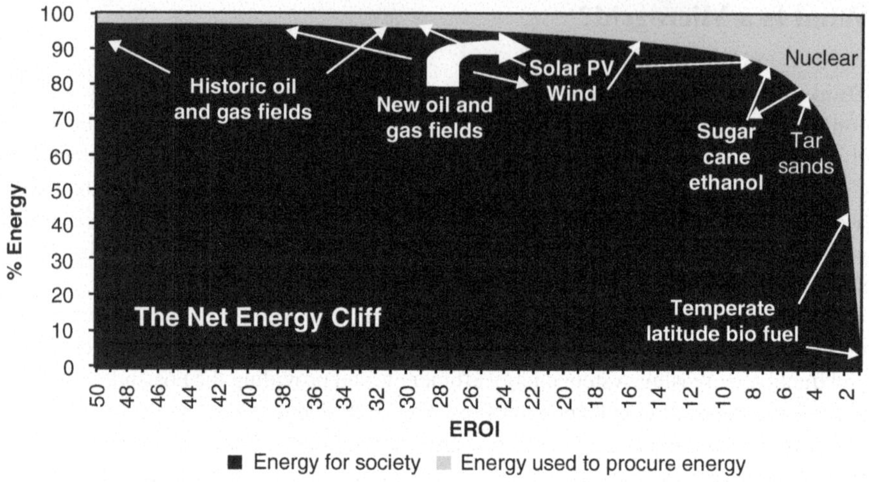

Fig. 10.6 The energy cliff, once we begin to use lower grade sources (Mearns 2019)

PV installations, we will have less net energy available, and it will be available less consistently (Fig. 10.6).

As renewables become an increasingly large portion of our energy mix, implying a decline in system wide EROI, the changes to the grid will become more extensive, and those changes will have to be built during an era of increasing energy variability. The fossil fuel grid was aggressively and confidently built in an era of increasingly consistent energy supply and falling costs, as was most of the electrical grid.

Presumably, as we move into the future, the one-fuel and one-way electrical, oil and natural gas grids currently in place will be replaced by a much more complex grid, evolved into a multi-fuel system capable of creating different forms of energy, and distributing, storing, and receiving them. The efficiency of the sub-systems is a high priority in our current, simple systems, and it is very easy to measure. But going forward, as the inclusion of inefficient processes such as hydrogen and thermal storage increases, the ability of the overall system to minimize its system infrastructure costs and maximize its service levels will require a different approach to planning.

There are benefits to a more inclusive and complex grid. A mature renewable energy system would eliminate curtailment waste since there would be so much storage capacity. The efficiency of some parts of the system may be less than ideal yet the mix of solar PV, hot water, hot air, and cogeneration might still offer higher efficiency than that of many fossil fuel systems we use now. On a systemic level, low-grade heat energy storage may make the adoption of renewable energy practical at northern latitudes where it might otherwise not be.

Descending down the face of the EROI cliff does not necessarily mean we will fall below the point at which a society is no longer able to produce toilet paper. It means we have to be much more thoughtful about our lifestyle and energy choices. The effectiveness of the grid will have a large bearing on the overall level of EROI at which our society can stabilize.

Timeframes

The timeframes necessary to build up the various components necessary to make a successful transition, span months to decades. It would take 4–6 weeks to upgrade an ordinary house and decades to upgrade the entire housing and commercial building stock of a nation. If we were intent on starting now, which tools would we have available?

- Residential
  - Conservation—technology is mature
  - Solar panels—technology is mature
  - Geothermal storage—technology needs refinement but workable now
- Industrial and commercial buildings—similar to residential
- Build Solar Farms—technology available
- Build Wind farms—technology available
- Replace ICE fleet with EVs—technology available
- Build Hydrogen storage capacity—technology not ready

There are 14.1 million residences in Canada (StatCan 2016). If the residential inventory is to be overhauled and upgraded over 40 years, that would mean adding a full suite of conservation, solar, and geothermal to each one, resulting in 2.5% of the stock or 350,000 homes a year becoming a full-functioning member of the renewable energy grid. At a cost of between $30,000 and $100,000 per home, the annual expenditure would be in the range of $11–$35 billion.

Every home and building would benefit from active and passive conservation measures. However, in looking at our housing stock, where it is located and how it is oriented, it becomes clear that putting solar panels on every roof will not make financial or energetic sense in the medium term. The cost of the required structural changes to re-orient many of the roofs would be crippling. This is where district heating and community energy projects come into play and possibly where the microgrid can have its greatest impact.

Building up Canada's capacity to manufacture the majority of its solar and wind infrastructure would take in excess of 15 years. Actually making the panels and turbines and installing them would require decades. Replacing the fossil fuel fleet with EVs could be done in as little as 20 years (assuming a 15 year vehicle life span) which would mean banning the sale of new internal combustion automobiles by 2025.

The renewable energy-based grid will no longer be just a standalone energy transmission system, but one integrated into society and functioning in concert with the environment. Energy flows and the grid will be much less than the "anonymous slave" as Andrew Nikiforuk would put it and much more of a part of our energy conscious lives.

"The Grid" can no longer be considered simply as a means of distributing electricity from the source to the point of consumption. It has to be seen as our energy network because, in the transition to renewable energy, the complex and interdependent mix of energy sources and sinks and storage will result in a system that has more in common with a living organism than an electrical circuit. The energy distribution system will not simply be an A to B means of transfer, but rather a system which enables the use of technologies which will allow us to maintain our level of progress through the process of transition to a renewable energy future. It will be mostly a future powered by electricity and one that will promote higher levels of efficiency, utilization of low grade energy and massive energy storage. Over the past 400 years of increasing fossil fuel use, human activity has become increasingly independent of, and disconnected from, the natural world. In the future, the grid will help re-establish the link that human society once had with earth's energy flows and resource stocks.

References

CGA (2019). http://www.cga.ca/gas-stats/

Dery S, Stadnyk T (2016) Recent trends and variability in river discharge across northern Canada

EnergyBrainPool (2019). https://blog.energybrainpool.com/en/physical-flows-vs-commercial-schedules-in-cross-border-power-trading/

Enviroblind (2019). https://www.enviroblind.com/rolling-shutters/faqs

Mearns E (2019). http://euanmearns.com/

Mesquita L, et al (2017) Drake landing solar community: 10 years of operation

Oil Sands Mag (2019). https://www.oilsandsmagazine.com/weekly-oil-inventory-report/2019/5/1

Prieto P (2019) From technologies to strategies email conversation with Pedro Prieto

Richardson JB (2019) Consulting Engineer

SEH (2019). http://www.solarelectricityhandbook.com/solar-irradiance.html

Solar Wall (2019). https://www.solarwall.com/

StatCan (2016). https://www12.statcan.gc.ca/census-recensement/2016/as-sa/98-200-x/2016005/98-200-x2016005-eng.cfm

Steffes (2019). http://www.steffes.com/electric-thermal-storage/room-units/

Ton DT (2012). https://www.energy.gov/sites/prod/files/2016/06/f32/The%20US%20Department%20of%20Energy%27s%20Microgrid%20Initiative.pdf

Valente A, Iribarren D (2019) End of life of fuel cells and hydrogen products

Yamba FD et al (2011) Climate change/variability implications on hydroelectricity generation in the Zambezi River Basin

Chapter 11
A New World for Public Policy

Abstract Governments have laid down the commercial boundaries and facilitated the fortunes of the industries which built the successive foundations of Canada's development from fur to fossil fuel. Industries which have ascended to prominence and leadership cling to their positions of power and privilege tenaciously. But the conditions which produced the need for resource exploitation and simple growth last only until the nation has matured and can turn its gaze from what made it great to what will allow it to maintain its greatness. Canada and the Americas started their rise through the massive exploitation of natural systems. But success in the future will depend on the ability of human society to sustain natural systems over the long term rather than drawing them down to the maximum over the short term as conventional (neoclassical) economic doctrine dictates. Commercial economics has a time horizon limited to the decision-making of commercial interests, and this, as well as the objectives, differs radically from the timeframe of natural systems and nations. In order to make good decisions and develop coherent policy, governments will have to be able to adopt clear goals based on new metrics. A critical step will be the replacement of neoclassical economics with biophysical economics in our universities and high schools.

The history of government in Canada has been one building a new nation out of a seemingly limitless body of virgin resources while displacing the original indigenous population from the best lands. Governments took a direct hand in all facets of infrastructure development from transportation to electrical utilities. The expansion of railroads and farming across the country and the rise of industry built the nation and expanded markets simultaneously. Government worked with productive industry to expand the output of real goods. This improved the economic welfare of most people and also built the strength and resilience of the nation by expanding the breadth of its economic base. Going forward, the nation will benefit from the reduction in energy and material consumption, and leadership will need to alter its approach accordingly.

All of the components of a viable nation were in place in Canada by the early 1900s. Rail lines tied the country together, most of the prime agricultural land was in production, and the population base was large enough to assure strong social

J. E. Meyer, *The Renewable Energy Transition*, Lecture Notes in Energy 71,
https://doi.org/10.1007/978-3-030-29115-0_11

infrastructure and a growing and well-distributed, vibrant industrial base. Further population growth was not needed to maintain a viable nation beyond the nine million citizens of the early 1920s.

The increasing sophistication of agricultural practices, the rapid rise in energy efficiency as well as the new capabilities of oil, and industrial capacity allowed the population to double to 18 million by 1960 at the same time standards of living and well-being were surging. It was a painless growth. Up until 1920, the interests of nation building and market building were one and the same. From 1920 to 1960, although the nation had been built, it undoubtedly gained strength as the population grew alongside agricultural production and in unison with the expansion and diversity of the industrial base.

From the late 1800s to 1960, energy became increasingly abundant and even cheaper while resource bases and industrial capability continued to expand rapidly. Individual well-being in terms of health, education, and material wealth increased dramatically for most people in this period, and by the early 1960s, Canada could boast the second highest level of equality in the world (Solt 2019).

Canada now stands as a mature, developed country with its environmental and resource limits fairly clearly defined. It is socially progressive with an extensive social safety net in place, and it at least pays lip service to the concept of egalitarianism. The population is well educated and socially mobile. But the building blocks were solidly in place by the early 1960s at which point government policy changed from building a nation to expanding the economy, a subtle shift, since the two had marched in lockstep to that point in the country's development. The economy continued to grow but individual well-being did not grow with it.

History and Current Reality

The new market expanding policies were based on population growth, initially through the natural baby boom of the 1950s and early 1960s supplemented by baby bonuses and higher immigration to almost exclusively mass immigration from the 1990s onward. There was no nation building component in these policies, just market making for the extremely profitable sectors of development, construction, and banking. The last six decades of rapid simple growth has resulted in unprecedented levels of debt at all levels of government and the quintupling of personal debt. In addition there exists a large infrastructure deficit which has resulted in declining service levels in such areas as public transportation. Per capita income has declined outside of the government, finance, and construction sectors, and the vaunted Canadian world equality ranking has now fallen number 2 in the world into the mid-20s. The industrial base lost hundreds of thousands of highly productive, well-paid jobs through free trade agreements.

The assumption (or mantra) that growth lifts all boats has proven to be untrue for the last two generations of Canadians. The promotion of Canada as a young country that can continue on the path of growth forever may still be mantra chanted by media corporations daily, but its message is now broadcast from glass towers through large, congested cities sprawling over what was once prime farmland. The legacy costs of the heady boom days of mining, energy extraction, and fisheries are coming into focus as the grades of resources decline across the board and the social and cleanup costs mount.

Every stage of Canada's development has involved different contributions to the building of the country and its commercial markets. Wave of elites rose to prominence on the strength of their rapidly expanding sector, only to fall back as they were in turn edged aside by larger and more rapidly growing sectors of the economy. Some industries like the fur trade were absolutely critical to the building of the nation yet involved very few people and did not massively expand the local market. Hence, the fur trade would be awarded a 9 for nation building and a 2 for market building. Others like farming and railroads laid a solid foundation for nation building while also dramatically increasing domestic markets and increasing population size to push the country through the threshold of viability. The various stages of Canada's development are laid out in Table 11.1 with their contributions to different stakeholders broken down. Nation building benefits all, whereas market building benefits a limited portion of the population.

Finally, population growth past a certain point is unnecessary for national viability and even becomes a liability as the land is depleted and the resources available per person declines. Canada has been in a mild stage of declining per capita resources for decades (as has the world), but the growth of the commercial economy has masked this basic reality.

Assuming Canada transitions successfully to renewable energy and approaches sustainability, the commercial market may well stabilize or even shrink dramatically. Asset inflation will be eliminated. Hence the transition to renewables will be of great benefit to the fiscal health of the nation and to most citizens, yet it may come at a very high cost to the size of the commercial market. Lower house costs

Table 11.1 History of nation building vs. market building (author)

Development stage	Nation building	Market building
Fur trade	9	2
East coast fisheries	8	3
Timber	6	4
Farming	10	8
Railroads	10	10
Industry	8	8
Energy and mineral extraction	6	4
Finance	3	4
More housing after 1920	0	10
More population growth after 1920	0	10
More consumption growth	0	8
Transition to renewable energy	10	0

and smaller, if any mortgages will be a boon for most individuals and families but a huge loss for the financial community as well as speculators.

Looking Beyond the Past

It is natural to want to continue the policies which have been so successful, but for the past two generations, Canada has confused the size of the commercial market and its dynamics with national welfare. Certainly the industries which benefit from a growth market depend on its continuance. The strain of development on the land and social structure has been showing for decades, but national priorities have been set aside as commercial market growth became entrenched as the only national metric of success. In this commercial market-making environment, individual and national interests and the strength of the nation suffered but the real estate and financial markets boomed.

Wikipedia defines the term "**Robber baron**" "as a derogatory metaphor of social criticism originally applied to certain late nineteenth-century American businessmen who were accused of using unscrupulous methods to get rich. The term robber baron derives from the *Raubritter* (*robber knights*), the medieval German lords who charged nominally illegal tolls … without adding anything of value, but instead lining their pockets at the cost of the common good (rent seeking)."

In Canada, the predatory practices of a small fragment of the business class exploited their control of media, banks, and, inevitably, politicians to engage in highly profitable but socially unproductive enterprises which saw a huge transfer of wealth from the general public to a narrow elite. In the late 1700s the American radicals rebelled against the prospect of taxation without representation, and certainly no nation will stand for long-tolerating "invisible taxation without benefit" which is essentially the end result of asset inflation and increasing debt which are both driven by population growth policies.

Canada, as a society, has yet to differentiate between an increase in the size of the commercial market, now commonly and mistakenly referred to as "the economy," and national welfare and the common good. Until we clarify what constitutes the real product and real wealth creation, rather than cash flow and monetary transfers, we will be locked into policies promoting market growth while the health of the nation continues to decline. Markets always react, they are always working, but they only produce long-term positive outcomes for society, if they are structured to produce those positive outcomes rather than just short-term profits for a few.

From its inception to approximately 1960, colonial and Canadian governments enacted rational and coherent nation building policies. By 1960, the country was well and truly built, but powerful interests could only reap enormous profits if the commercial economy continued to get bigger. From 1960 onwards, the focus shifted to promoting irrational market growth against the clear interests of the majority of the population and leaving progressively less capability to transition smoothly to renewable energy. All previous elites in Canada's history from fur trading monopolies to timber barons, to industrial titans and oil tycoons have contributed to the overall welfare of the nation as it progressed through the stages of its development. They drove the real wealth creation process, but many of our current elite class have accumulated wealth through asset inflation rather than merely taking a share of the real wealth they created. On a national policy basis, it is critical to identify which processes create wealth without increasing debt and which simply transfer wealth from one hand to another building debt.

This matters because if government maintains its sights on maintaining the growth trajectory of the nation's formative years, it will lock itself into a position that is completely incompatible with the actions needed to assure a successful transition to renewable energy. This transition is absolutely crucial to evolving into a sustainable society. Going forward, there will be no end of work or opportunity on some scale, but in terms of material and energy consumption, it will be impossible to maintain the commercial market dynamics which were driven by an ocean of cheap fossil fuels to exploit and endless farmland to pave. The intensive exploitation of Canada's resources has been a once-in-a-nation step, and the fossil fuel era has been a once-in-a-species event. We should use what is left of these gifts to invest in the renewable energy future that will inevitably follow fossil fuel decline.

Government policy must be one of adapting to the new reality of measured energy availability. Many countries have been in this mode for generations, but for Canada, the USA, and Australia, it will be a new mindset. It will also represent a changing of the guard as the power elites of what was essentially a simple growth economy are replaced by the new renewable energy oligarchs. "Solargarchs"? This process will also likely see a return to prominence of the real wealth producing sectors of manufacturing and farming as international trade flows decline with energy budgets and national focus returns to nation building/strengthening.

However, if government policy remains fixated on maintaining business-as-usual for as long as possible, the nation will not walk into the renewable energy era, it will be dragged into it and have to play catch-up. No nation can continue to attempt to live the fossil fuel dream and remain prosperous. Not in the long term, but right now, we are operating on the principle that long-term survival is bad short-term business.

Policy makers need to set goals and measure options in real physical units and cease mistaking the ascendance of the finance economy for real product that can be classified as increased real national output. We have to stop mistaking effort or action for productive work. We cannot mistake the acquisition of large amounts of wealth by a few as a viable path toward wealth accumulation by all, especially if wealth acquisition comes as the expense of someone else's increased debt. The items below represent huge flows of money but not one iota of real wealth creation.

- Bitcoin, blockchain
- Housing inflation
- Stock inflation
- Currency trading

Canada's commercial economy has grown more rapidly than any other developed country save possibly that of Australia over the past five decades. This growth has largely been on the basis of population growth since real per capita incomes have largely stagnated.

Increased population size and material consumption are incompatible with sustainability and energetically impossible in the renewable energy era with any known technology. Until government policy stops trying to ride two different horses in two different directions, implementation of rational energy and environmental policies will continue to fail. The difficulty for governments is realizing their horses are headed in different directions and that they need to switch to the new horse in a way that minimizes social and economic disruptions.

Of course, there will be intense pressure from the power groups for whom business-as-usual is not only highly profitable but is essential to the continued existence of their business models. There will still be a demand for house construction and mortgages in the sustainable era but large subdivisions and housing inflation will be things of the past. There will be a very large market for energy system installations and for energy renovations, but these will not have the scale or the margins that have made large developers wealthy. Nor will the new energy economy likely lead to the placing of splashy multi-page advertising spreads in newspapers. Banks will be faced with a smaller number of mortgages for smaller amounts. Without housing inflation, speculators will simply have to look for new, and hopefully productive, lines of work.

These groups are very focused on their own success and have a very clear understanding of where their money comes from. Political donations and advertising expenditures are simply a cost of doing business for them. Policy makers will come under intense pressure to maintain and even expand the policies of simple growth, namely more people and more resource exploitation, and the pressure will come in backroom discussions and in the stories carried in the corporate media. There will be continued promotion of the myths which justify growth and the business-as-usual mentality of "if we just had a little more, all of our problems would go away…" that currently drives decision-making.

Myths and Misses

Making a successful transition requires keeping the end goal of a fossil fuel free economy front-of-mind. In the process of transition there will be setbacks, but there will also be opportunities. One such opportunity presented itself in the financial meltdown of 2008 when the Big Three North American automakers were in distress and had to undergo financial restructuring. GM received $60 billion in loans from the Canadian, the US, and Ontario governments. Instead of using this opportunity to develop and produce more energy efficient automobiles, the Canadian and Ontario governments signed off on a plan to produce the Camaro, a 40-year-old pony car which advanced the technology base of workers, suppliers, and the country not one whit.

Clearly, governments were in crisis mode at the prospect of massive losses of high paid jobs, and the top and probably only priority was to prevent a cascade of plant closures and possible financial chaos. But the opportunity to restructure the automotive industry for the production of more efficient vehicles was lost and $60 billion was used to maintain business-as-usual.

Shortly thereafter, during 2011 and 2012, gasoline prices reached record levels and just about everyone was talking about downsizing their vehicles to something more efficient. But instead of taking that opportunity to maintain prices and channel the increased revenues into efficiency incentives, gas prices were allowed to fall to near record lows by 2016, and the vehicle-buying public upsized to much larger SUVs and trucks. We now have the largest and heaviest private vehicle fleet in history, making the transition just that much more painful for consumers as they deal with higher energy costs which will be locked in for the life of their nearly new vehicles.

Myth #1: Populating the North

Northern Canada is largely "empty" of people, and this is often misconstrued as a growth opportunity which would see cities and towns, on the scale of those in Southern Ontario and the Lower BC mainland, constructed all over the "empty" north. But this has not happened, nor will it ever, given the high thermal demand and low thermal potential of the region. Aside from the lifestyle restrictions of very long, cold, and dark winters and very short and not very warm summers, the energetic realities of the north make it one of the most expensive regions on earth in which to live, never mind being one of the most environmentally sensitive.

The obvious considerations of domestic heat and long transport distances come readily to the southern mind, but there are many unique issues which may not. For instance, currently there are two 100 kW diesel boilers in the community of Kuujjuaq, in northern Quebec used solely for heating the drinking water distribution system. The water source is 4.8 km away from the village and in order to prevent it from freezing, these boilers must be kept running most of the year. If they both ran all year, the energetic cost would be 1.7 million kWh simply to prevent the drinking water pipes from freezing. That is enough energy to power 200 efficient homes in the south for a year.

Aside from simple resource economics, the northern most populations of the Arctic nations are typically the least healthy in the nation. Despite large subsidies to offset the high material good and energy costs inherent in northern living, the toll taken by long dark winters and short summers is glaringly displayed in health statistics for Canada's Nunavut. Here the rates of TB, cancer, and suicide are many times the national average.

The Arctic Human Development Report by Joan Nymand Larsen and Gail Fondahl study observes that "Greenland and Nunavut—with over 85% of the population Indigenous, there is a wide gap in health status between these regions and

Denmark and Canada." Clearly more needs to be done to deal with the problems, but just as clearly, the extreme north is not a health-friendly environment for large numbers of people.

Current policy in the north is to improve the lives and health of the native groups while maintaining them on their ancestral lands. The financial and energetic costs of this will be whatever it takes, but heavy food and energy subsidy programs should not be used as incentives to bring people to the north as some sort of continuance of the colonial policies build up the more hospitable south in the era of energy abundance. Living in the north requires a great deal of energy and resource support from the south. There may be some mining developments which will justify these expenditures, but population growth in the far north will never be anything but a resource sink for the rest of the society.

Whether it is populating Canada's north, the Sahara, the Himalayas, the Gobi desert, or the Antarctic, growth proponents need to address the energy and resource constraints involved. Large areas of the planet have remained unpopulated for thousands of years, during which time hundreds of waves of desperate migrants have set out looking for a better life or any life at all. If these areas held habitation potential, they would already be full. Space does not equal resources, and no space will sustain a human population if the energy available does not exceed the energy required for survival.

Myth #2: Infinite Substitutability

A concept in neoclassical economics that means that no matter what needed material is depleting or otherwise becoming unavailable, another material can always be found to takes its place. It is a relatively recent hypothesis based on a faith in human creativity and technological prowess. It assumes the available of infinite energy (Cairns 2000). For example, copper is clearly finite, but aluminum is abundant. Aluminum can replace copper in some electrical applications, but it does so only with a large cost in efficiency and efficacy.

Myth #3: A Larger Economy Can More Easily Spend Its Way Out of Shortage

The theory is that a larger economy can afford to spend more on energy. But as depletion progresses, by using increasingly less rich resources, there will be smaller amounts of energy per capita available. The commercial economy is nothing more than energy flows. Declaring that the bigger it is, the more it will be able to spend on energy, completely misses the point that a larger economy demands more energy and draws down nature's supplies more quickly. Again this is a theory born from the commercial economic theory that has no basis in physics.

A direct analogy would be that of a region with a population of one million people experiencing famine. The theory suggests that by adding another five million people, food production would increase. That might well be true, but the severity of the famine would increase as the amount or food available per person would decline even further. It is impossible to outgrow scarcity. Well-being is a function of per-capita measures, not the size of the economy.

Myth #4: Endless Growth Is the Only Future

Naturalists dream of sparkling water burbling over rocky ledges and granite pebbles and flowing through verdant pastures and towering forests. Social progressives imagine communities in which healthy children run down streets lined with trees past well-kept, modest, mortgage-free homes whose doors are never locked.

Commercial market fixated individuals dream of other things. They dream of bigger, constantly growing markets, more demand, and ever higher profits. They appear not to dream beyond the realm of the commercial economy. Hence we have the dreams of media corporation employees espousing the need for the Canadian population to triple to 100 million as soon as possible. A book by one media employee calls for a policy of even more rapid population growth for the world's fastest (or second fastest) growing developed country. The author has suggested that a "bigger" Canada would then be able to project power around the world in the form of aircraft carrier strike forces. A lofty goal indeed if true enlightenment needed to be distributed through the barrel of a gun.

Even further, Jeff Bezos, of Amazon fame, foresees one trillion humans occupying the solar system (NBC 2019). What else would the world's most successful retailer dream about other than an infinitely large market?? Not human consciousness in computer memory, but living, breathing, walking around and wearing things out, consuming human beings. These one trillion people would be housed in large spheres in space which would be built by construction robots using material from the moon and other handy planets. There would be a million of these spheres housing one million people each. And, of course, since everything would be 100% artificial, with all inputs and costs quantified, every action, every breath would be monetized, adding to the size of the GDP and the wealth of the people prospering from flows of money.

If the earth's rate of population continues as it is now, at 1% annually, the one trillion number would come up in a scant 500 years; 1400 years beyond that, the number of spheres required to accommodate the population would itself be one trillion, which might begin to crowd spheres out past Pluto. If the rate of growth were doubled to 2% annually, we would hit one trillion people in 250 years and one trillion spheres in another 800 years. A dream market if there ever was one.

Think of it in leaps of roughly 1000 years. Humanity goes from the Viking landings in North America in 1000 AD to the end of the fossil fuel era in 2100 to populating the solar system with one trillion small moons full of people by 3000 AD. What a market dynamics! (Table 11.2)

Table 11.2 Population growth over seven doublings

Year	World (or solar system) population (billion)	Canadian population
2020	8	37
2090	16	75
2160	32	150
2230	64	300
2300	128	600
2370	250	1.2 billion
2440	500	2.4 billion
2510	1000	5 billion
Number of people added per year in 2511	14 billion	71 million

Population growing at a 1% rate doubles in 70 years (author)

There is never any end in demand for growth from the people who clamor for it. The world's current rate of population growth is 1% annually. Growth advocates want this continued in perpetuity. At a 1% growth rate, numbers double every 70 years producing the above progressions. In a mere seven doublings over 500 years, the same time frame that would take us back to Columbus, global population tops one trillion, 120 times the current level and is growing at 14 billion per year. Over the same period, Canada grows to five billion or about two thirds of the current world population and is growing annually at a rate of 25 cities of Toronto annually.

Growth advocates have no end in sight, no means of getting off the escalator, just business-as-usual until the system fails. The planet's environmental systems are failing at our current level of population and consumption. Ever higher levels of population dictate the adoption of some form of technology which allows us to survive with no natural systems functioning.

Eternal growth, or attempting to grow beyond sustainable limits, is something that would profit very narrow interests very handsomely, but societal collapse is always the final dividend. The reason we are transitioning to renewable energy and conserving earth's depleting resources is to avoid collapse. New business models will find opportunity in the markets which thrive in socially stable and environmentally balanced communities.

Metrics Have to Represent Both the Goal and the Means to Achieve It

The fog of GDP metrics and the debt and cash flows of the financial markets all serve to mask the reality that all physical bills eventually come due. Ultimately we cannot consume what we do not produce and selling off assets while inflating their value and printing money are no replacement for a broad-based, efficient, and productive economy.

We have no more right to consume happiness without producing it than to consume wealth without producing it.
Candida, Act I (1898) George Bernard Shaw

In the age of cash flow and increasing debt, the above quote may not have as much resonance as it did when real product was easily identified and fiscal balance was an assumption. But if the transition to renewable energy is to be successful, national policy will have to be grounded in physics rather than printed currency.

Attempting to include the cost of natural assets in National Accounts to better inform the GDP metric is a blind canyon. National policy should not attempt to express itself by means of the commercial marketplace. Rather than engaging in the futility of attempting to monetize natural processes and social welfare, national policy should be created on the basis of maintaining natural assets and setting social goals within those physical restraints. The marketplace should then be left to operate as efficiently as only it can within those limits and standards. Then GDP can go back to ably performing the roll it was designed to fulfill, that of a marketplace and tax measurement tool to be used in the national interest rather than representing the national interest.

Clear Goals

If our goal is to reduce our carbon emissions in the range of 80%–90% in the next 30–40 years, and ultimately to become 100% fossil fuel free, then we not only have to rearrange the means by which we produce energy but also the way we live our lives. Fossil fuels created the illusion that we could consume many times more than what the earth can sustain. Many decarbonization discussions center around continuing what we are doing by replacing fossil fuels with green energy. As previously noted, there will be many efficiencies inherent in the adoption of electric power but producing enough at the right time to maintain fossil fuel era consumption growth levels is extremely problematic.

Stated another way, we cannot get there from here yet.

Finding something to replace fossil fuels is one thing, and building a society without fossil fuels is something very different. Attempting to replicate the excesses of the fossil fuel age, where 3000 kg trucks are used as grocery-getters and 400 horsepower sports sedans are used for urban transport will not be energetically possible. Personal vehicles of the sizes above may well use two fifths of a kilowatt hour to travel 1 km, whereas smaller, more task-focused vehicles can do the job using less than one fourth of the energy.

Rather than just harvesting energy in one place and consuming it in the other, which is the fossil fuel template, energy harvesting would be designed into every building and daily 100 km commutes would be things of the past. In all, there would be more effort spent on tools than toys, and a more utilitarian approach to activities but the overall result would be much healthier living and a much higher degree of environmental balance. A willingness to let go of past practices is at the heart of the ability to set clear goals.

Physical Reality: The Missing Link

Few decarbonization discussions go past setting targets using broad brush, monetary tools. They deal with numbers on a page and then apply tools used in conventional economics to develop policy. But they do not get down to the point of quantifying the infrastructure, let alone characterizing the monetary and carbon costs needed to bring the projections on paper to real life fruition. They do not spell out the physical transformation that must take place. Below are a few simple hardware estimates. They are not based on the renewable energy necessary to replace fossil fuels of the current economy but on a ground up transformation of the way we will harvest and use energy in the future. They give an indication of the lifestyle we can realistically hope to achieve with a very aggressive buildup of our renewable energy-generating capability and of the complete overhaul of our buildings and transportation fleets.

In these simple examples, we take the best of technology from the real-world experience of Drake Landing, Art Hunter, and John Howe, add considerably to them through best practices, and complete energy integration into both district heating/microgrids and the national grid. The reader must keep in mind that even at this late stage in the energy/climate conversation, there are no real data because there are no developmental building sites into which a complete array of energy conservation and energy harvesting technologies have been built. Oil and natural gas are still used for limited critical application, and the ghg emissions stay within the range of 10–20% of 2019 levels. In other words, there would be an 80–90% cut in emissions and vastly longer lifespans for remaining fossil fuel reserves.

It must also be pointed out that the very large infrastructure investments made, the physical build, will be done mostly using fossil fuels. The ability to make such large changes, so quickly, over 30 or 40 years, depends on the burning of stored energy, rather than the use of our current trickle of renewable energy flows.

In Canada and other northern nations, the number one problem in the adoption of renewable energy is energy storage over the winter season. This level of storage simply cannot be achieved using electrical batteries now, and there is no technology on the horizon to suggest it will be practicable in the foreseeable future. Hence, low-grade geothermal energy storage is the only option, and fortunately, it appears to be quite viable in that singular application. In the very far north where soil temperatures are below freezing, even geothermal storage may not be viable, but for most of the country, it holds promise. The second problem, of course, is that generating energy at Canada's latitude from solar sources is both less productive and more variable than in more southern regions. Transitioning to renewable energy will be a challenging process in any part of the world but is increasingly so the further north a country is located.

Buildings of the Future

The illustration below outlines the major energy systems which can go into buildings in the near future that can work in cold latitudes. The outer surface could be described as "an active skin," "smart skin" for the marketing oriented, because it will actively be absorbing energy, repelling it, or holding it inside. In the example below, collectively speaking, buildings will generate all of their annual electrical energy and will transfer temporary surpluses to the grid or to an internal electrical battery or EV or convert it into heat for geothermal storage. Production peaks and demand peaks will be absorbed or supplied by the grid.

With two geothermal masses, it will be possible to have both heating and cooling storage. As cooling requirements cease in the fall, both masses would be used for heat and in the spring, one could again be used for cold storage. Building tens of thousands of kWh of storage capacity into every building makes renewable energy much more viable in northern countries. The system efficiencies involved, and the grades of the energy, are vastly different from those of the fossil fuel era which is why direct comparisons of current energy use and projected energy reduction targets are largely meaningless. They assume that current patterns of energy use will continue and therefore ask the wrong questions (Fig. 11.1).

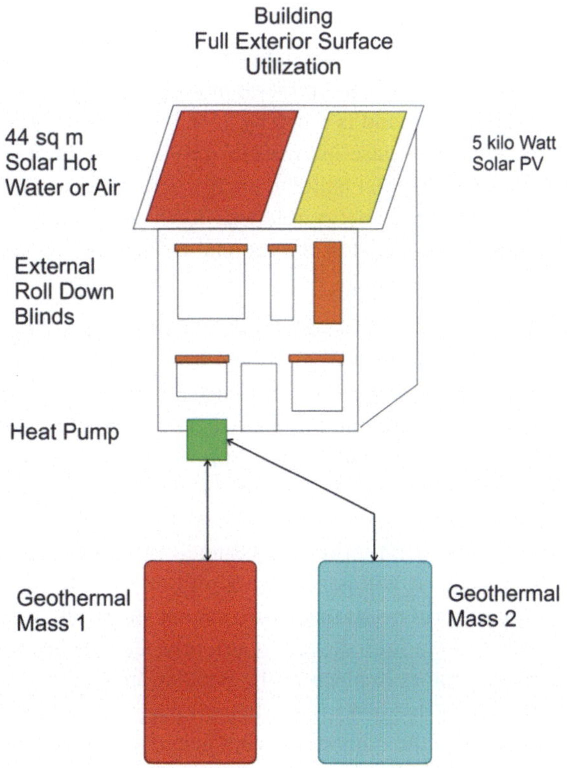

Fig. 11.1 House optimized for energy harvesting and conservation (author)

Renewable Energy Infrastructure from the Ground Up

It is not enough that buildings be made to be much more efficient than they have been in the past; in the future, they need to be active contributors. They need to be able to harvest energy and actively store it as well as leverage the energy they do use. In the example national energy system below, all buildings are at least energy neutral. Additionally, residential buildings have enough of an electricity surplus to power the private passenger vehicle fleet. With a 10 kW array on a residence, the household's surplus energy could power an efficient electric vehicle 30,000 km annually. Hundred percent of the flat roof surface of commercial buildings is covered with either solar hot water or photovoltaic panels. Commercial buildings are assumed to be nearly energy neutral and have no spare capacity to export energy.

Additional assumptions compared to 2017:

- Commercial energy demand is reduced by 80% due to on-roof solar PV and hot water.
- Industrial demand is increased by 33% and is discussed further below.
- Agriculture, fishing, and forestry sectors' energy consumption remains the same and remains fossil fuel based.
- Public administration demand is down by 50% for same reason as commercial.
- Construction demand is down by 30% due to less new build and more upgrading.
- Mining and gas demand down by 70% due to shutdown of oil sands and reduced conventional oil and mineral mining. The sector remains fossil fuel driven.
- Pipeline demand is down by 75% but still needed to deliver natural gas.
- Sale of retail gasoline goes to zero as ground transport fleet is 100% electric.
- All told, the total national energy demand goes down by 63%.
- Grid supplied energy totals 63 kWh per person per day. Self-generated energy from residential and commercial rooves is not counted.

This produces Table 11.3:

Using the Drake Landing data, solar hot water 29 panels were used for each residence, and if 17,300 W pv panels were added to create a 5 kWh array, there would be a total of 46 panels per residence taking up 70 m^2 or 700 ft^2 of roof space.

There is an infinite number of energy supply mixes which can be adopted but only one is shown below. The infrastructure build to generate the 466 billion kWh is as follows if we use a mix of 30% natural gas, 20% solar, and 50% wind (Table 11.4).

A 2 MW wind turbine is assumed to have a capacity factor of 35% in Canada which means it will generate 5.2 million kWh annually. If we make up 50% of the energy the grid needs that hydro cannot supply, then we would need to build 44,381 of these 150 m high devices. Currently, Canada has the equivalent of 6000 of these wind generators in place so the population will have to be multiplied by a factor of 8. If wind is deemed to be a better energetic investment than solar PV in different regions, then the number of wind generators would have to grow accordingly.

Table 11.3 Annual energy demand through grid (calculations by author from national statistics)

Total annual demand through the grid	848,000,000,000 kWh
Deduct Hydro 2017 output as the base load	382,000,000,000 kWh
Therefore wind, solar, and natural gas need to make up	466,000,000,000 kWh

Table 11.4 Infrastructure needed

		30% NG, 20% S, 50% W
Commercial building PV panels + hot water	per capita	9
Solar farm PV panels	per capita	19
Residential PV panels and hot water	per capita	17
Total panels	per capita	45
Roof surface needed per residence	(m^2)	70
Total number of solar panels nationally	PV + hot water	1.7 billion
2 mW Turbines (150 m tall)		44,381
Number of 1 gW natural gas plants at 70% capacity		23
Number of 1 gW natural gas plants at 50% capacity		32
Number of 1 gW natural gas plants at 20% capacity		80
Total annual ghg emissions	mega tons	72 mt
Percentage of 2019 ghg level	of 700 mt	10.3%
Percent reduction from 2019	ghg level	89.7%

The power production timing of these two renewables is very different, and the capacity factors, in Canada, are also wide apart at 12% for solar PV and 35% for wind. The capacity factor for solar hot water is a question mark at the moment but is likely several times higher than solar PV. Another difference between solar panels and wind turbines is that solar panels can be deployed almost anywhere and, most importantly, can be deployed on the wasted space of building roofs. This puts them within meters of the energy points of use and makes the generation of low-grade energy (heat) practical.

Wind turbines are typically "green field" projects, ones that are installed in forests, fields, and possibly shallow water meaning that there is little rebuilding of manmade infrastructure to perform. Solar panels, on the other hand, will require extensive modification of house roof structures with commercial flat roofs requiring much less radical alteration.

For comparison, a 1 gW natural gas plant has close the same output as one Candu nuclear reactor. The Pickering Nuclear plant has four Candu reactors which gives it about the same output as the Niagara Falls hydroelectric facility.

The number of natural gas plants needed would depend on the combined ability of the hydroelectric plants and the natural gas plants to cover the peaks in demand. The smaller the peaks, the smaller need for more NG plants as they would be used at a higher capacity level. If the plants can run at an average of 70% capacity and demand peaks can be filled with the remaining 30% headroom, then we would need only 23 plants.

If demand peaks were very dynamic, then we would need possibly 80 plants running at a low average of 20% capacity but able to produce five times that amount when needed. Storage will moderate demand peaks, but it remains to be seen by how much. In general though, high renewable penetration of renewables is much better enabled by high thermal storage and the leverage offered by heat pumps. A geothermal heat pump in non-heated earth (say 10 °C) might expect to have a COP or 4:1 (coefficient of performance 1 W in, 4–10 W out). If the earth is storing heat and is sitting at 30 °C, the COPs will go dramatically higher. The core temperature of the Drake Landing reaches a maximum of 70 °C.

The above example assumes the current Canadian population of 37 million. Once the population falls to the 20 million level, it is probable that hydro can completely replace natural gas as the quick reaction base load, leaving only those fossil fuels used in critical applications for which there is no electrical substitute. Further, natural gas use could also then be reserved for its vital roles of fertilizer and plastic production. But if population is pushed higher, more fossil fuels would be required in addition to larger solar and wind energy infrastructure. Natural gas reserves will be drawn down faster leaving less fast reaction base load capacity to support renewable sources. Increasing the population level either assures the failure to transition to renewables or, in the best case, it means achieving sustainability at a very low level of material comfort and social sophistication.

The EROIs for these complex systems remain a large and uncalculated question. The much higher efficiency levels of the new, ground up renewable energy system, may allow a higher standard of living at a higher latitude at lower EROI levels than may have been thought possible previously. In any case, a full energy accounting program needs to be in place early on in the transition process, and it should enable the most productive directions to be taken while perhaps avoiding some false starts.

The China Factor

Despite massive cuts across the board in this example, manufacturing energy consumption is increased by 33%. Why? Because domestic manufacturing stands an excellent chance of resurrection in a renewable and localized world. Over the past four decades, not only has the western world exported jobs and pollution, it has exported the energy consumption needed to make the goods it imports. Global world trade is currently taking place on the back of China's enormous coal consumption. The energy embedded in Chinese manufactured exports amounts to 3.2 billion barrels of oil annually.

The number of 300 W solar panels needed to replace the energy output from those 3.2 billion barrels would be 14.8 billion (at a 14% capacity factor). This would be equal to a daily energy budget of 11 kWh for each person in China; a huge amount in a world of precious energy.[1]

Once the transition to renewables has been accomplished, will it be desirable for China to continue large-scale exports if it is even physically possible? This is one issue which appears to be blindingly absent from Chinese papers. Is it in China's interests to continue to be the worlds' factory? It is hard to see what the benefit would be if it means choosing to reduce the amount of energy available for use by China's people. Will China need to trade for raw material other than what it already controls in Asia, the Middle East, and Africa? China's planners must certainly have considered the question, but it does not appear to be addressed in public documents.

China has a huge and well-developed domestic market and is extending its control of resources in Asia via the Silk Road and also in Africa through corporate farms and railroads. Certainly this requires some exports to pay for imports but will they need to export as much as they have?

The energy needed to move goods across the globe, while not immense, as ships and rail are very efficient, is still considerable and in a world where clean energy will likely be a precious commodity, a critically strategic commodity. In fact, few nations will be in a position to be net exporters either of raw energy or of energy embedded in manufactured goods. And manufacturing is not a one-way street. Currently, recycling material is often sent back to the countries which are producing the goods. Hence the manufacturing base in many developed countries will broaden once again and the material and energy inputs into domestically manufactured goods will likely increase.

Overall, we can expect material goods consumption to decline sharply as higher quality and greater upgradability of goods increase lifespans. But if foreign producers reduce their output of manufactured goods, not only will Canada and other developed nations have to make much more of the goods they consume, they will have to both produce a large part of their own renewable energy infrastructure and maintain it. This is critical. Right now most developed countries are planning to import their solar PV panels at Made in China prices with no domestic energy costs involved. It is quite possible that most countries will have to produce most of the goods they consume, and this will be done paying domestic rates for wages, energy, and taxes.

---

[1]Reference: 433 kg CO@ from 1 barrel of oil, 2014 China net emissions embedded in trade = 1369 mt

1369 mt = 3.2 billion barrels of oil

(reference: https://www.carbonbrief.org/mapped-worlds-largest-co2-importers-exporters)

(footnote: 3.2 billion barrels of oil = output of 1700 kWh = 5.44 trillion kWh or

1 × 300 W solar panel has output of 263 kWh at 10% capacity factor and 368 kWh at 14%.

5.44 trillion/368 = 14.8 billion solar panels or 10 solar panels per person in China.

5.44 trillion kWh = 3800 kWh annually per person = 11 kWh per person per day.)

Power Shifts: Provincially

Political power generally rides an economic horse. Historically in Canada, the provinces with the largest populations and biggest manufacturing bases have held the most power, namely Ontario and Quebec. In the last several decades, the energy powerhouse of Alberta has ascended as the economic health of Ontario has declined. This is not to say it is not a powerful province, but the economic basis for its power is in decline, and the consequences of being an energy have-not province will play out in some form in the coming decades.

How do things change for Ontario in a renewable energy world? In 2017, Ontario generated 153 tWh of electricity. Once it shutters its nuclear plants (scheduled for 2037), it will have to find another 87 tWh of energy.[2]

Its neighboring provinces of Manitoba and Quebec plus more eastern Newfoundland have a lot of water falling off rocks and generate annual surpluses of 10.7 tWh, 4.9 tWh, and 28.4 tWh, respectively. Even if these provinces were willing to direct all of their surplus to Ontario, instead of their current customers, that still leaves Ontario with just two thirds of the electrical power it has now.

Possibly, Ontario can quintuple its wind output to 60 tWh and increase its solar by a factor of 10–30 tWh, but it will likely only be able to draw from the neighboring border states if solar pv plants in the US Southwest begin to generate large surpluses. As the US shuts down coal power plants over the next several decades, the availability of energy for export may be questionable.

Although manufacturing is deeply rooted in Ontario, this may change due to the draw of reliable and relatively competitively priced energy in the provinces where it is produced. This might be enough to siphon off a portion of the goods production

---

[2]
- In 2017, Ontario generated 152.0 tWh of electricity (Fig. 11.2), which is approximately 24% of total Canadian generation. Ontario is the second largest producer of electricity in Canada and has a generating capacity of 40,123 MW.
- About 90% of electricity in Ontario is produced from zero-carbon emitting sources: 58% from nuclear, 22% from hydroelectricity, 8% from wind, and 2% from solar. The remaining 10% is primarily from natural gas, with some biomass and diesel. Ontario's electricity generating capacity is primarily located in the southern portion of the province with significant hydro generating stations located in eastern Ontario in the Ottawa River basin and northeastern Ontario in the Moose River basin.
- Three nuclear stations with a combined 13,500 MW of capacity provide the bulk of Ontario's baseload generation. Bruce Power on the east shore of Lake Huron is the largest, with eight generation units and a capacity of about 6600 MW. It is one of the largest nuclear power plants currently operating in the world.
- Ontario has over 66 hydroelectricity generation facilities and a total hydroelectric capacity of 8872 MW.
- Ontario leads Canada in wind capacity. About 4826 MW of wind capacity was added between 2005 and 2017.
- About 98% of solar capacity in Canada is installed in Ontario. In 2017, solar in Ontario had a total capacity of 2291 MW.
- Ontario Power Generation is the largest utility in Ontario's competitive electricity market, with over 16,000 MW of capacity (NEB 2017).

sector's highly paid jobs and taxes. The cost and availability of energy is a basic mainstay of social and political power and as energy availability shifts in different regions, there will eventually have to be shifts in political power.

Different agendas set by different governments with no overall authority, as is the case in Canada, have a debilitating effect on national performance. In Canada, the difference in the interests of the provinces with regard to fossil fuel as an income source is almost as great as the differences in the ease with which the different provinces could be expected to transition.

Alberta has an immense stake in the fossil fuel industry but has very little hydroelectricity. It does have significant wind and solar PV potential, compared to the rest of Canada. In contrast, Manitoba and Quebec have next to no fossil fuel resources while being quite rich in hydroelectric capacity. Only Newfoundland has a significant stake in both. In the medium term, it must be pointed out that Alberta's large natural gas reserves would allow it to both continue to export the gas and construct generating plants to replace oil and coal throughout its economy. Given its population size and potential energy base, Ontario will be a net importer of energy, no matter the type.

Gasoline Taxes

Tax structure will also shift as fossil fuel consumption declines. On average at present, Canadians pay 45 cents of tax per liter of gas and 39 cents per liter of diesel making up 33% of the pump price for gasoline on average and 30% for diesel. Federal and provincial governments will collect an estimated $24 billion in fuel taxes in 2018 (Taxpayer 2019). For passenger vehicles, the current gas tax of 45 cents on 37.2 billion L resulted in $16.7 billion of revenue. The distance driven was 349 billion km.

Once the automotive fleet has been fully electrified, the electricity necessary to propel the same number of total kilometers will be close to 52 billion kWh. The average cost for a kWh in Ontario currently works out to about 20 cents in all. The tax needed to cover the loss in gas tax revenue would be 32 cents per kWh. But that would mean taxing all electricity, since it will be impossible to distinguish the energy input into personal vehicles. Furthermore, in most cases, drivers will be recharging at home or at the office from the power being generated by solar panels on the roofs of the buildings. The electricity may well not even pass through any grid meter.

Hence, to cover the cost of road building and maintenance, either a flat annual fee will have to be charged per vehicle or vehicle owners will have to pay on a per distance basis. Right now that would work out to 4.8 cents per km. If taxes were to be levied on a flat charge per vehicle basis, each of the 19.7 million light vehicles in Canada could be expected to pay $850 per year as a mobility tax. High mileage drivers would benefit at the expense of those who drove less.

The road tax conundrum presented by EVs is also being studied in the United States, where electric car owners in Illinois are facing an increase in their annual registration fee from $17.50 to $1000.00 (FoxNews 2019). But does it need to be a conundrum?

Tax policy is usually designed to be fair, but in the case of the shift from fossil fueled vehicles to EVs, fairness is exactly the opposite of what is required. The sooner EVs replace our current fleet, the faster carbon emissions will drop. This cannot happen quickly enough. In order to maintain road maintenance revenues and encourage the rapid retirement of the internal combustion fleet, ICE vehicles should pay an additional $100 licensing fee which goes up by $100 each year for 10 years, assuming the annual revenue target is $1000 per car. At the end of 10 years, EVs begin to pay a $100 license fee increasing by $100 annually until the $1000 is met. At the end of 20 years, there will be no ICE vehicles on the road, and the EVs will be carrying the entire road maintenance tax burden. During the first 10-year period, ICE vehicle owners will have an increasing incentive to convert to electric as the would be paying road tax as part of their gasoline cost in addition to an escalating licensing cost of up to $1000 a year while the EV driver is paying no tax at all.

There will also be shifts in tax flows as energy production and manufacturing centers move around the country. If the oil sands are shut down, it will mean the stoppage of close to billions of dollars in annual subsidies from government spent, as well as the loss of income tax from a large number of highly paid workers, concentrated in and around Fort McMurray, Edmonton, and Calgary. Will wind, solar, and manufacturing provide enough new jobs to fill the void left by the shuttered oil sands? This speaks to the need to diversify industry and broaden the economic base to increase resilience in the face of large, sudden, and unpredictable changes to vulnerable sectors.

Tax policy will cross jurisdictional boundaries with cities exercising little control over emissions policies, but still able to create EV friendly infrastructure. Robert Hoffman, president of What-If Technologies, outlines the considerations as follows: "fuelling infrastructure for non-emitting vehicles as such infrastructure must be provincial, national or even international, remembering that as much as half the personal use vehicle kilometers occur outside municipal boundaries." But "cities can influence and control building codes and the development of transit and cycling infrastructure, although they depend on higher levels of government for funding." Further, "the penetration of non-emitting vehicles into commercial fleets such as taxis that are licensed municipally and city owned vehicles but overall, municipalities lack jurisdiction and the tax base to take actions needed to meet emissions targets within the time frame."

Currently, a carbon tax is a fairly broad tax since so much of the economy is fossil fuel driven. It will have more focused effect as an increasing spectrum of the economy electrifies. Where specific lagging sectors or problems become apparent, additional specific consumption of taxes could be levied to encourage more beneficial consumption patterns. For example, if good health is a national objective, it makes more sense to tax cigarettes, alcohol, and potato chips heavily while leaving broccoli and carrots tax free. Taxes are always resented, and ones designed to be discriminatory will face the greatest resistance, but the tax mechanism is one of the most effective means of modifying behavior short of outright bans and restrictions. If cutting carbon emissions is a central national goal, taxes will be used extensively to drive the needed cuts forward.

Power Shifts: Elites

Political power will shift away from development and finance to the new wealth creators of resurrected manufacturing and clean energy. In the tradition of industrial leaders like Henry Ford, Elon Musk, and Bill Gates, real wealth producers may once again become leading powerbrokers in place of the anonymous bankers and developers who have driven government policy toward simple growth for two generations after the Second World War. Although declining, oil will be a powerful voice for decades as will natural gas until they both deplete, over many decades, to insignificant levels.

In a successful transition, where progress is measured in physical metrics, it is realistic to expect leaders to arise from the sectors of the economy which deliver real benefits to the general population and are part of the transition. It is difficult to see how a successful transition would be led by individuals whose economic interests are embedded in a continuance of business-as-usual.

Real Options

In short, we cannot go back to the 1700s when some 20,000 Canadian colonists used their steel axes and saws to chew through immense stands of virgin forest. That policy will not work for a population of 37 million on a forest resource base, substantially displaced by roads, towns, and shopping malls. At 5 bush cords of firewood per residence per year (17.5 m^3), we would need to burn 245 million m^3 of wood to heat all of our 14 million residences. In 2016, 156 million m^3 of wood was harvested in Canada (Natural Resources Canada 2019). Going back to firewood, for primary residential, not commercial, heat besides being incredibly inefficient and polluting would require a 160% increase in wood production to a level highly unlikely to be sustainable. It would also move the source of woodsmoke equal to 25% of our forest fires into the center of our major urban areas.

We cannot rely on fossil fuels because they are running out and their use is killing the planets biosphere. Moving forward with much higher levels of efficiency and clean sources of energy is our only option.

Perhaps, one day, we will build small black boxes which generate immense amounts of clean energy or we will develop the ability to efficiently transform low-grade heat into electricity. But until some currently unknown, sustainable technology arrives to restore the energy bonanza of the oil age, we are left with the option of doing the best we can, as quickly as we can, with renewable energy and its conservation.

Moving Forward

People will drive change if given the opportunity. Certainly there will be a sizable contingent of nay-sayers and foot-draggers, but there exist a large number of energetic and motivated people very interested in making progress on the renewable energy transition. This group only needs the tools and the removal of obstacles to make rapid progress.

Governments should be enabling early adopters and developers through the provision of incentives and the removal of red tape. They should build the information networks enabling easy access to research and scientific expertise. There needs to be a central council on energy that provides low cost access to experts and testing and an expedited National Research Council clearing house for research projects. Regional learning and research centers can be energy templates to encourage optimizing a diversity of technology and region-specific solutions.

Communication and Pitfalls

Society cannot afford the setbacks like the willful misrepresentation of "curtailment" and cost of energy fiascos in Ontario we have experienced stemming from the failure to explain the inevitable hitches in new policy. Nor can progress be made while Yellow Vest protests, essentially triggered by a 4% hike in diesel prices, in France constitute the means of public discussion. Higher prices and mistakes and inefficiencies are inherent in any change of direction, and side effects like curtailment in the early days of renewable energy will decline rapidly and eventually go to zero in a mature system. In the middle to late 1800s, railways built Canada, but the debt accumulation their construction required almost bankrupted the government. For decades it would have been possible to get a full night's sleep while lying on the tracks without fear of being run over, so infrequent were the trains in the Prairies. Yet the trains had to come before the nation could develop and the tracks had to be laid ahead of the trains' arrival. Similarly, charging networks will enable the rapid adoption of EVs because without the necessary enabling infrastructure, they cannot rapidly become a mainstream alternative to the internal combustion engine. Policy makers have to be able to make a coherent case to the public.

China has a clear advantage in this kind of massive transition. The government controls communication to a large degree and has managed to encourage a high level of innovation while still maintaining a state-directed economy. But does a change as profound as the shift to renewable energy require state planning and control of both policy and people to the degree practiced in China? Can the traditional western democracies change successfully without going onto what is, in effect, a war footing where the society is solely focused on one all—important objective? The answer may be yes, if the transition can begin soon enough to allow adequate time for existing economic structures to adapt. If strong measures are put off and time becomes short, then central control will have to become stronger as the potential failure to transition looms larger. The later the transition is left, the more rapid the change must be. Conversely, the sooner changes are implemented, perhaps the more arbitrary it seems, as the crisis, "if there is one" is so far removed for many people. However, in terms of climate change, public awareness is increasing rapidly, and early initiatives are both vital and easy to implement.

The national conversation has to improve beyond the current state of polarized and commercialized information filtering and manufactured news. Mark Twain said, "if you don't read the daily newspapers you are uninformed. If you do read them, you are misinformed." Success in transitioning will require a better informed, expert intensive, minimally sensationalist, more open, and less factionalized national discussion.

The Energy: Dollar Disconnect

National options must be framed by physical reality, in energy, environment and resources. But individuals live in a dollar environment, and policy makers will have to bridge this disconnect by some means to make sure that energy policy is made transparent to the public as a necessary and beneficial program. Easing the pain of transition will involve incentives to conserve and possibly short-term income supplements.

Long-term subsidies are a trap however, as they bake inefficiency into the system and rob it of its ability to innovate and adapt. When subsidies come to an end, as they inevitably do, the fallout takes the form of higher than necessary levels of pain and social disruption. Consumption always needs to be directly connected to real costs but costs have to be distributed manageably and gradually. Unfortunately, fossil fuel subsidies across the globe amounted to some $400 billion in 2018. That means there was $400 billion less of a reason to become more efficient, and it also means there was $400 billion less to spend on conservation and renewable energy infrastructure. The report from the International Energy Agency (Matsumura 2019) indicates these subsidies were put in place "to meet social policy objectives" (read: take pressure off the sitting governments) and illustrates the disconnect among social, energy, and environmental policy worldwide.

Public policy cannot be set completely by technical metrics, it has to be within the range of acceptability to the public, and here timing and perception become key factors. Does the public mind respond to unusual weather and reports that extinctions are increasing dramatically or does it only respond to direct impacts such as food and fuel shortages and local catastrophic weather events? According to scientists, what we do now will determine the events we experience 30 years or more from now. Will the public accept sacrifices now to offset future disruption? And what degree of sacrifice?

Governments will have to undertake "privation policies" which will only pay off decades down the road during which time conditions may well worsen, providing little evidence that the sacrifices are worthwhile or even working. When can we declare "Mission Accomplished"? Probably never, and from some quarters the cries of "Are we there yet?" will be unrelenting.

Winston Churchill's "blood sweat and tears" speech, which is remarkably relevant to our current situation, took place when the United Kingdom was fighting a losing war on several fronts, its cities were being bombed and the homeland was under imminent threat of invasion. Visualizing the issues at hand was not a problem for the British people (Churchill 1940).

Also, the implication was that after the crisis, normalcy would return. That is, after the challenge to an imminent and well-defined threat was met, life-as-usual would return. But the transition to renewable energy will see no return to fossil fuel norms because we will be transitioning to a sustainable society, a normal that western societies, at least, have not experienced for millennia.

The transition might appear to be similar to evacuating a functioning luxury cruise ship and getting into lifeboats in a rough sea, but policy makers have to make it clear that, in the foreseeable future, the cruise ship will no longer be viable transportation as it will have disappeared beneath the waves.

Energetic Equality

Social stability is a prerequisite for successful transition. Social stability can be assured either by informing and involving the public well and by assuring a high level of fairness or social stability can be enforced by iron-fisted measures that crush dissent and leave no other choice but complete compliance with government directives. The chart below displays the energy intensity of personal income levels by country. Mostly, this indicates that people in northern countries spend more a larger portion of their income on energy than do those further south.

It also displays the disproportionately high percentage of income spent on energy by the lower-income groups in northern countries. One notable exception is that of the Netherlands which can be considered fairly northern, yet its expenditure on energy appears to be, if not Mediterranean, quite low. The other very notable feature is the evenness or egalitarian distribution of relative spending on energy through the spread of income groups. The policies that led to this outcome would themselves be worthy of detailed study (Fig. 11.2).

Lower income groups are more sensitive to energy cost increases. They spend more on the basics, and they have less financial room to move. Social friction will be increased by the rising costs of energy, particularly if these increases are felt more by lower income groups. Since energy costs disproportionately impact the poor, then they should be the disproportionate beneficiaries of conservation incentives. If the pain is too great for some individuals, then subsidies or supplements can be added to their gross income, but real costs should never be obscured and consumption never encouraged or subsidized.

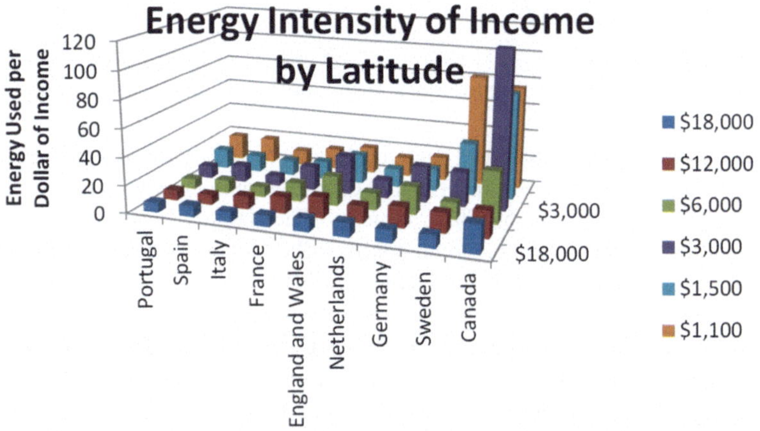

Fig. 11.2 Energy intensity by country and by income group (Unger and Thistle 2013)

The collapse of Syria (Ahmed 2017), the Yellow Vest riots in France and the backsliding in Ontario demonstrate the necessity of a well-explained and egalitarian approach to the energy transition. Egalitarianism rides a fine edge between being seen as a "free ride" and punishing individuals in poor circumstances. Francesca Dobbyn, a social worker in Ontario, notes that putting too much pressure on people results in their making poor decisions. She terms this the "tyranny of the moment" and "tunnel vision," whereby poverty creates conditions in which people have to deal with everything in a short time period, and it costs much more than taking a longer view. Researcher Jack Alpert would explain this using chaos theory.

Dobbyn finds that the shame associated with poverty is debilitating and that most people want to pay their bills have independence and be respected. When the threat of winter hydro disconnect due to late payments was put under moratorium, the hydro utilities experienced fewer bad debts. Also, when poor people were given more money, they smoked and drank less. The message seems to be less stress means better decisions, and this undoubtedly applies to people of all incomes as well as to governments.

Decisions by government will always be seen as arbitrary by some, no matter how much context they are given and how well explained and transparent is the process. Like taxes, which are always too high, and services, which are never good enough "for the taxes I pay, we should be getting much more!" Clearly, a coherent and comprehensive approach is necessary to the social policy aspect of the energy transition because a workable path to transition will not be arrived at when there is chaos in the streets.

A national interest approach to national policy making is very different for the Canada of the last 60 years, but not so different from that of the nation building period. Nor is it a different process for those countries which have a comprehensive, long-term view of public welfare, a shortlist of whom would include Norway, Denmark, China, Germany, Singapore, and Japan. Names not on the list would be post-Lougheed Alberta, Venezuela, Ontario, North Korea, and the UK.

What Is the Solution?

It is far more likely we will settle on a sustainable path that requires continuous effort, than arrive at a nirvana-like solution. There will probably be never-ending tension between human demands and the renewable resource and energy flows nature can provide. We have to be prepared to adjust continuously, as the planet and the universe are very dynamic environments. The natural world under stress is becoming much more dynamic, but humanity will undoubtedly become much more efficient with the resources it has available for it. But moving beyond fossil fuels, it is critical to distinguish between low-grade energy and high-grade energy. In a renewable world, high-grade energy will be mostly electric, and it may be expensive to store, but it is very easy to transport. Low-grade energy, namely heat, is next to impossible to transport, but it may well be easy to store. Low-grade energy and high-grade energy are used for different tasks. These distinctions have to be built into planning.

For southern, sun drenched latitudes, it is possible to go a long way further toward the complete decarbonization of society in a fairly short period—maybe two or three decades. For northern countries, the path is much steeper and progress harder to come by. Given the energy demand and the lack of solar power in the periods when demand is highest, huge amounts of storage will be required. The solution, if there is one, is recognizing that we will always be on a path rather than expecting to arrive at a destination.

Leadership

Energy transition means social transformation in the way we live, the way we work, the way we trade, and in the changing of the guard of the power elites. We will have to cease activities which are generally regarded in the halls of power, as economic mainstays.

In sum, there are initiatives which run counter to current policy, and some which are very complementary. Improving public transit, aggressive energy conservation programs, district heating programs, increasing recycling, and re-building manufacturing infrastructure do not conflict with business-as-usual interests. The following list of initiatives will be necessary components of a successful transition but are more or less directly in conflict with BAU:

- Stop population growth
- Stop urban sprawl
- Close oil sands
- Reduce conventional oil output
- Generally reduce consumption
- Broaden base of economy and increase local production

However much of the implementation of these initiatives would be of benefit for most people, for much of the power structure and media in western, high growth countries, the above list reads like a recipe for economic suicide. For a narrow portion of those economic elites, it certainly is. But for the vast majority of stakeholders, given enough time and well thought out transition policies, economic Armageddon can be avoided.

Governments must use the proper tools and not be swayed by estimates of staggering costs with their implication that, because of high costs, survival is not worth discussing (LSE 2018). Transition is going to happen and government must hold onto the course which sees it succeed to the highest possible level, as smoothly as possible. It should be made clear to all that a healthy commercial market can only exist in a stable society. Policy makers need to adopt real physical metrics because only a grasp of what is occurring in the physical world can prevent the chaos of the moment, whether physical or financial, from derailing rational policy implementation. Any nation which does not have a clear understanding of its relationship with energy and the environment is doomed to suffer a litany of seemingly incomprehensible shocks and setbacks.

Governments are loathe to pick commercial enterprise winners, but they must be able to at least pick a direction for their incentives for innovation and conservation to work. Backing research and test facilities are much more surefire than backing specific untried technologies. Wide-ranging test programs, expensive as they may be, allow the most promising technologies to be identified. They can then be brought up to the level of inclusion in real-world projects, most likely in leading communities which experiment and develop new processes. At this point in 2019, the technologies which have demonstrated clear promise are heat pumps, building surface energy control, EVs, solar and wind, and the projects which integrate these for use in district heating, micro-grids, and main grid integration.

The policy process will have to be open enough to enable experts to speak information to power and to the public.

Consistency of support over decades for renewables and conservation is vital. As noted by the International Energy Agency's Executive Director, Dr. Fatih Birol, in May 2019, "*Energy investments now face unprecedented uncertainties, with shifts in markets, policies and technologies. But the bottom line is that the world is not investing enough in traditional elements of supply to maintain today's consumption patterns, nor is it investing enough in cleaner energy technologies to change course. Whichever way you look, we are storing up risks for the future.*"

Risks and Instability

Change is best carried out in a stable environment since shocks will impede the success of the transition. Shocks are, by definition unpredictable, in both time and nature, but some can be anticipated. For instance, reserves of relatively easily available oil will last for decades. But that does not mean their prices will not be subject to wild fluctuation. Once past their peak, major conventional oil fields may decline at 3–7% annually in fairly predictable fashion. Fracked oil fields decline precipitously, in the order of 70% annually. If new tight oil plays do not continue to match the output of existing fields, oil production in the United States may fall rapidly and lead to price and possibly economic instability. Tight oil contributes less than 10% of the world's total production but a rapid decline in that sector could unsettle financial and trade mechanisms.

The shocks from climate change can stretch across all sectors and regions. If the climate acceleration thesis is correct, we can expect further challenges for our infrastructure and budgets going forward in ways that are impossible to anticipate. Whatever the shocks are, governments cannot allow the transition programs to continue to be funding rollercoasters producing sputtering efforts and intermittent results. More complete data, working physical models, and comprehensive mathematical models will help to develop better planning. But resolving is also critical. Leadership is doing with what you have and pressing forward, rather than waiting for the cavalry to charge over the hill. The fantasy of just-in-time solutions appearing as needed cannot be allowed to displace solid planning and scheduling commitments. Unlike the timeframe associated with Winston Churchill's famous nation-rallying speech, it will be decades before we will be able to declare victory in the renewable energy transition.

References

Ahmed N (2017) Failing states, collapsing systems; biophysical triggers of political violence. Springer, Cham

Cairns J (2000) Sustainability and the future of humankind: two competing theories of infinite substitutability

Churchill W (1940). https://www.youtube.com/watch?v=fKBR0igDPfM

FoxNews (2019). https://www.foxnews.com/auto/illinois-1000-electric-vehicle-legislation

LSE (2018) Costs of cutting ghg emissions London School of Economics

Matsumura W (2019) Fossil fuel consumption subsidies bounced back strongly in 2018

Natural Resources Canada (2019). https://cfs.nrcan.gc.ca/statsprofile

NEB (2017) Canada's Renewable Power Landscape 2017 – Energy Market Analysis. https://www.cer-rec.gc.ca/nrg/sttstc/lctrct/rprt/2017cndrnwblpwr/prvnc/on-eng.html

NBC (2019). https://www.nbcnews.com/mach/science/jeff-bezos-foresees-trillion-people-living-millions-space-colonies-here-ncna1006036

Solt F (2019). https://fsolt.org/

Taxpayer (2019). http://www.taxpayer.com/media/2019-GTHD-EN.pdf

Unger R, Thistle J (2013) Energy consumption in Canada in 19th and 20th Centuries. Consiglio Nazionale Delle Ricerche, Rome, Italy

Chapter 12
Steps Toward the Other Side of the Transition

Abstract The ability of society as a whole to identify its own best interests, as distinct from the interests of smaller groups with influence, based on either power or their ability to be heard, will determine its ability to absorb and process the information it needs to be successful. Using real physical measurements to expand society's view of the planet to beyond the scope of the commercial economy and determine our options will give us a clear idea of the best direction to choose because they give us the clearest understanding of history and why events unfolded as they did. Attempting to monetize the stock and flows of earth's natural systems and value them at levels which reflect both natural and human priorities is an exercise in futility. It is much more effective to use biophysical measures to define the parameters within which the commercial economy can perform efficiently. Conserving as much of the potential of earth's biosystems as possible and developing human potential to its fullest, whatever the unforeseeable impacts of the increasing dynamics of the planet's climatic and resource systems will produce the best outcomes. Our steps must be measured in physical units representing biophysical reality.

Europeans came in numbers to the North and South American continents 500 years ago. They succeeded on the ability of their iron tools to generate the energy necessary to spread out and dominate the land and the native populations who had been living in relative balance with nature. Now that we can see the limits of the New World's once virgin resources, we find ourselves aspiring to be in the place of relative balance once occupied by previous indigenous societies.

The scientific toolkit we will carry forward will be much more advanced than that of past societies because of the learning potential unlocked by fossil fuels. This energetic rocket launch allowed human populations and consumption to explode in Canada and across the world. But putting aside the immense waste, vain-glorious consumption and environmental damage accompanying this boom, levels of health, education, social equality, and science made huge strides on the back of the increasing availability of energy.

Canada was a treasure trove of natural resources, but in the millennia before fossil fuels, it remained largely uninhabitable, certainly for populations of any size. The development of Canada thus benefited more than most countries from fossil

J. E. Meyer, *The Renewable Energy Transition*, Lecture Notes in Energy 71,
https://doi.org/10.1007/978-3-030-29115-0_12

fuels which enabled a sufficiently large population to exist and thrive amid initially rich resources. Now that we have succeeded at building a viable nation, how will we maintain that success when the reasons for our high level of achievement are removed? We need to ask ourselves first, what are we trying to achieve?

Are we embarked on a mindless journey of simply using up all of the resources available and then moving on like a modern-day slash and burn subsistence culture? Are we sitting back and seeing where business-as-usual will take us, or are we aiming for a society living in balance with nature? A near worst case scenario has business-as-usual greenhouse gas emissions increasing global temperatures by an average of 5°C by 2100 producing a 2 m rise in sea levels. The probability of a 2 m rise is low in that timeframe, but the probability increases over time to near certainty as glaciers melt more quickly on a warming planet (Bamber and Oppenheimer 2019). The impacts of this level of warming, however, would go far beyond sea level rise and include huge and negative impacts on global agriculture, biosystems, and human infrastructure via increasingly extreme weather causing floods, forest fires, and economically disruptive weather patterns.

Fantasies and Dreams of Perpetuating a Moment in History

Despite our collective difficulty to date in getting out of the blocks on effective carbon emissions reductions, most countries, and certainly most citizens, believe strongly that the causes of climate change must be addressed. In doing so, we will have to let go of both current expectations and past fantasies. Yes, during peak fossil fuel era, we could have grown bananas on Baffin Island. It would have been technically possible. But we did not do it and we no longer seem to fantasize about projects like that or of using nuclear bombs to develop the oil sands. The wildest fringes of our flights of fancy seem to have been sanded off by a growing awareness of the reality of our climatic and resource base situation. Maybe we are gaining some environmental and planetary perspective!

Fantasy still exists however in the acceptance of the basic premise of business-as-usual and the assumption of continuous growth. Perhaps "fantasy" is the wrong term since the life we live today would have been regarded as wild fantasy 500 years ago. But certainly, if growth does continue, it will be made possible only by technology which does not exist today. The availability of large quantities of raw energy gave us access to resources which were previously beyond reach but now, both the sources of our energy and the resources we use, fossil fuels, to exploit are in various stages of decline. Even if we did discover a viable replacement for fossil fuels, in the medium to long term, would that be enough to overcome the decline in grade of the resources critical to an advanced society?

From this dim realization stems the discussion of mining asteroids and other planets. The energy required to do so, it seems, has not been a part of the discussion because even the crudest calculation would show the energy costs per kilogram of any returned ore would be astronomical. Experimentation and research will continue, but until humans move off the planet, it will likely remain cheaper to mine terrestrial resources, no matter how degraded, than to launch miners and mining gear into space and bring back the resultant minerals.

But fantasy has a way of flowing past such seemingly insurmountable barriers. Developers dream of endless waves of roofs and gleaming glass structures washing over farmland and forest. George Bezos, the most successful retailer in history, dreams of 1 trillion consumers. Media corporation employees appear to dream of a Canada of 100 million people by 2100 and growing forever to swell circulation numbers and ad revenue.

A book, "Maximum Canada," written by one media worker, promotes policies which directly align with the interests of media corporations which are the growth of the local market, and strong demand for the products of their largest advertisers. One can imagine the recommendations for national health policy made in a book written by employees of a tobacco company or by representatives of the fast food industry. In the case of "Maximum Canada," the author has suggested the following:

- A large population will facilitate the building of the aforementioned carrier strike groups.
- Increasing population will cure congestion and urban sprawl.
- A growing population will cure the problem of never having enough workers to fill low paid jobs.
- A higher population will increase income equality levels.

The corporate headquarters the author works out of are in downtown Toronto, the recipient of a large portion of Canada's immigrant stream, the highest rate of immigration in the developed world over the past six decades. The Greater Toronto Area has the worst commute times in North America and is spreading over farmland faster than any other city in the OECD. The Canadian equality level has fallen from the second highest in the world in the early 1960s to the mid-20s in 2018 due, in substantial part, to housing cost inflation and stagnant wages. Canadian productivity rates are among the lowest in the OECD over that same period. Since 1990, the proportion of people working in minimum wage jobs in Ontario has quintupled. Hence the perpetual provincial government budget deficits. Since the highest rate of growth in the western world is obviously failing to produce the desired results, one might wonder what rate of growth is high enough to make endless growth actually work the way its supporters claim it can.

A more thorough review of the book, including its good points, was done by Roderic Beaujot and Lise Patterson (Beaujot and Patterson 2018). However, its core theory, apparently endemic to the media industry, of bigger-is-always-better, fails to take humanity's history of resource collapse and population cycles into account. It also lacks the ability to distinguish between the size of the population and the health of the people, the size of the commercial economy and per capita incomes and between total tax revenues and fiscal balance. It is unable to connect the dots of any physical process by relating the visible symptoms of a problem to its root causes. Can the public really expect media corporations, which are very self-aware private businesses to act against their own self-interest, no matter what our expectations of neutral and comprehensive reporting might be?

Media corporation revenues are tied to the size of their local markets and the financial health of their largest advertisers. This builds in a huge bias to promote business-as-usual, which is to say, more growth. But in Canada, they have been given an even greater incentive to promote BAU by direct government subsidies of more than $100 million annually. This brings into focus the issue of subsidies which should not be used to extend current practices beyond their useful lives but rather to ease and hasten the transition to more sustainable policies and business models. Subsidies mask the real cost of whatever process to which they are applied and should only be granted to emerging processes which show promise in delivering long-term benefits. Subsidies are suitable for upgrading and transition, not for allowing existing interest groups to carry on in the manner to which they have become accustomed.

Whether for the fossil fuel industry, media corporation, developers, or other groups with the influence to divert tax money to their own cause we need a full and honest accounting of subsidies. What we pay now and what legacy costs will burden future generations should be a matter of public record, to assure public funding goes toward the long-term common good, rather than working directly against the public interest.

Given the commercialization of media and the message control exerted by powerful interests, the extreme political polarization of the national conversation leaves very little room for what could be termed "free speech" or real policy discussion. Humanity is facing its greatest threat since agriculture began; with one of its most powerful tools, the ability to collectively communicate, discuss, and learn greatly diminished.

A Failure to Communicate

Science and context are largely absent from the national conversation. Scientists have the information but are reluctant to step outside of their own fields of expertise. Journalists, and the media corporations they work for, are clearly unable to absorb technical information and present it to the public with any context. Many journalists may want to provide as full a reporting as possible but lack the technical education, the time, or the journalistic freedom to do so. Therefore it falls upon scientists, the people with the information to begin with, to step out of their comfort zones and develop the ability to speak information and context to both the media and to power.

The media are the custodians of the national conversation, and ultimately, no matter privately or publicly owned, any worthwhile, fully fleshed-out and inclusive national conversation will have to pass through the hands of professional communicators. This process is not working now as witnessed by the extreme unwillingness and inability of our elected officials to address even the most basic core concepts involving energy, resources, population, and climate. When these now seemingly toxic topics are mentioned, they are given short shrift in isolation with no hint they are deeply interconnected. Scientists, engineers, farmers, hunters, and people with solid expertise in any field involving the physical world do not appear to be able to pass through the filter media corporations have erected by means of their control of the public discussion.

The opposition to replacing growth with sustainability as the nation's top priority will be fierce. As Roger Boyd, who prospered in the finance industry before turning his attention to the environment and the need for downsizing put it, "There is so much financial wealth, and high paid jobs, dependent on future growth I just don't see them ever voluntarily accepting the huge financial losses that degrowth would entail." His book "Energy and the Financial System: What Every Economist, Financial Analyst, and Investor Needs to Know" looks at the implications for the financial system of tackling our environmental issues.

Like a glint of light in our 1 million year history, humanity's moment with abundant cheap stored energy and boundless resources is almost over. Our future will now resume in working with natural flows of energy of which there is a great deal. At the moment, the technology we possess is capable of harvesting a very small portion of these flows. But we do have some capability to harvest and store raw energy and our ability to use that energy efficiently is growing.

In the long-term future, not only will Canada not be able to use its own fossil fuel energy, it will not be able to profit from the goods and technology built by nations subsidized by their cheap, abundant energy. Coal will no longer drive the production of cheap Chinese products now nor as it did the advancement of German and British technology in the early days of the industrial revolution. Cheap oil will no longer serve as the rocket fuel that drives American NASA programs or the advanced learning that really built the USA. Certainly the age of science need not come to an end and we will be able to hold on to what we have learned, but the huge surplus of time and work potential that fossil fuels have given us will no longer be available.

What Are We Really Trying to Accomplish?

If humanity can avoid catastrophic collapse and maintain the health of the planet, there is no limit to what we can accomplish over time. Conversely, if we continue to make all-or-nothing bets on high growth and rapid resource exploitation, sooner or later our collective luck will run out no matter the gains we have made. Regional population collapses have happened repeatedly throughout human history, but any future large climate and energy shocks will be felt worldwide. Now that much of the world has developed to the point of providing adequate food, shelter, health, education, and material comfort for its citizens, there is no reason to take huge risks to provide even higher material consumption or further build commercial markets that benefit the few. Those whose business models depend on providing "more" would disagree, but they are a small, albeit still powerful group.

Are we going to create a healthy planet that will support a balance of humanity and the natural world in perpetuity that may eventually lead to exploration of the galaxy, or will we maintain our current level of growth and have no choice but to attempt to leap off a dying planet within 200 years in a fleet of artificial environments housing millions each? If we can make that choice, all of our other decisions fall into line much more easily.

Terraforming

Humans have been building and modifying the surface of the earth for thousands of years. Nowhere is this more evident than in China from the Sea of Japan inland to Beijing. Flying in from the coast to the Chinese capital presents nothing to the eye but transformed features. Re-directed rivers, rows upon rows of tall apartment buildings dominate the landscape over the nearly 200 km where 100% of the land surface has been processed by humans to suit their own needs, completely displacing nature.

How much industrialization of the Canadian landscape in the form of wind and solar farms will Canadians tolerate? Will we simply close our eyes and build our urban sprawl out among the energy farms and hope that somehow, natural systems are not really necessary. That is what we must believe if we continue to pursue endless growth. And then why stop at population of 10 billion or 1 trillion? If growth is necessary, there is no end to the expansion of our numbers until nature finally calls a halt. Believers in endless growth are convinced not only that we can replace all natural systems but that doing so is inevitable. On the other hand, an environmentally balanced world with a high degree of technology is entirely possible and that can endure forever.

It is time to look around us and understand what our real options are. Despite 15 generations of heady development, Canada, the land, is much less healthy than when we found it, with a good deal less of the rich resources it once had. The nation was built on cheap and abundant energy and, if we are to protect our environment, in a matter of decades, 80% of that energy will be unavailable. Not cutting fossil fuel use now will only buy use another two or three decades before we have to. We need to pick a course. Which vision, which economy, which social order, and which leading elite are we going to choose?

Canada's circumstances are unique. Looking past the vain-glorious consumption levels to which we aspire, it still takes a great deal of energy to live in a northern region or in any country with a severe climate. Renewable energy flows, although large, are still very diffuse and intermittent while demand remains consistently high. Canada will never be Central Mexico, the Nile Valley, Okinawa, or even Southern Europe in terms of its ratio of energy available to energy demand. If we choose balance over growth, eventually we can achieve equilibrium with earth and its other inhabitants where healthy people will enjoy living in healthy communities amid a healthy environment.

Stepping Stones of the Transition

The first stepping stone is a solid understanding of the realities of our situation. All nations have unique options and must arrive at their own best solutions. We cannot expect to apply the same technology that worked well in other regions with different conditions. Canada has to understand its options as a northern nation and all nations have to recognize their vulnerability to disruption during the period of withdrawal from dependence on fossil fuels. The physical reality presented by the graph below should be kept front-of-mind by all northern nations (Fig. 12.1).

As mentioned in Chap. 10, thermal storage may be the most promising way to deal with the huge problem of seasonal storage in northern countries. The issue of seasonal storage is an immense problem and a deal-breaker for renewable energy if no way around it is found. And, of course, failure to transition for northern countries ultimately means failure of the nations themselves. Hence the need to revisit this key issue. Geothermal storage may have the potential to allow us to dodge the fatal flaw inherent in building our future based on intermittent energy.

Geothermal energy storage offers the scale necessary to address seasonal storage with the levels of efficiency high enough to make it practical. The technology is at a workable level now. It is simple. It uses energy that is relatively cheap to harvest. It has its limitations, but currently, there is no other seasonal storage solution on the horizon. Thermal storage can be used only locally and only for the production of low-grade energy, heat, but it can also accept low-grade energy for storage where previously we had only been able to store high-grade (fossil fuel or electrical) energy. There is a great deal more low-grade energy available to us than high grade, and it costs much less to acquire.

There are four different approaches to large-scale thermal storage, and their details can be found in the paper "Design Aspects for Large-Scale Aquifer and Pit Thermal Energy Storage for District Heating and Cooling" produced by an

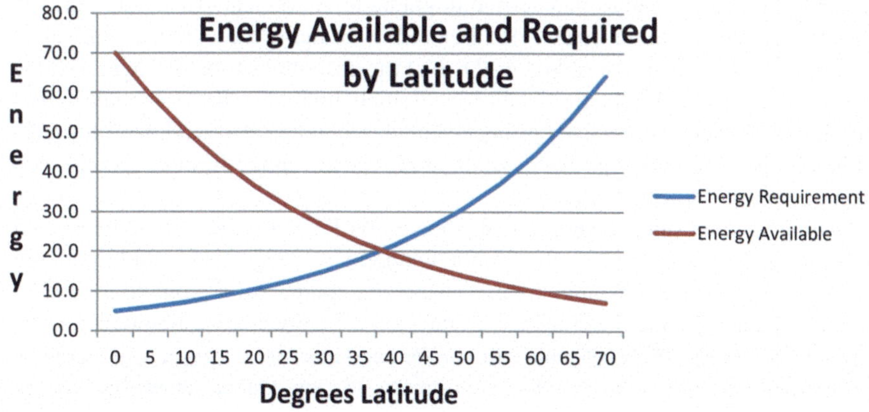

Fig. 12.1 A key relationship, energy required vs. energy available (author)

Table 12.1 Types of geothermal mass heat storage (IEA 2018)

	Tank thermal energy storage (TTSE)	Pit thermal energy storage (PTES)	Borehole thermal energy storage (BTES)	Aquifer thermal energy storage (ATES)
Description	Concrete or steel tanks sit on insulated pad and covered with insulation and earth	An excavated pit is lined and filled with water and gravel	Vertical holes are bored 30–100 m into the ground and U-pipes are inserted in each one	Aquifers are water-filled caverns of permeable sand, gravel
Heat storage capacity per m^3	60–80 kWh	60–80 kWh	15–30 kWh	30–40 kWh
Temperature range	5–95 °C	5–95 °C	−5 to 90 °C	2–80 °C
Cost	\$30–\$60 per m^3	\$30–\$60 per m^3	\$30–\$60 per m^3	\$30–\$60 per m^3

international group of contributors including Natural Resources Canada. The paper lists four types of large-scale underground storage and their characteristics (Table 12.1).

Drake Landing in Okotoks, Alberta, uses the borehole method, and the area covered by the field of boreholes is 35 m in diameter giving their geo heat battery a volume of approximately 20,000 m^3 of earth. The efficiency of the system increases with size with the benefits tapering off once over the 10,000 m^3 level. Borehole systems would seem to be the simplest and cheapest method, but, at this early stage, there is no comparable efficiency data. The Drake Landing BTES efficiency is 45% round trip.

At a cost of \$50 per cubic meter and with a heat storage capacity of 20 kWh, the cost per kWh of storage would be \$2.50. For the same 13 kWh capacity as a Tesla PowerWall, the cost would be \$33. The PowerWall costs approximately \$10,000, but it stores high-grade energy which can be used for any purpose. Thermal storage can only be used for heating. Also, the PowerWall, being an electric battery, is close to 100% efficient, whereas geothermal might be 50% efficient. This would double the effective cost of the storage to \$5.00 per kWh, bringing a geothermal equivalent to a PowerWall in at a cost of \$65 or 0.6% of the cost of electrical storage. And the geostorage recyclability factor would be extremely high as well. Geostorage is best applied to district heating-sized projects, but clearly it can also work well for individual homes and buildings since geothermal systems, in the form of ground loop heat pumps, are in wide use now.

Are there downsides to geothermal storage? Will it cause the earth to heave and subside as it warms and cools over a temperature range that could span see 50 °C–60 °C changes? Would any expansion and compression promote long-term water seepage reducing the efficiency of the system? Only more testing will demonstrate how resilient these systems can be. We need answers because, as it sits now, geothermal storage stands as the centerpiece of any viable renewable energy strategy.

Economic Structure Going Forward

For a period of several decades during the rise of manufacturing, it appeared Canada was maturing beyond the "hewers of wood and drawers of water" level. Free trade agreements and the Dutch Disease impacts of the oil bonanza however have pushed the economy back into the resource/construction and debt pattern as the manufacturing sector declined. Manufacturing constituted 23% of Canada's economy in the 1965 but only 10% in 2015. The only major developed country with a lower percentage is that of the UK. The broad-based economies of Germany and China have manufacturing sectors which make up 23% and 31% of their commercial economy respectively (RBC 2017).

As Canada transitions to a sustainable and renewable energy society, the manufacturing sector is likely to increase its share of the economy as mining, oil, banking, and FIRE (finance, insurance, and real estate) retreat from their fast-growth economy levels. Economic stability will likely increase as the boom and bust sectors become far less dominant and as debt levels decline.

Broad-based economies with much more localized production will improve social stability and reduce energy consumption and all manner of pollution. A great deal of the plastic floating throughout the world's oceans comes from inefficient and dirty production processes in less-developed countries struggling to be the lowest cost producers for goods consumed in developed countries. As production of goods shifts back to developed countries, an increasing percentage of the material consumption will be produced under stricter environmental and labor laws.

Global trade is driven by Chinese coal. This pattern might change dramatically in the coming decades given the depletion of China's coal reserves and the constraints inherent in a shift to renewable energy (Mohr 2015). The map below illustrates the transfer of embedded carbon emissions in international trade and is a clear indication of manufactured goods trade flows. We have studied on the possible impacts of a stoppage or slowdown in the Gulf Stream. But what happens when the flows of carbon based energy stop or slow down significantly? (Fig. 12.2)

Countries with a very high net embedded energy imports may find it difficult to increase their domestic energy production and manufacturing capability in the event that large manufacturing exporters reduce their output. Canada has close to an embedded carbon emissions import/export balance, but this is due to the large amounts of fossil fuels it exports. Net of these, Canada, like most developed countries, is a substantial importer of embedded energy and carbon and will have to offset this by growing its domestic manufacturing capacity. The size of a nation's renewable energy capacity will be a major determinant of its ability to produce the goods which it consumes. Canada is fairly well positioned in this regard but countries like Switzerland and the UK may face greater challenges (Fig. 12.3).

Clearly the economic structure of many countries will change during the transition, and the governments must put stability before economic growth to avoid the setbacks which have plagued early transition efforts. The ideal circumstance for any country undergoing structural changes and seeking to become more efficient is

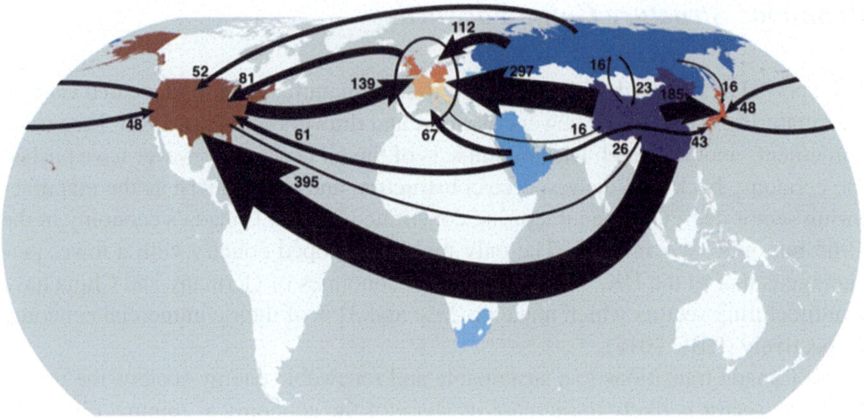

Fig. 12.2 Imported and exported embodied energy and ghg emissions from carbonbrief.org (Caldeira and Davis 2010)

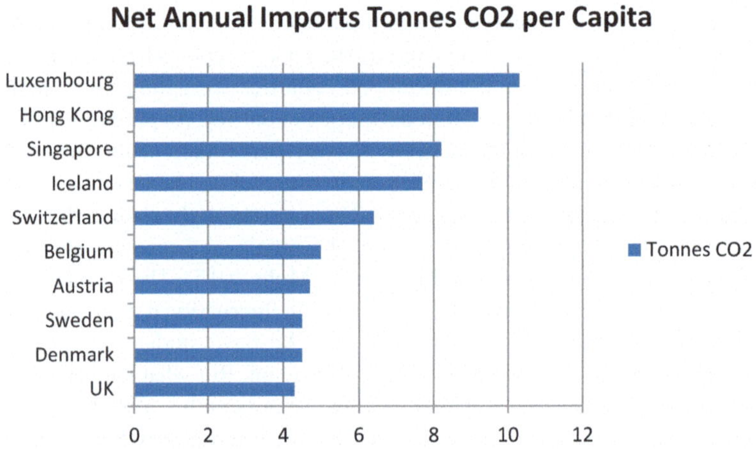

Fig. 12.3 Large per capita net embodied CO_2 importers (Davis 2010)

some degree of labor shortage. A tight labor market will assure the quick absorption of dislocated workers and the investment to make them continuously more productive. A nation awash in cheap labor will find the process of retraining and reemploying highly paid workers displaced by industry downsizing, particularly in one-industry towns, to be extremely difficult and expensive.

Countries with a history of developing the creative and productive potential of their people to the highest degree will have an easier time implementing employment transitions than ones with governments which see their function as merely building markets and finding ways to supply existing business models with the labor they demand at the wages they want to pay.

Models: What to Believe

Productive policy needs to be based on full-context and detailed analysis. Mathematical or quantitative models offer the capability of making wide and deep analysis explicit and transparent to all stakeholders. Modeling is absolutely vital to the process of transition, but many studies by reputable researchers differ wildly in their assumptions, datasets, and, therefore, conclusions. It is safe to dismiss any long-term study based on dollars. These snapshots have no little relevance to the physical world now and certainly none in the world 10, 20, or 30 years out. Studies assuming rapid technology advances or breakthroughs may be interesting but cannot be used for planning purposes. Many older studies assume the results from one region are transferrable to another region. Some studies still use nameplate capacities or optimum results to inform their conclusions which are consequently optimistic or even unrealistic.

There is now a growing wealth of datasets for installations all over the world which give a realistic view of the capacities of installations subject to storm damage, failures, theft, and poor design. Studies based on this data, which means relatively recent ones, can be very useful. Full-energy accounting is approaching the point of becoming mainstream as both governments and corporations improve their data gathering and analysis. No matter how good the data, studies which simply extend past trends into the future are more likely to mislead than inform, since the road ahead is beginning to twist and looking in the rearview mirror will no longer be a reliable means of forecasting.

Modeling is rapidly evolving as this comment from Robert Hoffman illustrates. "We started off doing classical static input-output modelling which focuses on flows among industries and sectors measured in dollar value units using linear and proportional relationships between inputs and outputs. Over time it became clear that these relationships would be better based if flows were measured in physical quantity units. It also became clear from Jay Forrester's system dynamics modelling that it is necessary to keep track of stocks and their evolution over time in order to escape the dysfunctionality of comparative statics. In essence, we combined the compositional richness of input-output modelling with the time structure of system dynamics."

The scale of the study should always be kept in mind so its conclusions are not automatically extrapolated over the entire society. The latest study on the costs of replacing fossil fuels completely in Canada's electricity system, "The cost of decarbonizing the Canadian electricity system" (Dolter and Rivers 2018), looks at the question in considerable detail and is a rich source of information. It attempts to estimate the cost, in dollars, of replacing coal and natural gas generation with wind, solar PV, and hydro capacity and using pumped storage and upgraded grids to mitigate renewable variability. Readers should note that 22% of the electricity generated in Canada is fossil fuel, and electricity is currently 19% of our final energy consumption. This means the study addresses the effort required to decarbonize only 4% of Canada's total energy consumption.

Time and timing matter in building our policy. Researcher E. Krause (Krause 2019) points to the benefit modeling provides in policy sequencing as a framework for understanding specific mechanisms and processes that support a ratcheting-up of policy by helping to identify and overcome barriers. Her thesis examines the policy mechanisms used by the governments of Alberta, Canada, and Westphalia, Germany to develop just transition processes through a policy-sequencing framework. The Canadian Just Transition Task Force and the German Commission for Growth, Structural Change and Employment were both tasked with finding the best means for closing out coal mining in Alberta and Westphalia while supporting the affected workers and communities through the process. Transition brings both opportunity and disruption, making it essential for even-handed and comprehensive government policies to be in place before the large-scale greening of our economy begins in earnest.

Energy Consumption Changes with the Type of Energy Generated

The example below shows the finish to start differences between the energy consumption during the fossil fuel era and that of the renewable energy era. Most of the energy forecasts done today substitute green energy for fossil fuel energy using our existing pattern of energy consumption. But the energy consumption pattern in a renewable energy society will be very different from the current one. Not only will the electrification of all processes offer significant efficiency gains, the processing and conversion losses of fossil fuel primary energy are avoided (Fig. 12.4).

Here, for the sake of demonstration, we assume the house and vehicle are still getting all of their energy from the grid. The source of the energy would be a mix of solar PV, wind, and hydro. The distance driven every year is cut in half from the long-commute and drive-to-the-corner-store days of maximum oil. The house uses

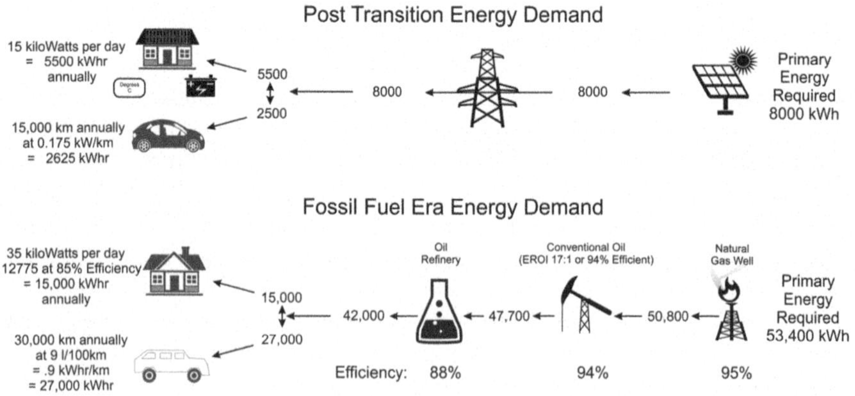

Fig. 12.4 Total energy demand in fossil fuel economy vs. electrified economy (author)

both geothermal storage and a heat pump. In the previous calculations in Chap. 11, it was assumed that residences would not be net users of grid energy and would also be able to generate enough surplus energy to power a vehicle for the year. In this example, we show both the house and the vehicle running off grid power.

The energy demand for the green era house is less than half that of the current era house. The energy demand of the electric vehicle is only about 10% of that of the internal combustion engine vehicles we drive today. No transmission losses are used for either example, and effectively, the electricity from the source goes directly to the end use. Not so in the fossil fuel system where natural gas is used to drive the oil extraction sector, and the oil must be refined and the resulting gasoline burned in the vehicle engine. There are significant losses of energy at every stage and, of course, the efficiency of the internal combustion engine is only about 35%.

Added together, the system demands become radically different. The primary energy needed to drive our current energy system is 53,400 kWh, but the renewable energy system needs only 8000 kWh of primary energy, a reduction of 85%. Our current models do not take this fully into account and thus overstate the amount of energy we will need to produce in the future. If we had used the worst-case scenario, using coal as the primary energy to drive the extraction of bitumen from the oil sands, losses would have been quite a bit higher and the comparative advantage of the renewable energy system just that much higher as well. This raises the possibility that a society could prosper on lower primary energy EROIs than previously assumed.

But the challenges of providing energy in the extreme north are still very substantial. A very recent study has shown soil temperatures of barely 1 °C at a 30 m depth compared to the 10 °C at more densely populated latitudes (Miranda 2019).

Evolving Metrics and Models

Our current fossil fuel-based energy model still features fairly high EROIs and very inefficient consumption. It could be compared to the open hearths of primitive houses which had access to large amounts of energy but used it very inefficiently. But we have learned to develop efficient heating systems and better designed and insulated buildings, increasing comfort while reducing fuel consumption. In a post fossil fuel world, EROIs will be much lower but, at least in several important areas, such as ground transportation and building heating, efficiency levels will be much higher.

EROI is a ratio on the energy production side and COP (coefficient of performance) is the ratio of energy input to energy output on the consumption side of heating buildings. These ratios look the same, but there is a critical difference in the way they are calculated.

Since the consumption side of the energy equation is currently populated with devices which are less than 100% efficient, the thinking to this point has been that there could not be an energy level in play that would effectively elevate the COP above 1:1. In other words, consumption energy would produce no more useable energy than it contained, yet a high COP, as we have seen from heat pumps and geothermal storage, will allow the "mileage" we get from consumption energy used for heating to be multiplied several times.

Wooden stoves might be 50% efficient, oil furnaces 65%, and the latest natural gas boilers can get over 95%. Electric heaters are de facto 100% efficient since they only produce heat, and there is no exhaust or incomplete combustion involved.

Heat pumps and geothermal storage change the game however. A geothermal heat pump can produce four times the energy in heat that is used to run it for a COP of 4:1. The EROI would be 4:1 as well since they measure the same thing. This is the ratio of heat produced to energy input at a normal ground temperature of say 10 °C. If heat is stored, boost that temperature to 50 °C, then the return on the energy used by the heat pump can exceed 20:1.

So now, if the EROI of a solar PV system was 8:1, and the energy from that system was used only to run a geothermal storage system and heat pump with a COP of 20:1, the system EROI would appear to be 160:1, taking us back to the glory days of early oil! But there is one very large caveat regarding the coefficient of performance metric that differentiates it from the EROI metric. EROI counts all of the energy that went into the construction and processing infrastructure of a particular fuel, while COP is based on only the direct energy that goes into driving the process, like the electricity used to pump the glycol through the system. It does not include the energy required to build the entire system and maintain it.

In that respect, EROI and COP are very different, and the 160:1 ratio would fall dramatically if all of the energy used to create the hot water and storage system were included. Despite that, it appears the energy lever of geothermal storage is a powerful one. The question is to what degree will this affect our ability to live with low EROIs? The answer is "significantly." Reducing the severity of the problem of seasonal energy storage for northern communities is a critical consideration in the transition to renewable energy for many of the countries in the world. Given the energy multiplier effect of heat pumps and geostorage, we need to develop EROIs spanning the entire heating process, from generation of energy to the resultant energy that is ultimately consumed.

An electrified, renewable energy society will use much less total energy to do the same work as a fossil fuel society due to the efficiency and lack of conversion processing in the electrical network. The new processes and opportunities which will be part of the renewable energy network have to be taken into account in the modeling process, which to this point has been largely based on our current fossil fuel template.

EROI of Energy Resources in Regional and Resource Base Context

The energy returned on energy invested is a key national metric, but similar EROIs will have different social implications at different latitudes and in regions with different resource bases. A healthy society may be maintained with very low EROI energy sources if it is located in a region that requires no heat for buildings or irrigation for crops. If that region has abundant natural resources, it will demand very little energy to harvest and process those resources. The reverse is true for a society located in a harsh environment with sparse and low-grade resources. Also, the type of fuel is relevant to a discussion of EROIs since it is likely that 10 kWh worth of electricity will be able to do more productive work than 10 kWh worth of gasoline because electrical systems are so much more efficient than ones which burn fossil fuels. EROIs have to be developed using local data and evaluated in light of local conditions. They have to be applied with their full context spelled out.

Open Questions

Can high-gain heating systems and efficient electrical systems overcome the constraints of much lower primary energy availability inherent in the post fossil fuel world?

Will lithium battery recycling ever become sufficiently efficient to allow it to be used as a large-scale, long-term solution over centuries? Possibly, if the development work being done at Lithion (Lithion 2019) meets the target of 100% recyclability, it will be able to put to rest the concerns put forward by John Howe on the necessity of battery recycling.

Will we eventually be able to perform all vital tasks with electricity? Currently, heavy outdoor energy demands as found in mining, snow clearing, agriculture, marine transport, and space exploration cannot be filled with battery/electric drive systems. Large-scale air transport falls into this category but is not absolutely critical to societal continuance.

People with a Point to Make

Many people have worked for decades on different aspects of the energy/human relationship. Here are but a few of the thousands of personal initiatives driving progress on these interconnected issues.

- Charles Hall
  - Developed the critical metric of energy returned on energy invested and built up awareness of the declining productivity of our energy resources. He also created awareness of how critical energy is to society and developed the tools necessary for us to give structure to its study. Author of "Energy and the Wealth of Nations" and many more books and co-founder of the International Society for BioPhysical Economics with Kent Klitgaard and Jessica Lambert.

- Robert Hoffman
  - And his partner Bert McInnis (1940–2016) established WhatIf Technologies to model public policy in real units, the stocks and flows of a physical wealth creation process. In setting the distortions of dollars aside, they are able to offer insights into systems as diverse as flows in the catchment areas of Australian rivers to municipal renewable energy strategies to modeling the infrastructure and processes necessary for countries to manage the energy transition process for municipalities. They work in real units to determine real options.
- Jun Yan
  - Established the superiority of management of using energy metrics rather than currency for decision-making in an energy firm. Essentially an energy costing system was used to review the efficiency of the production system (Yan et al. 2019).
- William Rees
  - Developed the concept of the ecological footprint. A metric designed to put ecological overshoot into perspective and drive the point home on a personal level.
- Norman Borlaug
  - Won the Nobel Prize for his work contributing to the Green Revolution and pointed out in his acceptance speech that the Green Revolution was not a solution to world hunger, but rather a stop gap measure that would buy the world time while leaders dealt with the fundamental problem of human population size and growth. He made clear that technology was no long-term savior.
- David MacKay
- Provided a detailed reference of how society (specifically the UK) generates and uses energy and also what the realities of transitioning to renewables would entail. He called out "greenwashing" with basic arithmetic and put energy issues into 3D infrastructure perspective with an "adult conversation" on the physics of the task ahead. (MacKay 2009)
- Long-time activists Alec Adams and Mike Nickerson
  - The millions of activists around the world who have worked through their lives to advance awareness of climate change, resource depletion, and the impossibility of ever increasing growth are voices in the wilderness no longer. Most people now get it. Whether through dogged organizational efforts like Adams or through published works like Nickerson, they will never have the fame of the David Suzukis or David Attenboroughs of this world, but they are just as vital to making the transition to a sustainable society a reality.

International Cooperation

The establishment of broad-based national economies flies in the face of the current version of globalism, but globalism as a means to increase commerce may ultimately give way to a version of globalism that promotes sustainability. These are very different objectives, but the channels of communication between nations to arrive at mutually beneficial agreements would be the same. Green globalism would represent a dramatic change in direction but not necessarily a dramatic change in international relations. And without strong international cooperation, it is hard to imagine the world or even more than a few individual countries succeeding in the transition to renewable energy. This new version of globalism would be more of a cooperative climb to the sustainability threshold versus the current commercial race to the environmental bottom.

Strong, resilient, broad-based economies would more easily meet international standards and be in the position to offer much more open green energy technology sharing. The trade in manufactured goods and energy may be restricted in the national interest, but some trade in energy could leverage higher levels of penetration of renewables. Take the case of hydro-rich Norway. Their abundant hydroelectric power is seen by other European nations as a fast reaction baseload they could put to good use. But Norway's hydroelectric capacity is not large enough to both supply its home market and act as a buffer for even a significant part of Europe's grid outside of Denmark.

Norway has not been motivated to develop much of its considerable wind potential due to its existing hydro capacity, but if its wind capacity were to be increased, it would be able to provide a much higher level of buffering capacity for Europe. This would allow other nations to leverage their renewable penetration to a higher degree resulting in a disproportionate drop in ghg emissions. Possibly a similar situation exists in the Southwest US where the solar capacity factors are almost triple those of Canada and large solar farms could supply Canada during the winter.

Energy Budgets and Efficiency

The post fossil fuel energy budget will be dramatically lower given the lower EROIs of renewable energy sources, but the energy system will also be a great deal more efficient. Eliminating 80% of the energy supply will mean overall consumption will be reduced but given the efficiency improvements and conservation measures likely in place by the time fossil fuels disappear from the energy menu, lifestyles will not necessarily see large negative impacts. Net energy or usable energy will be a much higher percentage of the energy budget than was the case with fossil fuels which involved multiple processes, all with some degree of energy loss with the final use step involving very lossy, inefficient combustion (Fig. 12.5).

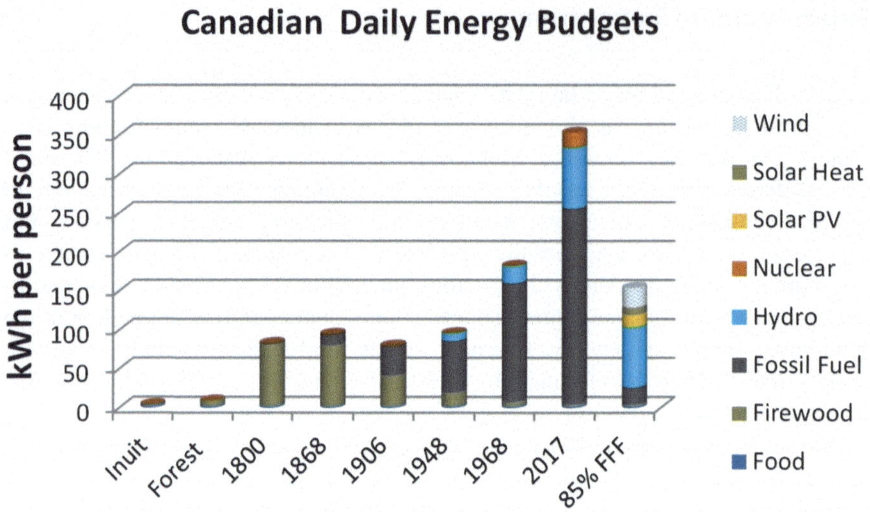

Fig. 12.5 Per capita energy budgets peak with maximum fossil fuel and decline (StatCan 2019; Beaujot and Patterson 2018)

Efficiency has increased dramatically from the 1750s, when wood was the primary energy source. Despite the huge increases in energy availability during the fossil fuel era, process efficiency increased continuously even if its benefits were squandered by the end user. The electrical era will see much higher levels of efficiency than previous eras, and increasing efficiency, both in the use of energy and in its harvesting, is something that science can deliver consistently going forward.

Nuclear energy, given our current level of technology, will not be a long-term solution to reliable and low-cost (high EROI) energy supply since it requires a mined resource, uranium or thorium. These minerals will ultimately deplete past the point of energetic viability just as fossil fuels are in the process of doing now. Nuclear can be a very useful stepping stone to 100% renewable energy if its clearly identified issues can be successfully addressed. Unquantifiable risk is perhaps its greatest question concerning nuclear power. It is something that renewable energy does not suffer from. Legacy costs with fossil fuels are large in the case of conventional oil and gas and extremely large in the case of the oil sands. But the biggest legacy cost of fossil fuels is the alteration their use has caused in global weather systems.

For the most part, hydro, wind, and solar seem to have minimal downstream costs, but there can always be unintended consequences. Charles Hall, originally involved in studies of fish, once opposed the construction of power plants on the Hudson Rivers because of the negative effects on fish stocks. The additional power plants were never built, and power was supplied by a new dam in northern Quebec. But there were consequences in the form of mercury buildup and the poisoning of the local people, something that had simply not been foreseen, making a power plant on the Hudson, perhaps, the lesser of the two evils.

The next chapter in mankind's legacy must integrate both the economic and biophysical world. The future of humanity is depending on it!—J.G. Lambert

Stable, Physically Illuminating Metrics Needed

The need to measure our resources in physical units and use those metrics to guide national policy is critical given the tremendous distortions inherent in using commercial market metrics to represent the health of society and the various capacities of the resource base. As the impacts of climate change begin to be felt, the oscillations reverberating through the commercial economy and society can only be captured by a suite of physical metrics. These will represent a consistent view of reality, whereas the gyrations of the monetary/trade/commercial markets will cause commercial economic metrics to provide even less reliable guidance than they do in a stable environment.

Money is the lifeblood of the commercial economic system. Energy is the lifeblood of every society and every living being as it allows us to process and absorb the resources around us. Money should be used for commercial economic transactions and planning and physical units of energy and resources should be used for national policy planning. Nations are not markets. Cash flow is not asset management in the financial world much less the biophysical world. The physical world can only be properly represented in physical units. Attempting to use tortured fiat currency metrics to represent the natural world completely clouds the greater picture. We live in a physical world, and our endeavors that have been monetized and valued for the purposes of building a commercial system are merely a small subset of the energy and biological flows that make life on this planet possible.

Science isn't offended by stability, commercial economics is.

Metrics of growth—imagine a country whose current consumption is 100 environmental units while sustainable consumption is 33 units meaning it is in a position of environmental overshoot. It would need to reduce its footprint by 2/3 or 67% to achieve balance. If population doubles, then per capita consumption cuts must be 83% i.e. twice as high. If population triples, cuts must be three times as high, and sustainable per capita consumption will be only 11 units. When it comes to environmental assets and conservation, economies of scale are negative. The environmental realities of boosting population anywhere, and especially in northern climates, are grim. In short, the continued pursuit of growth guarantees complete failure.

Coherent National Decision-Making Needed

The ability to implement clear policy in the face of great challenges is a critical function of government. In Canada, our failure over the past 25 years to meet or even remotely approach our environmental targets is spelled out in clear physical numbers. This failure is based on an overlap of federal and provincial responsibilities and a lack of will and a disconnect between policy makers and technically competent advisors. A report on the CBC website by Jason Proctor (CBC 2019) noted that the federal government has the power to authorize a project, but provincial governments have the authority to regulate, and all levels of government are allowed to make laws on the environment. According to one provincial judge quoted in the article, "The kinds of environmental concerns at play in 2019 didn't exist in 1867 (when the constitution was adopted). And divisions over the way to handle them aren't likely to vanish anytime soon—no matter what the courts decide."

Successful countries are able to put everyone at one table and come away with a binding decision based on national goals. All levels of government must be clear on national goals and the power and limitations of commercial markets. The EU, China, Norway, and others have been able to do this but clearly, to date, Canada and the United States have not. Splintered political decision-making is often exploited by entrenched lobby groups to delay and deflect public policy that runs counter to their narrow interests. The tobacco lobby worked against province-wide smoking restrictions in Ontario, hoping to have decisions made on a municipality-by-municipality basis. This would have allowed them to apply their well-funded and sophisticated lobbying machine against the smaller and less sophisticated local councils bringing the process of de-smoking to its knees. So far, powerful groups have been successful in continuing to have the pursuit of simple growth maintained as the core national goal. As this is being written, the administration of President Trump is seeking to reassign responsibility for environmental standards to individual states, which would be an absolute godsend to carbon-intensive industries.

Until sustainability replaces ever higher consumption as the business-as-usual condition, an effective transition to renewable energy cannot occur, and this will require nationally binding legislation to be in place. As Valorie Allen makes clear in her book "Growing Pains," we are living on a planet in distress, and an awareness of that should inform every government policy. We need to sort out our priorities and our options through an adult conversation based on physics and our collective view of what a society should be.

Science is a relative newcomer in human history and occupies a rather unique position. It is well-respected generally, but wields little economic power, and therefore cannot have the direct control over public policy which groups spending large amounts on political donations and media advertising possess. Science can represent the health of the planet in cold, hard numbers and three branches, environmental science, atmospheric science, and energy science, are sending the same message: continuation of business-as-usual, however much commercial economic sense it might make in the moment, will assure physical disaster in the near future.

In the era of abundant cheap energy, it made little sense to apply science to public policy. Going forward, with our energy, climate and resource options becoming increasingly constrained, science, with its long view and comprehensive overview of planetary systems, can offer positive guidance that commercial interests simply cannot. Public policy, to date, has been formed by the balance of commercial interests and the moral suasion of advisors representing the broader public good. Science adds a third axis to the scale of public policy by representing physical realities or limits to what is possible. It also offers time horizons, the extent of which, have never been available to past policy makers.

A cohesive society would feature high levels of public trust in its institutions. The Edelman Trust Index polls people in 26 different markets and nations with regard to their level of trust in their NGOs, businesses, government, and media. They broke respondents down into two groups roughly the top 25% of income earners and the remaining 75%. Institutions in rapidly developing countries generally enjoyed the highest trust levels with China at the top. Media and government are the two least trusted institutions with NGOs being highest and business second highest. There is a level of disparity between trust in institutions for upper 25% income groups and lower 75%. Canada had a 20-point spread (and growing) between the elites and the general public which was the second highest spread in the world behind the UK at 24% (Edelman 2019). A cohesive society trusts its institutions, and given the stresses which will be associated with both adapting to climate change and transitioning to renewable energy, a low level of social cohesion speaks to the difficulty a nation may experience in managing change in the coming decades.

We have barely touched upon what is probably the most important step in this process: the preparation of our young people to be able to effectively assess the situation before them and to make effective decisions. At this time, more or less the largest major in our colleges and universities is economics, finance, and related fields. All of these fields are essentially devoid of an understanding and use of science and indeed are generally based on assumptions that would not pass the sniff test of any natural science (e.g., see Hall and Klitgaard 2011). But what economics does well is to provide a seemingly rational approach to making decisions, even if the basis upon which they are made is fundamentally flawed. An approach that melds both the natural and the social aspects of economics is urgently required.

Fortunately such an approach, called BioPhysical Economics, is available and has a coherent textbook to that effect (Hall and Klitgaard 2011). While it cannot be expected that most economists will drop their conventional textbooks, there is a strong rationale that scientists, engineers, economists, and others who do believe in the basic physical concepts can begin their own conversation. A form of economics that is not based on the assumptions of growth, debt, and destruction of nature will be consistent and aligned with the other courses in natural sciences the students are taking. In this way, we can set the stage for the transition that we must make, if we are to survive as a society and perhaps as a species.

No End to Transition

The need to change and adapt will not cease once we have burned our last drop of oil. Fossil fuels have enabled us to develop the tools we need to prosper, but we need to have the will to use them as well as the will to consume less than what we know the natural world can sustain at any single point in time.

Leadership tools:

- Extensive energy accounting data
- Full toolbox of physical unit metrics
- Sophisticated modeling
  - Open access to all interested parties
- Awareness that if we do not push hard in one area, we will have to push harder in others.
- Responsible media
- All parties at the table and in the conversation
- Clear national goals
- Coherent strategy and the long view
- Resolve

Short list of positive steps:

- Stop population growth, whether by birth or migration, in every country, particularly in the high consumption ones
- Upgrade energy conservation in all buildings and begin installation of energy harvesting capacity in all buildings
- Convert to EV fleet as soon as possible
- Implement long-term wind-down of oil sands
- Build up infrastructure to install geothermal storage on a national scale
- Coordinate district heating plans with plans for consolidation of communities
- Incentivize localism for food, manufacturing, and consumer goods
- Convert commercial globalism to green globalism based on knowledge sharing and international environmental standards
- Set aggressive standards for local recycling and national waste reduction

These are the low hanging fruit which will deliver the greatest progress toward the goal of sustainability for the least amount of effort, but even with their full implementation, success is far from assured. Ultimately, we need to balance where and how we live with energetic and environmental realities. As David MacKay noted, nations have a number of levers to push to achieve the de-carbonizing of society. We will all have to push hard on the levers available to us to get the job done. When the efforts in one area fall short of expectations, we will have to push harder on the remaining levers to reach our goal in the remaining time.

Well thought out and aggressively implemented programs are going to be required to ease society away from fossil fuels. Climate and resource dynamics however will not likely allow humanity to enter a blissful state of rest after the trials and tribulations of engineering our conversion to renewable energy systems are over. Change is the one thing that will probably never change for human society, but once we have converted to renewable energy, we will have acquired a deep cultural understanding of the earth's workings as well as a large inventory of tools to enable us to sustain ourselves regardless of the challenges ahead.

Words from a Sustainable Past

It is common to romanticize the relationship between indigenous people and the land before the arrival of Europeans but the lives of these people were tough, short, and often violent. It is not something most would aspire to revisiting, but these cultures undoubtedly had a connection to the natural world lacking in modern society. Tribes on the west coast of North America lived in a very favorable environment compared to their cousins on the plains or in the north or the northeastern forests. They had abundant fish and forest resources, and the climate was relatively moderate. Chief Sealth led a tribe in the Washington state area, and since he lived the life of his ancestors, he was able to appreciate the impact European settlement would have and possibly its end result.

Chief Sealth

Every part of the earth is sacred to my people. Every shining pine needle, every sandy shore, every mist in the dark woods, every clearing and humming insect is holy in the memory and experience of my people.

The white man is a stranger who comes in the night and takes from the land whatever he needs. The earth is not his brother but his enemy and when he has conquered it he moves on. He leaves his fathers' graves and his children's birthright is forgotten.

All Things share the same breath—the beasts, the trees, the man. The white man does not seem to notice the air he breathes. Like a man dying for many days, he is numb to the stench … What is man without the beasts? If all the beasts were gone, men would die from great loneliness of spirit, for whatever happens to the beast also happens to man.

All things are connected. Whatever befalls the earth befalls the sons of earth …The whites too shall pass—perhaps sooner than other tribes.

Continue to contaminate your own bed, and you will one night suffocate in your own waste. When the buffalo are all slaughtered, the wild horses all tamed, the secret corners of the forest heavy with the scent of many men, and the view of the ripe hills blotted by talking wires, where is the thicket? Gone.

Where is the eagle? Gone.

And what is it to say good-by to the swift pony and the hunt, the end of living and the beginning of survival.

(Reference: Chief Sealth (1786–1866) was a Suquamish and Duwamish chief and the city of Seattle is named after him (Sealth 2013))

Over the past several decades much of western society has come to recognize that the earth is our partner rather than our enemy or slave. Yet our socio-economic structure still treats our planet like an expendable accessory, something to be consumed on the way to greater things, or just to keep the economic wheels from falling off for yet another day. But the simple fact is, decoupling our society from the embedded infrastructure of high consumption and endless growth can either take place chaotically over a short period by nature's hand, or relatively smoothly over decades through our own very intensive actions.

How well we manage the transition to renewable energy and sustainable lifestyles will determine how much of the best features of our lifestyle we will be able to retain. In addition, a stable, sustainable society will be well positioned to continue to learn, develop our human potential, and progress socially and scientifically.

Either we turn the focus of our highly developed consumption technology toward sustaining biosystems of the earth by regaining communion with our natural systems, or we risk suffering repeated failures and decline, as we futilely attempt to re-live the past glory days of fossil fuels and endless raw resources. Canada and the rest of the world are riding the crest of the two greatest natural endowments in human history, 10,000 years of abnormally stable climate and the wealth of learning enabled by mining fossil fuels. Managing a smooth transition to renewable energy will allow us to continue to both preserve the endowment of favorable climate and reap the benefits of continuous progress.

References

Bamber JL, Oppenheimer M (2019) Ice sheet contributions to future sea-level rise from structured expert judgment

Beaujot R, Patterson L (2018) Maximum Canada: what do Canadians wish to maximize? Canadian Studies in Population 45(1–2)

Caldeira K, Davis SJ (2010) Consumption-based accounting of CO2 emissions

CBC (2019). https://www.cbc.ca/news/canada/british-columbia/transmountain-carbon-tax-climate-change-law-1.5149179

Davis SJ (2010) The supply chain of CCO2 emissions

Dolter B, Rivers N (2018) The cost of decarbonizing the Canadian electricity system

Edelman (2019). https://www.edelman.com/sites/g/files/aatuss191/files/2019-03/2019_Edelman_Trust_Barometer_Global_Report.pdf?utm_source=website&utm_medium=global_report&utm_campaign=downloads

Hall CAS, Klitgaard K (2011) Energy and the wealth of nations: understanding the biophysical economy. Springer, New York

IEA (2018). https://www.iea-dhc.org/fileadmin/documents/Annex_XII/IEA_DHC_AXII_Design_Aspects_for_Large_Scale_ATES_PTES_draft.pdf

Krause E (2019) Sequencing a just transition: a comparative analysis of Alberta and North Rhine-Westphalia's coal phase-outs

Lithion (2019). https://www.sdtc.ca/en/projects/pilot-plant-for-the-recycling-of-lithium-ion-batteries/

MacKay D (2009). www.withouthotair.com

Miranda M et al (2019) Temperature-depth profiles measured in the Inuit community of Kuujjuaq, Northern Québec, Canada

Mohr SH (2015) Projection of world fossil fuels by country

RBC (2017). http://www.rbc.com/economics/economic-reports/pdf/other-reports/Manufacturing%20Trends-Feb2017.pdf

Sealth, Chief (2013). https://nativeamericannetroots.net/diary/1536

StatCan (2019). https://www150.statcan.gc.ca/n1/en/pub/11-516-x/pdf/5220018-eng.pdf?st=KwHbSHsv

Yan J, Feng L, Steblyanskaya A (2019) Creating an energy analysis concept for oil and gas companies: the case of the Yakutiya Company in Russia

Further Reading

Ahmed NM (2017) Failing states, collapsing systems. Bio-physical triggers of political violence. Springer, Cham

Al Gore (2007) The assault on reason: Earth in the balance. Penguin Press

Allen VM (2013) Growing pains: planet in distress

Banks F (2000) Energy economics

Barbero A (2007) The day of the barbarians: the battle that led to the fall of the Roman Empire

Bardi U (2014) Extracted: how the quest for mineral wealth is plundering the planet. Chelsea Green Publishing

Blain B (2010) Weaving golden threads: integrating social theory. Institute for Economic Democracy Press

Bolgiano C (2007) Mighty giants: an American chestnut anthology. American Chestnut Foundation/ Images from the Past, Bennington

Bonner W, Wiggin A (2009) The new empire of debt: the rise and fall of an epic financial bubble. Wiley

Borjas GJ (2016) We wanted workers: unravelling the immigration narrative

Brooke JL (2014) Climate change and the course of global history—a rough journey. Explosion of knowledge about climate. Cambridge University Press, Cambridge

Brooks A (2014) Musings from the oil patch: could the US LNG market be derailed before it starts

Brown LR (2009) Plan B 4.0. Mobilizing to save civilization. WW Norton, New York

Caplow T, Hicks L, Wattenberg BJ (2000) The first measured century. Illustrated guide to trends in America 1900-2000. AEI, Washington

Carter J (2006) Our endangered values: separation of church and state. Simon & Schuster

Cary JR, Tuljaputkar S (2001) Life span: evolutionary, ecological, and demographic perspectives

Cavalli-Sforza LL, Cavalli-Sforza F (1995) The great human diasporas: history of diversity and evolution. Addison-Wesley, New York

Chang J (2005) The unknown story MAO. Political biography

Clements J (2005) The Vikings: brief history of the last pagans or the first modern Europeans

Cline EH (2014) 1177 B.C.: the year civilization collapsed. Princeton University Press, Princeton

Cochrane RO, Fairbank CO (2000) Oil heritage tour of Lambton County: the birthplace of the Canadian Oil Industry. Oil Petroleum Institute, London, ON

Collier P (2013) Exodus. How migration is changing our world. Oxford University Press, Oxford

Daly HE (1977) Steady-state of economics: economics of biophysical equilibrium and moral growth. W. H. Freeman

David Sanborn S (2007) Smelling land: the hydrogen defense against climate catastrophe. Canadian Hydrogen Association, Westmount

J. E. Meyer, *The Renewable Energy Transition*, Lecture Notes in Energy 71,
https://doi.org/10.1007/978-3-030-29115-0

Davies G (2016) History of money: from ancient times to the present day. University of Wales Press, Cardiff
Day JW, Hall C (2016) America's most sustainable cities and regions. Surviving the 21st century megatrends. Copernicus, New York
Denevan WM (1992) The native population of the Americas in 1492: history of the Americas. University of Wisconsin Press, Madison, WI
Diamond J (1997) Guns, germs, and steel: the fates of human societies. W.W. Norton
Diamond J (2005) Collapse: how societies choose to fail or succeed
Dilworth C (2009) Too smart for your own: ecological predicament of humankind good. Cambridge University Press, Cambridge
Durschmied E (2002) From Armageddon to the Fall of Rome: how the ancient warlords changed the world. Hodder & Stoughton Limited, London
Ellen Hodgson Brown JD (2008) Web of debt: the shocking truth about our money system and how we can break free
Ferguson N (2008) The ascent of money: financial history of the world. The Penguin Press HC
Fioramonti L (2017) The world after GDP: analysis of problems with GDP
Flannery T (2002) Future eaters. Insight into current destructive path
Flannery T (2007) The weather makers: how we are changing the climate and what it means for life on earth
Fletcher I (2010) Free trade doesn't work. What should replace it and why. U.S. Business and Industry Council, Washington, DC
Fraser EDG, Rimas A (2010) Empire of food: feast, famine, and the rise and fall of civilizations. Random House
Friedemann AJ (2016) When trucks stop running: energy and the future of transportation. Springer, New York
Gauthier B (2011) Santa Claus is alive and well and living on Wall Street. Financial guide to Wall Street
Grubel H (2009) The effects of mass immigration: Canadian living standards and society. The Fraser Institute, Vancouver
Hall CAS (2011) New studies in EROI. Sustainability (Special Issue)
Hall CAS (2017) Energy return on investment: a unifying principle for biology, economics, and sustainability. Springer, Berlin
Hall CAS, Klitgaard KA (2011) Energy and the wealth of nations: understanding the biophysical economy. Springer, New York
Hargrove B (2009) Laying it on the line: driving a hard bargain in challenging times. HarperCollins, Toronto
Haywood J (2000) Encyclopaedia of the Viking age: story of the Vikings. Thames & Hudson, New York
Heinberg R (2007) Peak everything: waking up to the century of declines
Heinberg R (2015) Afterburn: society beyond fossil fuels. New Society Publishers
Homer-Dixon T (2006) The upside of down: catastrophe, creativity, and the renewal of civilization. Random House
Howe JG (2016) The end of fossil energy and per-capita oil. McIntire Publishing Services, Waterford, ME
Huffington A (2003) Pigs at the trough: how corporate greed and political corruption are undermining America
Jackson T (2009) Prosperity without growth: economics for a finite planet
Johnston DC (2008) Free lunch: how the wealthiest Americans enrich themselves at government expense (and stick you with the bill). Portfolio, New York
Karl TL (1997) The paradox of plenty: oil booms and petro-states. University of California Press
Kelly M (2001) The divine right of capital: dethroning the corporate aristocracy. Berrett-Koehler, San Francisco

Kennedy M, Lietaer B, Rogers J (2012) People money: the promise of regional currencies. Triarchy Press, Axminster
Kerr D, Beaujot R (2015). Population change in Canada: statistics, population and the environment
King R (2008) Atlas of human migration. Causes and effects
Levitt SD, Dubner SJ (2006) Freakonomics: a rogue economist explores the hidden side of everything
Lietaer B, Arnsperger C, Goerner S, Brunnhuber S (2012) Money and sustainability: report from the Club of Rome. Triarchy Press, UK
Livi-Bacci M (2017) A concise history of world population. Population changes. Wiley
MacKay DJC (2009) Sustainable energy—without hot air: various low-carbon pathways. UIT, Cambridge
MacNeil J, Winsemius P, Yakushiji T (1992) Beyond interdependence: the meshing of the world's economy and the earth's ecology. Oxford University Press, Oxford
Magnusson M, Palsson H (1965) The Vinland sagas: the Norse discovery of America. Penguin Books, New York
Malthus TR (1999) An essay on the principle of population. Tendencies to outstrip resources. Oxford University Press, New York
Mann CC (2005) 1491: new revelations of the Americas before Columbus. Knopf
McEvedy C, Jones R (1978) Atlas of world population history. History of world population
Meyer KE (2003) The dust of empire: the race for master in the Asian heartland
Nickerson M (2006) Life, money & illusion: living on earth as if we want to stay
Nikiforuk A (2014) The energy of slaves: oil and the new servitude. Greystone, Vancouver
Oreskes N, Conway EM (2010) Merchants of doubt: how a handful of scientists obscured the truth on issues from tobacco smoke to global warming. Bloomsbury Press, New York
Parker G (2013) Global crisis: war, climate change & catastrophe in the seventeenth century. Yale University, New Haven
Phillips K (2008) Bad money: reckless finance, failed politics, and the global crisis of American capitalism. Viking
Piketty T (2013) Capital in the twenty-first century. Economics
Pimentel D (2008) Biofuels, solar and wind as renewable energy: benefits and risks systems. Springer, New York
Pimentel D, Pimenel MH (1979) Food, energy, and society. Useful facts and tables linking over-population, energy requirements, and food production. CRC, Boca Raton
Pointing C (2011) A new green—history of the world: the environment and the collapse of great civilizations. Random House
Preston SH, Heuveline P, Guillot M (2000) Demography: measuring and modeling population processes. Wiley
Raven PH (2000) AAAS Atlas of population & environment. Analysis of relationships between human population and the environment. American Association for the Advancement of Science, University of California Press, Berkeley
Reinhard CM, Rogoff KS (2009) This time is different: eight centuries of financial folly. Princeton University Press, Princeton
Roberts P (2005) The end of oil: on the edge of a previous new world
Roberts WO, Lansford H (1979) The climate mandate: quality and character of human life
Robertson H (2010) Walking into Wilderness Toronto carrying place and nine mile portage: crucial land and water trails between Lake Ontario and Georgian Bay. Heartland Associates
Rubin J (2009) Why your world is about to get a whole lot smaller: oil and the end of globalization. Random House
Ruddiman WF (2005) Plows, plagues, and petroleum: how humans took control of the climate. Princeton University Press, Princeton
Russell C, Russell WMS (2000) Population crises and population cycles. Med Confl Surviv 16(4):383–410
Sellman RR (1957) The Vikings: Vikings and their impact on the Christian world

Simmons MR (2005) Twilight in the desert: the coming Saudi oil shock and the world economy
Simpson J, Jaccard M, Rivers N (2007) Hot air: meeting Canada's climate change challenges. McClelland & Stewart, Toronto
Smil V (1987) Energy-food-environment: realities, myths, options, London, Oxford University Press
Smil V (2017a) Energy and civilization: a history. MIT Press
Smil V (2017b) Energy transitions: global and national perspectives. Praeger, Santa Barbara
Sorenson JL, Johannessen CL (2009) World Trade and biological exchanges before 1492: trans-oceanic transfer of plants, diseases, and animals. iUniverse, Inc., New York
Soros G (2008) The new paradigm for financial markets: credit crisis of 2008 and what it means
Stanford DJ, Bradley BA (2012) Across Atlantic ice: origin of America's Clovis culture. University of California Press, Berkeley, CA
Stanton W (2003) The rapid growth of populations: histories, consequences, issues nation by nation. Multi-Science Publishing Co, New Jersey
Taft K (2017) Oil's deep state: how the petroleum industry undermines democracy and stops action on global warming
Takach G (2016) Tar wars: oil, environment and Alberta's Image. University of Alberta Press
Taleb NN (2007) The black swan: the impact of the highly improbably. Random House
The Trottier Energy Futures Project (2013) An inventory of low-carbon energy
Thirgood JV (1981) Man and the Mediterranean forest: a history of resource depletion. Academic Press, London
Thirgood JV (1987) Cyprus: a chronicle of its forests, land and people. University of British Columbia Press, Vancouver
Unger RW, Thistle J (2013) Energy consumption in Canada in the 19th and 20th centuries. A statistical outline. Consiglio Nazionale Delle Ricerche, Italy
Victor PA (2008) Managing without growth: slower by design, not disaster. Edward Elgar Publishing Ltd., Cheltenham
Wackernagel M, Rees W (1995) Our ecological footprint: reducing human impact on the earth. New Society, Gabriola Island
Weaver A (2008) Keeping our cool: Canada in a warming world. Viking
Weisman A (2007) The world without us: what would happen to the planet if humans disappeared
Woodward B (2006) State of denial: bush at war, part III
Wright RJ (2007) An illustrated short story of progress. Rise and fall of past civilizations

Index

J. E. Meyer, *The Renewable Energy Transition*, Lecture Notes in Energy 71,
https://doi.org/10.1007/978-3-030-29115-0

The manufacturer's authorised representative in the EU is Springer Nature Customer Service Centre GmbH, Europaplatz 3, 69115 Heidelberg, Germany. If you have any concerns regarding our products, please contact ProductSafety@springernature.com

Printed and bound by CPI Group (UK) Ltd, Croydon, CR0 4YY
30/09/2026
02225657-0011